Dutch Design

Dutch Design

A History

Mienke Simon Thomas

REAKTION BOOKS

Published by Reaktion Books Ltd
33 Great Sutton Street
London EC1V 0DX, UK

www.reaktionbooks.co.uk

First published 2008

Transferred to digital printing 2012

This translation was supported by grants from The Prince Bernard Fund and The Mondriaan Foundation.

Prins Bernhard Cultuurfonds geeft cultuur de kans

Mondriaan Stichting
(Mondriaan Foundation)

Printed and bound by University of Chicago Press

British Library Cataloguing in Publication Data
Simon Thomas, Mienke
Dutch design: a history
1. Design, Industrial – Netherlands
I. Title
745.2'09492

ISBN 978 1 86189 380 2

Contents

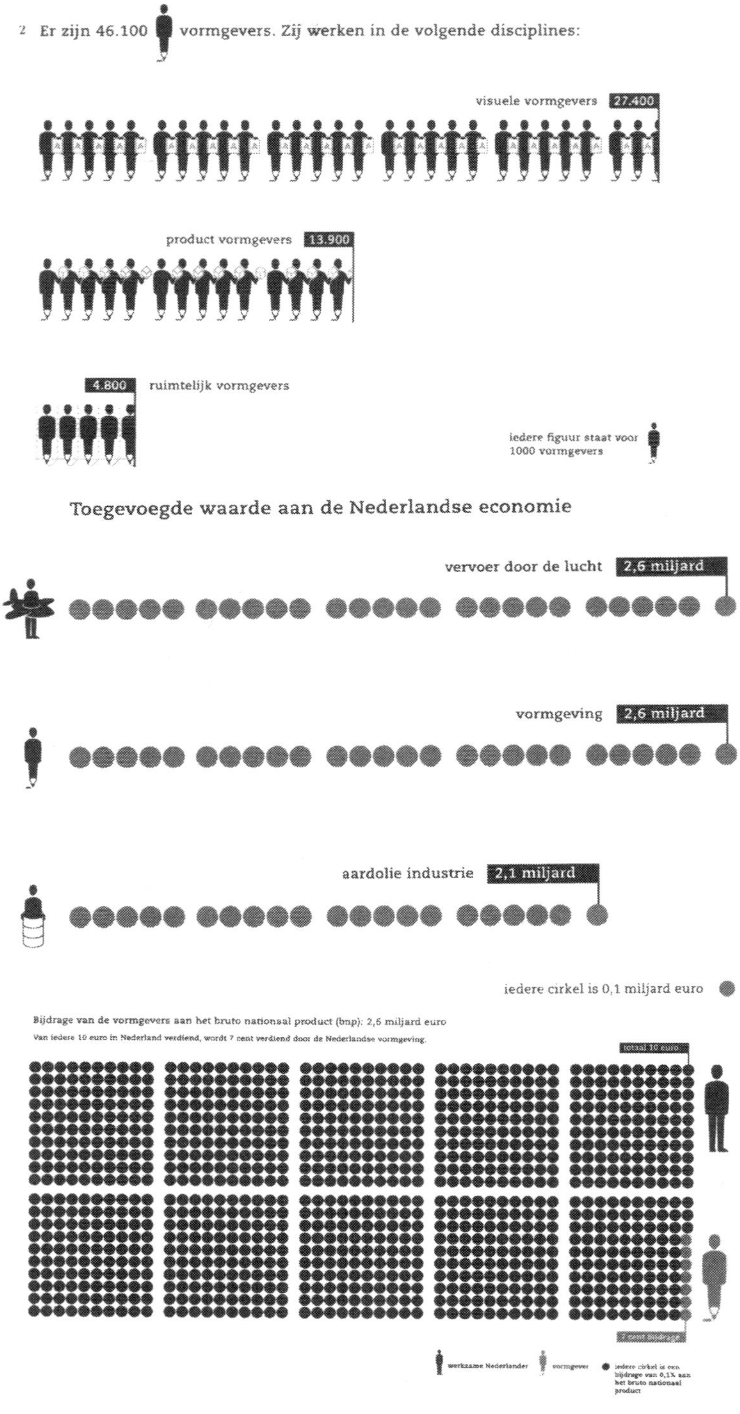

Studio Mijksenaar, visual statistics in the TNO Report *Design in the Creative Economy* (*Vormgeving in de Creatieve Economie*), for Premsela and the Ministry of Economic Affairs, 2005.

Introduction

In 2001 the Dutch government set up an Interim Advisory Committee on Dutch design to map out the infrastructure of design culture. The aim was to use this information as a basis from which it would be possible to make more specific recommendations on design policy in the future. The committee advocated more 'synergy' between the social, cultural and economic sectors involved in design, and the establishment of a new design institute that could offer guidance. It reasoned that the Netherlands has always enjoyed a design tradition in which great attention has been paid to social ideals and cultural values, but less to economic concerns. Four years later, in 2005, the last hypothesis was put to the test by the information research group TNO, which needed to know the precise importance of design as part of the creative economy. This exhaustive study produced remarkable results: the astonishing conclusion was that, when grouped together, Dutch designers were as important to the national economy as the profits accrued from air transport or the petroleum industry.[1] This made a very surprising outcome indeed if we consider the prevailing image of the 'thrifty' Dutch – with their supposed lack of ostentation and small-scale production system.

The way these two reports came about invited criticism. First, the Advisory Committee set up in 2001 was composed entirely of people from the cultural scene, who had a limited knowledge of economic affairs. In 2005, on the other hand, professional flower arrangers were assessed in the TNO study alongside industrial designers – a mismatch that many saw as detracting from the validity of the conclusions. In short, a scholarly, value-free analysis of design culture is an extremely difficult task, even using the most modern research methods. These reports proved that an assessment of the design sector depends to a large degree on the perspective, aims and sources at the researcher's disposal. This was no different in the past. There was, for example, a hidden agenda in 1878 when the senior official of the Ministry for Home Affairs, Jonkheer Victor de Stuers, and the State Commission he installed were asked to judge the state of the Dutch art industry.[2] The same held true in 1945 for the designers Piet Zwart and Paul Schuitema, who had just as many predetermined motives when they drew up their report on the future of industrial design in the Netherlands.[3]

These examples show that writing a historical survey of Dutch design culture can be a hazardous undertaking. The primary sources at our disposal usually throw light on just one side of the story. Even the secondary literature still in existence has its limitations, since until now design history in the Netherlands has mainly been the province of art and architectural historians. It is only natural that they have mainly described the history of design from an art–stylistic perspective. Only a small number of studies has approached design from a different angle, by, for instance, taking an interest in economic, sociological and political-philosophical views.[4]

In this book the central focus is on Dutch design culture in the twentieth century. This means that our attention will be fixed primarily on the cultural, economic and political-social context of design, and only in the second instance on the products and designers that figure within these realms. The main theme is the development of design in modern Dutch society. We shall look at the relationship between designers and manufacturers, at the artistic and moral mission designers thought they had to proselytize in the discussions they held on the subject in their specialist journals. The content and organization of the design academy courses will also come up for discussion, as well as the role of the Dutch government in providing subsidies and commissioning work from designers. Finally, we shall examine design criticism and – to a certain extent – the Dutch consumer's opinion about design.

The subject will be divided up into five themes that cover the subjects or issues that were foremost in people's minds when thinking about design, and as such provided the ideological framework within which designers carried out their work. The main thrust of these themes occurs in different eras and by dealing with them in chronological order we shall cover the entire century.

The first chapter addresses the theme of artisanal design, an issue that was of central importance at the beginning of the twentieth century, but crops up again regularly afterwards. In this chapter we shall discuss the strange paradox that during this period, despite increased industrialization, the interest of Dutch designers (then still called decorative artists) was mainly in producing products manually, with the Middle Ages providing an important source of inspiration. Even when design education was reformed and the Vereniging voor Ambachts- en Nijverheidskunst (VANK) was set up in 1904, it was initially handmade crafts that were the focus of attention. Thus around 1900 Dutch design was in a certain sense 'conservative', but it would, surprisingly enough, be proudly presented to the following generation as part of the developmental history of 'typically Dutch' Nieuwe Kunst (New Art). At the same time this traditional, crafts-based movement was of marginal importance for the growth of industrialization, and for innovation in a wider sense. Unopposed, modernization continued its course.

In chapter Two some light will be shed on the designers in the 1920s and '30s who made frenetic attempts to promote their opinion that design should be 'art'. All the same, some of them did begin to see at this point that collaboration between designers and industry was inevitable, and possibly desirable, but nevertheless for many of those involved the products resulting from this collaboration still had to remain 'art'. This was the opinion of many VANK members at the time and was also common among designers of the Amsterdam School, but was apparently also upheld by the more progressive artist-designers of De Stijl movement and members of the Bond voor Kunst in Industrie (BKI). In these circles their great longing for art and artistry continued undiminished. So for a long time, and in a certain respect up to the present day, they have recognized a fundamental difference between artistic products emerging from a collaboration between designer-artists and 'ordinary' industrial bulk goods. Only a few progressive designers, like Piet Zwart and Willem Gispen, had already managed to liberate themselves from these artistic aspirations before the Second World War.

An important theme that dominated Dutch design throughout almost the entire twentieth century was the need to make the world a better place through beautiful design: beauty and ugliness in the Netherlands have often been synonymous with good and bad. In chapter Three it is argued that the main thrust behind this issue is modernism before and after the Second World War. This 'Moral Modernism' concentrated on the virtues: simplicity, honesty and functionality. The politically committed architects of Nieuwe Bouwen (New Building) and the designers connected with them were motivated to aspire to what was morally classified as a 'good' form by adhering to these values. The same held after the war for architects and designers involved in the reconstruction of the Netherlands. The Goed Wonen Foundation is the clearest manifestation of this Moral Modernism in the 1950s and '60s.

Chapter Four deals with the way in which, from the 1950s onwards, the Netherlands brought the process of professionalizing design as a discipline to completion, and in so doing made industrial design a factor of real social importance. Design culture in the post-war reconstruction years was characterized by, at long last, the arrival of a flourishing industry, more and better design courses, enthusiastic designers and, above all, far more products made with the involvement of a designer. The Instituut voor Industriële Vormgeving (IIV), with its showroom in Amsterdam, played a major role in giving design culture the necessary exposure. Industrial design became an important part of the policy pursued by manufacturers of electric household appliances and was gradually adopted by the furniture industry too. It seemed as if the whole of the Netherlands was being redesigned in those years. There was evidence of this at Schiphol airport, in trains, at stations, on motorways, the money in our purses, in post office design and products, in supermarket design and packaging, and in department stores' designs and wares: well-considered modern design was filtering through on all sides. For that matter we must not neglect to mention that in getting the public to accept modern design an important role was reserved for a few large design studios, as well as stores such as Metz & Co. and the Bijenkorf, and later HEMA and IKEA.

In the last chapter reactions to the issues handled earlier come up for discussion. It then becomes clear how much some themes have constantly continued to dominate the design culture debate. In addition, we shall also see that in the last three decades of the twentieth century a number of designers and critics begin to loathe the 'perfect, but boring' Modernist design in evidence all around them. Running parallel to this reaction is their criticism of the over-commercial character of design and designers, and the total lack of concern shown by manufacturers for conserving the environment. This sparks off debates and counter-cultural or oppositional movements all over the place. At the same time, the dividing lines between design, fashion and art become more indistinct. New anti-design becomes internationally famous thanks to the generous, progressive subsidy policy pursued by the Dutch government. Thus Dutch design currently stands for critical, ironic and conceptual – in other words, intellectual – design. However, the question posed at the beginning of this book about the concrete economic importance of design at the start of the century could equally well apply to present-day, celebrated Dutch design.

The exhibition *Dutch Design Port* by Rotterdam's VIVID gallery at the International Contemporary Furniture Fair (ICFF), New York, 2007.

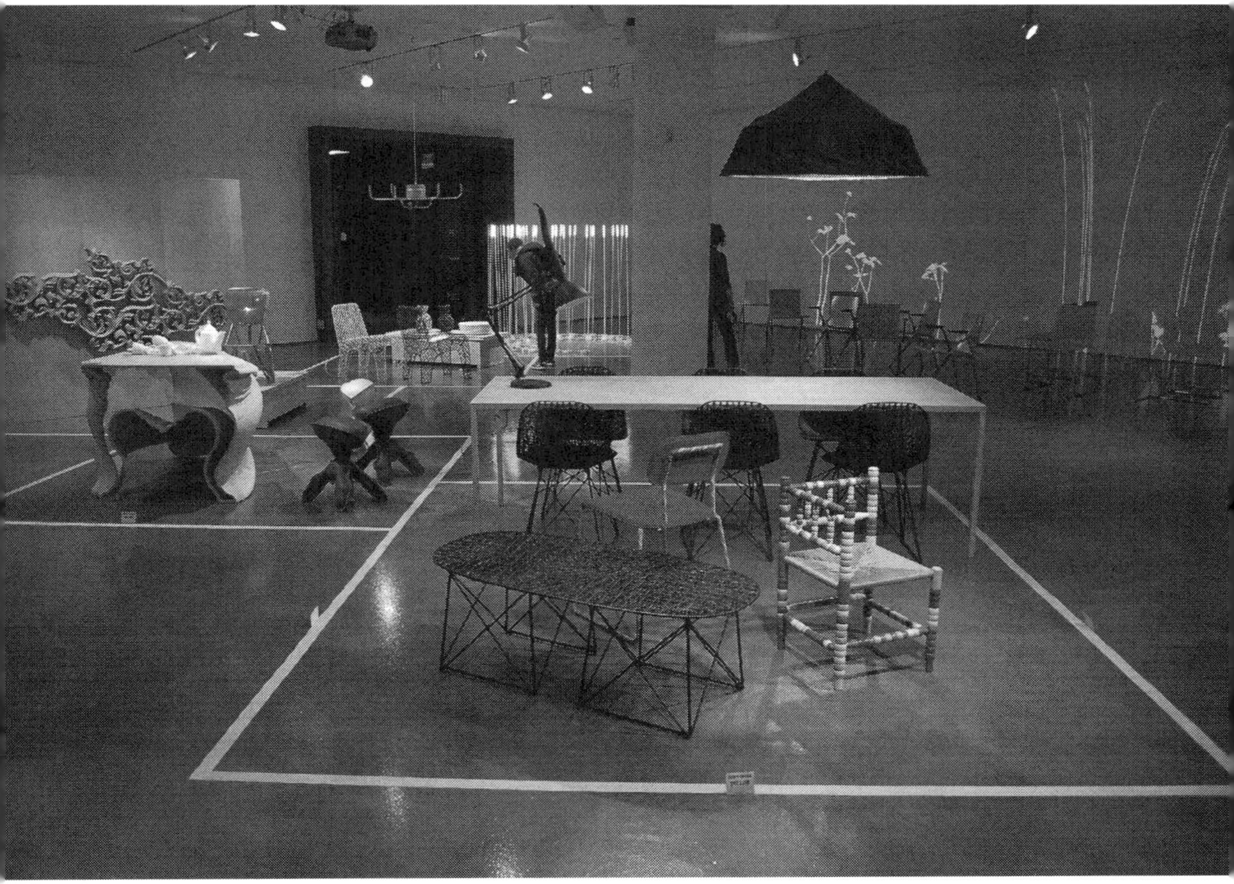

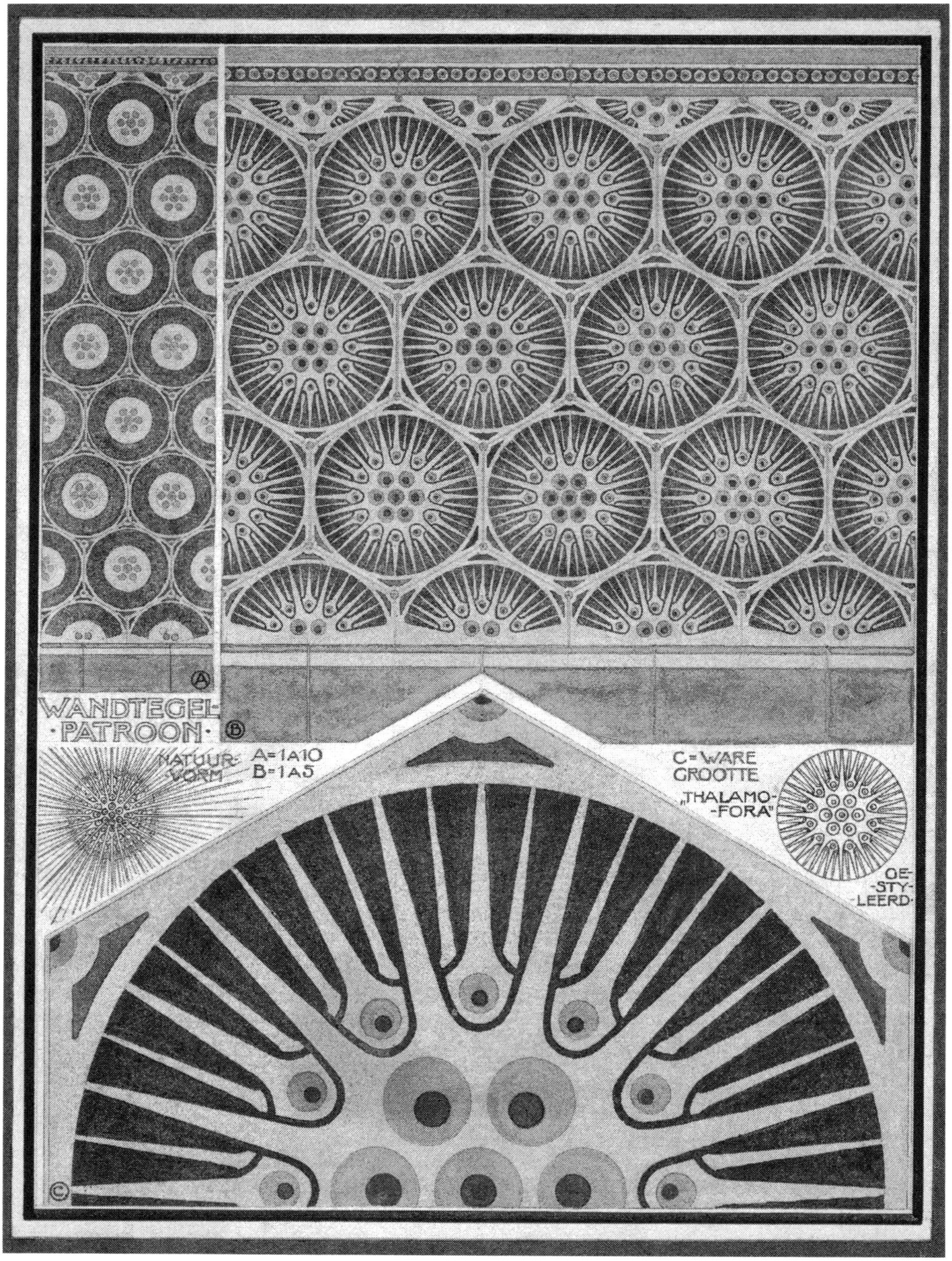

WANDTEGEL-
·PATROON·
NATUUR-
VORM
A=1A10
B=1A5
C=WARE
GROOTTE
„THALAMO-
-FORA"
GE-
-STY-
-LEERD·

1

New Art, Old Craft, 1875–1915

The Paris Exposition Universelle of 1900 offers a useful starting point for a view of Dutch design at the turn of the twentieth century.[1] The Dutch entry gives an idea of the products then considered interesting, representative and beautiful enough to be presented internationally. However, it is almost just as fascinating and revealing to consider what was not selected for this special occasion.[2] The organization of the entries was in the hands of a committee set up and funded by the Dutch government, comprising members of parliament, ex-ministers, a member of the Amsterdam Chamber of Commerce and the chairman of the Advisory Council of the Museum of Applied Arts, Haarlem, as well as the chairman of the venerable Pulchri Studio artists' society in The Hague.

The Netherlands was represented in Paris by no fewer than 559 exhibitors. While this may appear to be a large number, when set against a grand total of 83,071 participants this was in fact rather small. Despite this modest number, however, the Dutch economy was then flourishing. One should not forget that ever since the seventeenth century it had been based on trade. Around 1900 this state of affairs was even consolidated by the opening of the Dutch East Indies for exploitation by private enterprises and the growing coal and steel industries in Germany.

The 1900 Exposition Universelle was still organized along nineteenth-century lines in that every branch of industry in the widest sense of the word was represented. Thus exceptionally designed, artistically decorative and functional objects formed but a small part of the entry. Agriculture and livestock businesses were also represented with their products, even including a number of cattle and horses. Visitors in Paris could also study new developments in the shipping and fishing industries, get acquainted

H. P. Berlage, tile design based on Ernst Haeckel's *Kunstformen der Natur*, c. 1905.

with the results of the new and flourishing chemical and mechanical industries, or view products by the then rapidly expanding Dutch food and drink manufacturers, including the attendant packaging industry. And, not least, they could sample the results of the Netherlands' famous genever (gin) distilleries and breweries.

Among more than five hundred participants, only a few dozen displayed products categorized as 'industrial art' or 'applied art' – products that because of their extra attention to design, artistic decoration, costly materials and skilled finish put them 'above' everyday functional objects. These were mainly to be found in the 'Decoration and Furnishings' department, a section of the Dutch entry selected by a subcommittee that included, among others, Adolf Le Comte, who had formerly taught at the Polytechnic School in Delft, and E. A. von Saher, director of the School of Applied Arts in Haarlem.

In this department nearly all the space was reserved for entries from the Dutch ceramic industry. In addition to a few smaller pottery manufacturers, De Porceleyne Fles from Delft and the Haagsche Plateelbakkerij Rozenburg proudly showed their large and varied collections of modern decorative pottery. The same department presented colourful carpets, stained-glass windows, furniture, decorative silver objects and various base-metal items.

The entire Dutch exhibit was housed in a series of individual pavilions designed by Karel Sluyterman, lecturer in decorative art and theory of ornament at the Polytechnic School, Delft, who was assisted in this by The Hague architect Joh. Mutters. Sluyterman chose an exuberant, contemporary version of International Art Nouveau – the so-called Congo style. This imaginative blend of Art Nouveau and Exoticism came into vogue following the Exposition Internationale in Tervuren, close to Brussels, in 1897, where the Belgian Congo pavilion had been executed in this arresting style. At the committee's request, Sluyterman's remarkable design, including decorative batik fabrics, striking colours and contemporary lettering, had attempted to create uniformity among the somewhat disparate Dutch departments. The result evidently met with the approval of the international jury, which presented him with highest possible award for his design at the end of the exhibition.[3]

None of the leading industrialized companies from the Netherlands producing decorative or functional objects was represented at the Paris Exposition Universelle. Both the Dutch organizers and potential entrants obviously felt that products should be handcrafted, or at least partly so, in order to fall into the 'industrial art' or 'applied arts' category. An artistic product had to be unique and not mass-produced in a large factory. For this reason neither of the two largest ceramic factories in the Netherlands, The

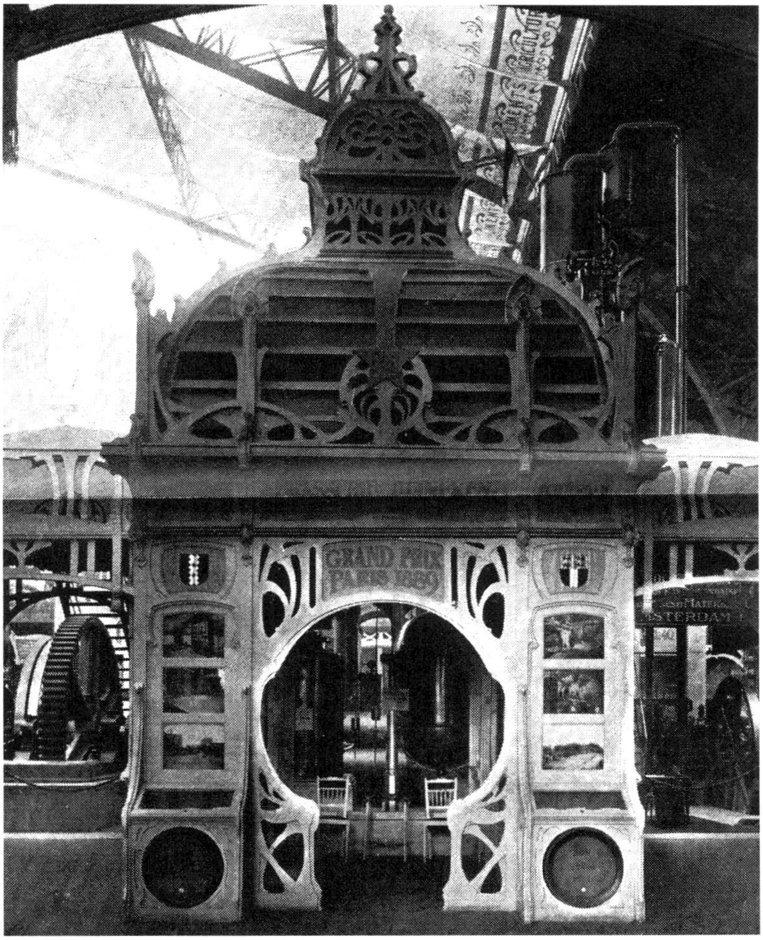

Karel Sluyterman, Heineken pavilion at the Exposition Universelle in Paris, 1900.

Sphinx (formally Regout) and Société Céramique, both in Maastricht, were present. At the time these two companies belonged to a handful of truly large industrial manufacturers in the Netherlands. With more than 3,000 employees, including many children, these firms, with the help of steam power, produced virtually anything to do with ceramics around the clock.[4]

The well-developed Dutch textile industries were also noticeable by their absence, including not only the wool factories and damask weaving mills in Brabant, but also the cotton textile factories in Twente, which were then among the country's largest industrial companies. Like the four leading calico printers in Haarlem, Leiden, Rotterdam and Helmond, they exported virtually all their production to the former Dutch East Indies.[5] Also absent from Paris were the equally large and important furniture

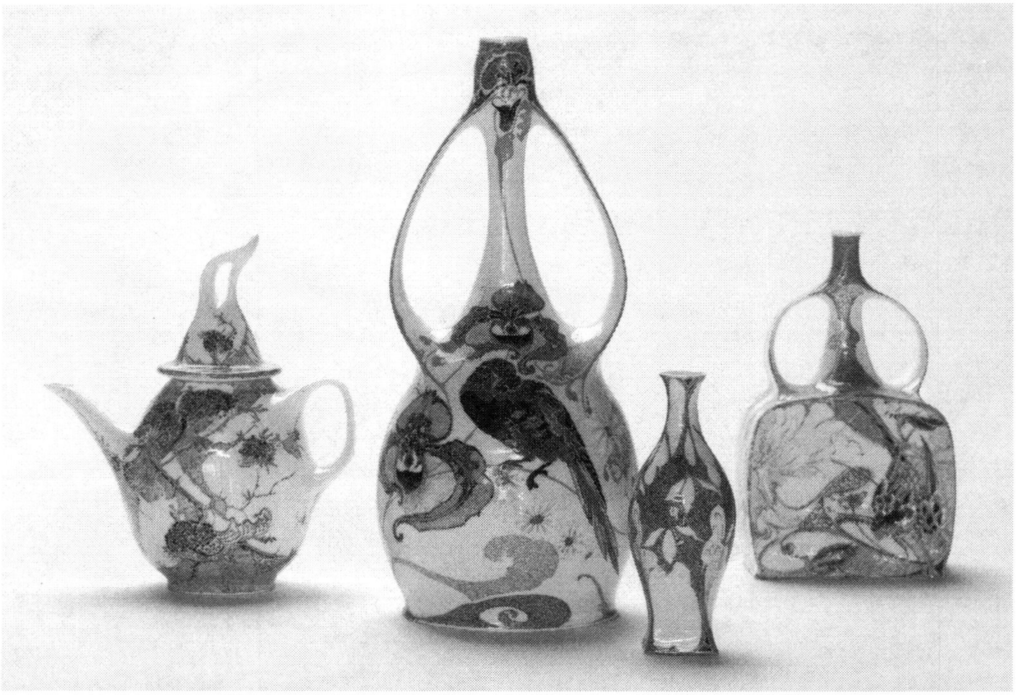

Jurriaan Kok (Haagsche Plateelfabriek Rozenburg), teapot and three vases, eggshell porcelain, c. 1900.

firms, such as Pander, Mutters and Eckhart, and the leading metalware firms, like Daalderop and DRU. All these manufacturers seem to have had little confidence in the commercial advantages to be gained in Paris and, despite the organization committee's urgings, they were not prepared to spend time and money on proper representation there.

Despite these omissions, a review in *L'Art décoratif* declared that 'Holland is presented at the Exhibition as one of the nations most active in pursuing a new style'.[6] Fifty years of official endeavours to take applied art to a higher level had reaped results. Thus the jury concluded with a certain satisfaction that, artistically speaking, the Netherlands could compete with the rest of Europe; even the President of France, who visited the Dutch exhibit on 30 May 1900, described it as a 'huge success'.[7]

Looking Back: Design in the Mid-Nineteenth Century

Half a century earlier at the Great Exhibition, the first international exhibition, held in Kensington, London, in 1851, it had been a different story. Time and again this exhibition has been seized upon to highlight the abominable quality of Dutch industry at the time.[8] It is indisputable that from the

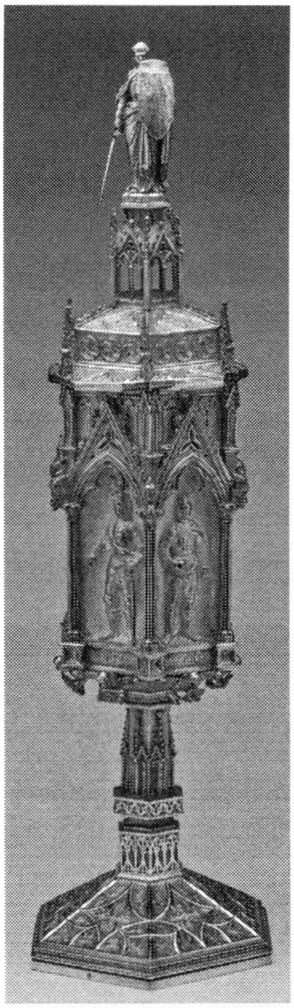

J. M. van Kempen, Utrecht, silver goblet decorated with representations of medieval ancestors of the Orange and Nassau Houses, 1847, shown at the Great Exhibition, London, 1851.

eighteenth century the once flourishing industry, artisan skills and favourable trading position of the Netherlands markedly declined due to the ascendancy of Great Britain. The abolition of the guilds in 1798, followed by the division of the Low Countries and the establishment of Belgium in 1830, meant that little now remained of this industry. Well-to-do Dutch preferred to obtain artistic, well-made consumer goods from abroad. Luxury furniture from France, Belgium and Germany was considered more appealing than that of Dutch manufacture.

While it is true that industrialization and modernization occurred more slowly in the Netherlands than elsewhere in Europe, recent research shows that developments there had their own specific character.[9] It is inappropriate to link industrialization solely to the introduction of steam power, as is often the case. For a long time the hundreds of windmills all over the Netherlands, as well as the smaller gas engines, were simply much cheaper and more efficient for most of the small Dutch factories. This places a different light on the batik decorative friezes designed by Karel Sluyterman for the 1900 Paris Exposition Universelle: windmills, represented in a decorative Art Nouveau style, were a generally accepted feature of Dutch industry at the time and had not yet become a hackneyed traditional symbol. The small scale and versatility of Dutch industry also gave it a flexibility that ensured that modernization would ultimately make its way there too.

In retrospect, the Dutch entry for London in 1851 was not representative of the situation in the Netherlands. It was not the stagnant industry but rather the lack of interest by the Dutch government that was the chief reason for the sparse representation. Prime Minister Thorbecke had handed over responsibility for Dutch participation to private initiative, with the result that only 115 companies were prepared to send products to London at their own expense. Unlike other countries, the Netherlands still did not consider a good international display of its national industry to be a government matter.

During the second half of the nineteenth century, however, the artistic quality of Dutch decorative and functional objects became a cause for concern among the cultural elite. Triggered by subsequent international exhibitions in Paris (1855, 1867, 1889), a second in London (1862) and others in Vienna (1873) and Chicago (1893), a debate had begun about the languishing state of Dutch design.[10] In official reports and cultural magazines the reason for this was sought in the immense lack of feeling for art, be it among employers, workers or consumers. Moreover, it was customary to point out how this contrasted sharply with the Netherlands' glorious past, particularly the 'Golden Age' of the sixteenth and seventeenth centuries. Thus, art and architectural history were brought to bear in an attempt to raise national awareness in the field of industrial art. In so doing the critics hoped that the 'industrious and down-to-earth Dutchman' would finally emerge and be a match for the 'inventive Englishman', the 'refined Frenchman' and the 'practical American'. That way the greatness of former times could undoubtedly be recaptured.

For these reasons illustrious Old Dutch applied arts were proudly displayed at the first international industrial exhibition in the Netherlands in 1877.[11] The organizers, who by now included government representatives, were convinced that the display of such fine old examples would boost contemporary industry and stimulate Dutch manufacturers and consumers' sense of national pride. At this event, titled *Exhibition of Art Applied to Industry*, the design and artistic standard of the exhibited products were pivotal, rather than the technological advances so prominent in other industrial exhibitions.

The responsibility for this concept lay with the newly appointed arts official of the Arts and Science department at the Ministry of Home Affairs, Jonkheer Victor de Stuers. This first Dutch 'arts' official was driven by an ambition to awaken an interest for their own past among the Dutch. During the last quarter of the nineteenth century he was at the forefront when all aspects of Dutch culture were being determined, including museum policy, art education and the preservation of historic monuments and buildings. His power was such that he had the casting vote in awarding national architecture commissions like the one for the Rijksmuseum and the Central Station in Amsterdam, for which the Gothic Revival architect P.J.H. Cuypers was appointed.

The architect J. R. de Kruyff was actually the most important figure in organizing the exhibition. He also designed the presentation. In a brochure published prior to the exhibition, he defined the concept of industrial art as 'containing those products of human endeavour, in which the imagination is harmoniously reconciled with the guiding sense of beauty, which extends to the production of domestic objects which industry brings forth to satisfy the numerous requirements of everyday life'.[12] But there were few examples of mass production or everyday items; the exhibition was more about luxury household goods, hand-knotted carpets, lavishly carved furniture made from gleaming, expensive types of wood, heavily ornate mirror frames and silverwork. Exceptions to this were the modest exhibits from the ceramics factories of De Porceleyne Fles in Delft and Regout in Maastricht.

More important than the exhibition itself were the jury report and the other publications that appeared in its wake. One government-appointed committee, in which De Kruyff again played a central role, wrote a report

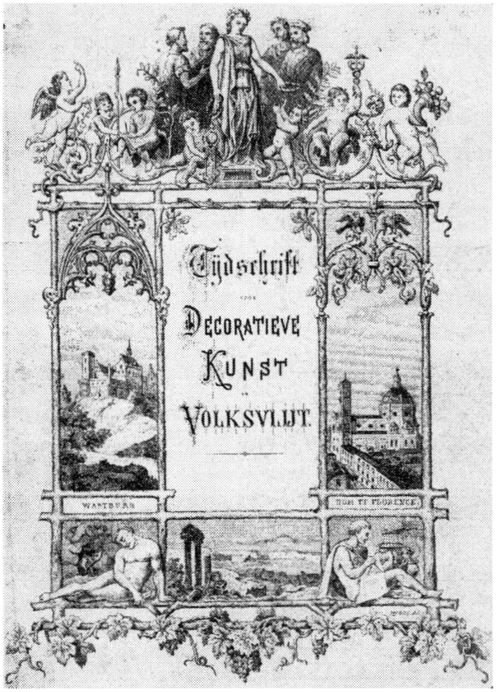

Cover of the magazine *Decoratieve Kunst en Volksvlijt*, 1875.

on Dutch industrial art in 1878.[13] It concluded that the situation was in general still 'depressing'. Fortunately, the critical committee members saw a few rays of hope. They considered the carpets of the Royal Carpet Factory in Deventer to be outstanding, while the furniture companies of H. P. Mutters and H. F. Jansen were praised for the diversity of genre styles. Yet the entries from the two ceramic factories were judged far below standard, with severe criticism of the decoration applied mechanically to the Maastricht wares. The 'depressing' results were then seized upon by De Stuers and other interested parties to implement several reforms in the Netherlands. A Museum of Applied Arts was founded in Haarlem and serious plans developed for new courses to be established.[14] Much use was also made of knowledge and experience from abroad.

Foreign theoretical treatises were also useful for a small group of Dutch specialists. In particular, Gottfried Semper's views, as expressed in such publications as *Der Stil in den technischen und tektonische Künsten* (1860–63), were initially critical in forming opinions in the Netherlands. After a visit to the International Exhibition held in London in 1862, the secretary of the Netherlands Society for the Trade and Industry, F. W. van Eeden, for instance, wrote a series of articles that prominently featured his knowledge of Semper's published works.[15] A decade later Van Eeden became the first director of the Museum of Applied Arts in Haarlem. Semper's conviction that the style – or design – of a product should be derived from its function, its material and the technique by which it was made had already become common knowledge by the 1870s.

Following writers like Owen Jones, Ralph Wornum, Richard Redgrave and A. W. Pugin, the study of historic styles became essential in the Netherlands. In 1884 Carel Vosmaer's translation of Lewis Foreman Day's *Everyday Art* (1882) appeared as *De Kunst in het Daaglijksch Leven*. Towards the end of the nineteenth century the Netherlands became acquainted with the more socially engaged design ideas of John Ruskin and William Morris. The major consequence of this was a steadily increasing appreciation of craftsmanship and a better understanding of the position of the industrial artist in society. The publications of the French architectural theorist Eugène-Emmanuel Viollet-le-Duc, which became known in the Netherlands chiefly through the architect P.J.H. Cuypers,[16] were to have just as big an impact as those by Semper and the English writers. As a result the Netherlands became familiar with new Gothic-based ideas about architecture and design. French, German and Austrian periodicals as well as sample portfolios were constantly scrutinized in the Netherlands during the last quarter of the nineteenth century. The first Dutch magazine on design was a version of the German monthly *Gewerbehalle* and first appeared in 1870 as *Kunst en Industrie* (Art and Industry). The first original Dutch periodical was the decorative art magazine *Tijdschrift voor Decoratieve Kunst en Volksvlijt* in 1875. Unfortunately, this spirited initiative from C.A.J. Geesink, the owner of an Amsterdam printing firm, who also made plans for a Netherlands Art and Industry Museum, folded after only two years.

Early Design Education

During the late nineteenth century the Netherlands was particularly interested in how the newly reformed education in art and design was organized abroad.[17] At first it seemed that education reform in the Netherlands was on a par with the growth of industrial production. Steady increase in mechanization, scale and division of labour had led, for instance, to the founding of the first technical school in Amsterdam in 1871. Now that it was increasingly evident that future workers could not be trained as well on the factory floor, special vocational courses had to be set up. Pupils ranging in age from twelve to sixteen were then taught, among other things, how to become carpenters, blacksmiths and painters. The second Dutch technical school to open its doors was in The Hague.

When the government committee on industrial art, under the influence of Victor de Stuers, argued in its 1878 report for improvements in education, the director of The Hague technical school, H. L. Boersma, wrote a lengthy reply in which he warned against the slavish adherence to traditional applied arts emphatically advised by the committee. Each era had its own characteristics and its own applied arts: by failing to recognize this, he felt the committee did not do justice to the requirements of industry. Boersma was also against the distinction the committee made between industrial designers and artistic crafts people, and the priority it wished to give to the former group. The director argued that Dutch industry was on such a small scale compared to neighbouring nations that the artistic development of crafts people should take first place.[18]

The first 'School of Design for Applied Arts' was founded in Haarlem in 1879 on the initiative of the Netherlands Society for Trade and Industry as a logical extension to the town's Museum of Applied Arts, which the Society had opened two years previously. Its first director was the architect Eduard A. von Saher, who had trained at the Polytechnikum, Zürich, and had been taught by Gottfried Semper. The combination of a school and museum was already to be found in various foreign museums, the earliest and most notable example being the South Kensington School and Museum in London. In 1881 a National School of Applied Arts was incorporated into the Amsterdam Rijksmuseum, where, alongside drawing, emphasis was placed on theoretical training. Here an influential teacher was Pierre Cuypers, who had initiated another type of training a few years earlier. During the building of the Rijksmuseum he had noticed that skilled stonemasons and woodworkers were few and far between, so in 1879 he set up a new training school for the purpose. This on-site building shed or workshop later became the Quellinus School of Applied Arts. Here the refining of practical traditional skills rather than drawing and theory was the main concern in the early years.

The objective of this and various other new schools of applied art inspired constant debate for the rest of the century, which led to the curricula often being modified and adapted. Since the prime intention, with the exception of the Quellinus School, was to train future designers for industrial design, the new schools were chiefly schools of technical drawing. For so many days a week pupils were supposed to work in a workshop or factory and then receive additional 'theoretical' instruction at school. Since the 'art' component, in the context of industrial 'art' and applied 'art', was virtually synonymous with ornament, teaching mainly covered the history of ornamentation and the technical drawing of well-conceived decoration.[19] Much attention was devoted to studying historic styles, including those from the East. The underlying principle of acquiring such knowledge was not to copy styles, but rather to establish a way of achieving well-founded new designs: 'Study the Old so that you will remember it and gain strength to begin afresh', as Cuypers wrote in fine Gothic lettering on the walls of the Rijksmuseum. Moreover, armed with

Petrus Regout & Co., Maastricht, jug with imitation marble, c. 1860.

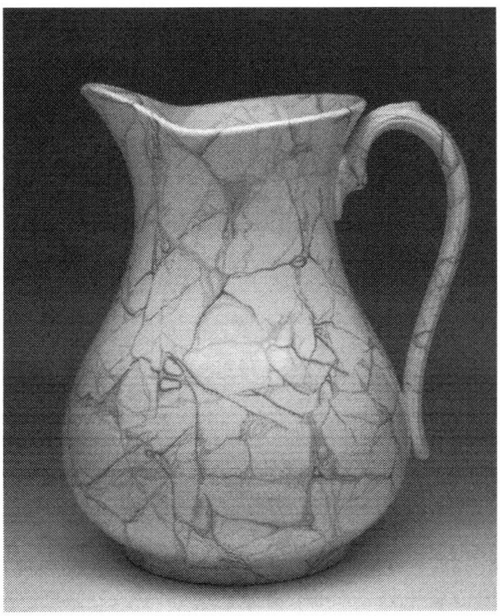

this improved knowledge about styles, the designers would be less likely to give in to the commercial malpractices of many manufacturers. Jacob de Kruyff, who became the director of the National School of Applied Arts, railed against the 'terror' of commerce and the various fads found in the industry of his time. Particularly objectionable, in his view, was the trend to imitate expensive materials in a cheap ersatz manner, like painting cheap wood to make it resemble costly marble or much rarer types of timber. He also roundly condemned as a 'fad' the popular use of naturalistic plants and animals as decorative elements.

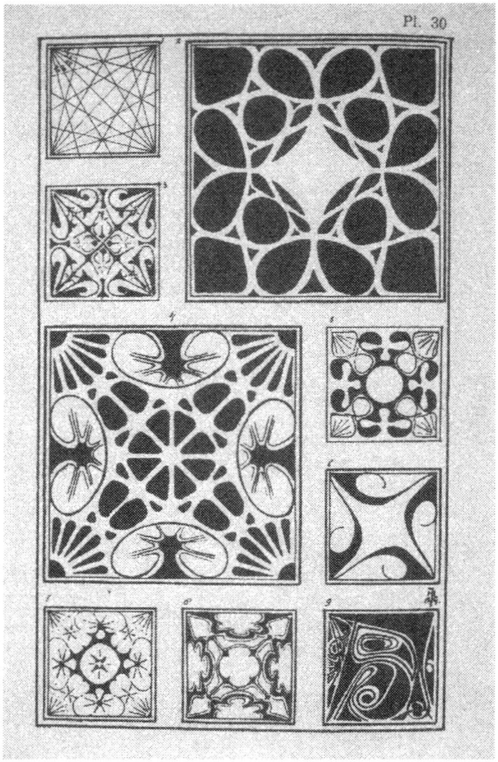

A page from J. H. de Groot, *Driehoeken bij Ontwerpen van Ornament* ('The Use of Triangles in the Design of Ornament'), 1896.

Indeed, stylization of flowers and plants was central to ornamentation training. Pupils were taught how to make nature more abstract and reduce it to simple, repetitive, decorative motifs. Towards the end of the century, however, geometric or systematic design began to permeate Dutch applied arts education. While patterns of straight lines or triangles had in fact almost always underpinned decorative design, the ideas about this, under the influence of the growing popularity of Theosophy, gained an entirely new relevance in the Netherlands.[20] According to Theosophy, mathematics and the laws of measurement and numbers had a divine meaning, while an almost mythical significance was ascribed in particular to the Egyptian isosceles triangle.

The architects Karel de Bazel and Mathieu Lauweriks, both of whom had trained with Cuypers, joined the Theosophical Association in 1894. They were so fervent about their discoveries that they even set up a special artists' section intended to serve as a 'temple' for studying and spreading the message of these 'revelations'. In this Vâhâna lodge classes in design were started in 1897. Every Friday night a few dozen pupils would gather for this purpose in a room at the old Hotel American in Amsterdam. By 1904 some two hundred artisans had followed the Vâhâna lodge's course in systematic design. In their turn, the artists trained there then taught in applied arts education. Consequently, the principles of designing according to geometric systems were widely disseminated in those years. The architect J. H. de Groot and his sister, the needlework artist J. M. de Groot, even put together in 1896 a small

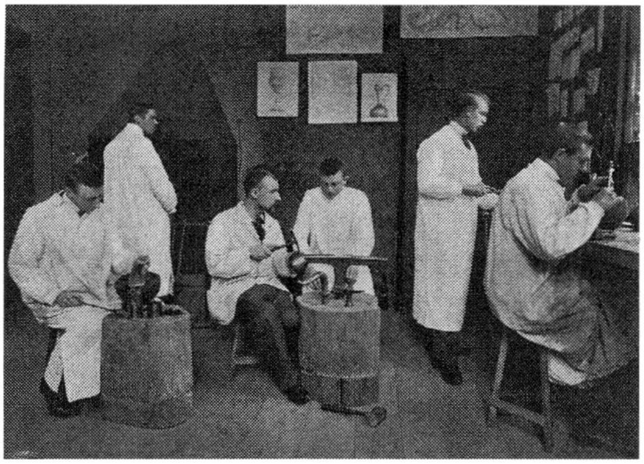

Metal workshop of Frans Zwollo, Sr, at the Haarlem School of Decorative Arts, c. 1905.

practical manual entitled *Driehoeken bij Ontwerpen van Ornament* (The Use of Triangles in the Design of Ornament), in which the new method was explained with the help of examples. This manual also showed that systematic design could be explained in a much less vague and esoteric manner. Using a triangle, a compass and a ruler, anyone could learn to draw the most varied new decorations.

Around 1900 applied art schools mainly combined the stylization of nature with design based on geometric systems. This led to the flat two-dimensional decoration considered characteristic of Dutch decorative art of the period. The finest and most typical examples of this are the batiks and damask designs of Chris Lebeau and the ceramic decoration of Chris van der Hoef and Bert Nienhuis.[21] The most spectacular results of combining nature and geometry in decorative art were achieved by the architect H. P. Berlage. Around 1900 he transformed illustrations of micro-organisms from Ernst Haeckel's book *Kunstformen der Natur* (Art Forms in Nature) into marvellous, almost purely geometric decorative designs for tiles, stained-glass windows, plates, wallpaper and even three-dimensional objects like lamps.

Meanwhile, decorative design lessons at the applied arts schools were no longer aimed only at future designers or draughtsmen in industry. More to the point, it was becoming apparent that this type of designing was becoming an objective in itself. An increasing number of pupils who were not already working and practising a traditional skill were enrolling at the schools. As a result, a few critics warned that future designers should be better aware of the purpose for which they were making the decorations. The architect Jan de Meijer complained about the 'dry affair' that killed the personality of the artists, and his colleague Willem Retera feared that this theoretical work would restrain their fantasy.[22] The term *sierkunstenaar* ('decorative artist'), initially slightly demeaning, now became a fashionable description of those artists who specialized in designing ornament, but who no longer possessed the skills to make the products themselves. Such a lack of practical skills was now seen as a shortcoming.

Subsequently, the schools of applied arts – besides the one in Amsterdam, a school was founded in 's-Hertogenbosch (1882), while new applied arts departments in existing art academies were opened in Utrecht (1886), The Hague (1889), Rotterdam (1902) and Groningen (1903) – introduced classes in a number of straightforward skills. Around 1900 it was possible to learn lithography, woodcarving and batik work, crafts in which it was possible to incorporate decoration without too many technical aids. A further step was later taken when workshops for making ceramics, metal objects and furniture were added. The precious metals worker Frans Zwollo was one of the most confirmed believers in this 'workshop' concept in applied arts schools.[23] About ten years later the potter Bert Nienhuis also became an influential proponent of this practical form of education.

While industrialization continued apace, design education paradoxically focused increasingly on the artistic and skills side of manufacturing. Around 1900 the 'artist-craftsman' emerged – a pattern designer, artist and craftsman all in one. So while the schools of applied arts had been founded in the nineteenth century to raise the standards of industrial products and had given an initial spurt to the evolution of the later industrial designer, this process changed course again in what can best be described as a conservative direction. For the time being the schools did not train industrial designers but produced textile artists, creative metalworkers and potters.

Paris 1900

If we now return to the Exposition Universelle in Paris and focus in greater detail on what was to be seen there, it is evident that many Dutch decorative products were the logical result of the developments outlined above. De Porceleyne Fles, the only ceramic factory in Delft in the eighteenth century to survive fierce competition from Britain, had patently taken to heart the advice of such as Victor de Stuers and De Kruyff.[24] From 1877 the company had successfully concentrated again on producing traditional blue and white tin-glaze pottery, for which Delft had become so famous two centuries earlier. At the same time the firm was experimenting with new glazes and firing processes, as well as more contemporary designs.

Partly as a result of the successful initiative in Delft, the Haagsche Plateelfabriek Rozenburg was founded.[25] The decision to take on the architect Theodoor Colenbrander as a designer of new forms and decoration turned out to be an inspired move. This idiosyncratic artist quickly helped Rozenburg to establish a reputation by designing a number of original and exciting decors and models. This groundbreaking work later earned Colenbrander the unofficial title 'Doyen of Dutch applied art', awarded to him in 1923 by H. E. van Gelder, director of the Gemeentemuseum in The Hague.[26] At Paris in 1900 the director of Rozenburg, Jurriaan Kok, was again able to present something with novelty value – a new paper-thin type of

T.A.C. Colenbrander (Haagsche Plateelfabriek Rozenburg), *Constantinopel* wall plate, 1886.

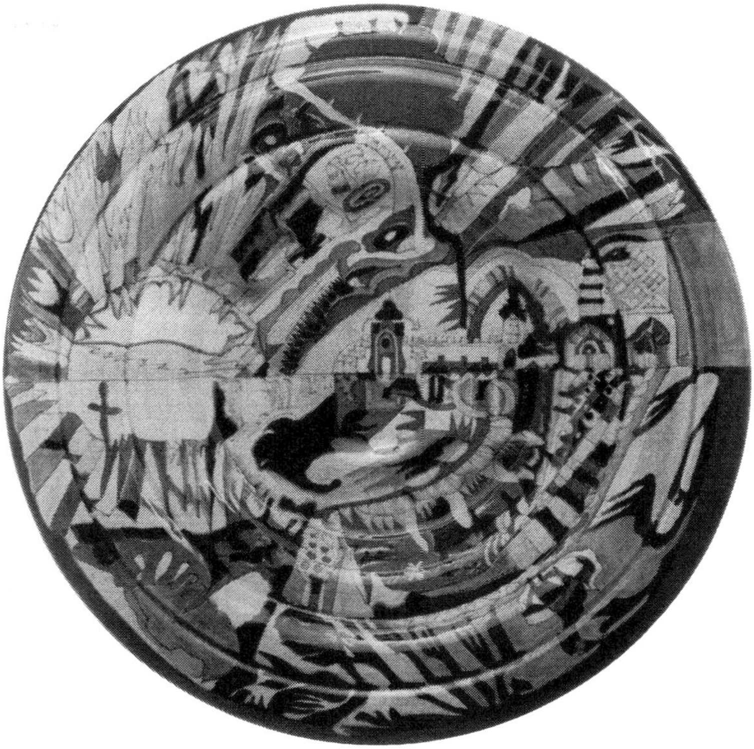

semi-porcelain called 'eggshell'. Partly based on Kok's own designs, Rozenburg had cast a major collection of new ware from this exquisite material, which was decorated with extremely refined, colourful, Japanese-inspired depictions of plants and birds. This new product was an overwhelming international and commercial success in Paris. After the young Queen Wilhelmina honoured the firm with the privilege to use the title 'Royal', Jurriaan Kok showed his gratitude by creating a specially decorated eggshell porcelain tea service for the wedding of Wilhelmina and Hendrik van Mecklenburg on 7 January 1901.[27] De Porceleyne Fles and Rozenburg's winning formula stimulated the founding of various new Delftware factories between 1890 and 1900, five of which submitted work to Paris. The influence of decorative design classes and the stylization of nature were clearly evident in the modern designs of this new Dutch pottery.[28] There were also products to be seen in Paris from a handful of earthenware factories in Friesland, where most of the traditionally designed, everyday kitchen and tableware in the country was still made. Around 1900 these companies were still just about able to ward off competition from British mass production.[29]

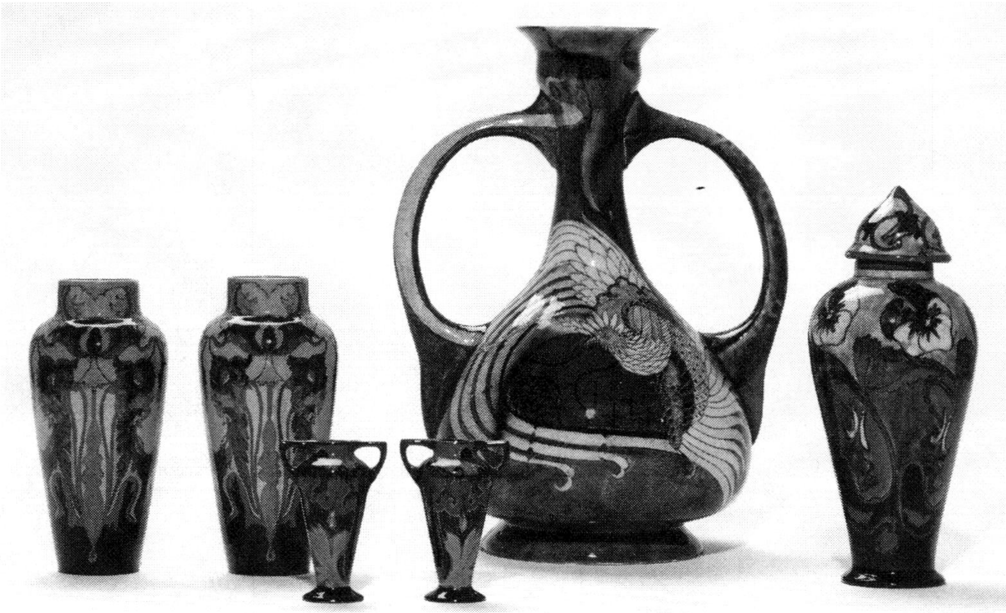

Decorative pottery: large vase decorated by C. J. Lanooij, 1907, from Wed. N.S.A. Brantjes Co. firm, Purmerend; and five smaller vases, 1900–05, from Plateelfabriek Zuid-Holland, Gouda.

The Royal Carpet Factory from Deventer regularly participated in international exhibitions.[30] Since the first half of the nineteenth century the firm had produced, as well as simple cow-hair rugs, luxury, hand-knotted Smyrna carpets with patterns inspired by Near Eastern carpets. Artistically, the patterns were very much in keeping with the increasing focus on colourful, exotic decoration, especially on textiles. In Paris the Deventer factory presented not only these popular designs, but also one or two new ones by Theodoor Colenbrande. Since about 1895 he had been creating patterns for hand-knotted carpets that, as with his pottery, resulted in something completely new. His colourful expressive designs had more or less the same structure as Near Eastern ones, yet at the same time were totally innovative in their free style. They rapidly caught on among the Dutch cultural elite: Willem Hendrik Mesdag, the influential and wealthy marine painter and collector of oriental art, furnished his grand home in The Hague with Colenbrander's carpets and started to collect his Rozenburg ceramics as well. Still more or less in its original state, this is now called the Museum Mesdag and is open to the public.

Interest in oriental textiles was also apparent in a growing fascination for the Javanese batik technique. This had become increasingly familiar in the late nineteenth century partly due to the strengthening of relations with the Dutch East Indies (now Indonesia). Dutch calico print-works were even

T.A.C. Colenbrander (Royal United Carpet Factory, Deventer), carpet in the Museum Mesdag, The Hague; remake of a design from 1893.

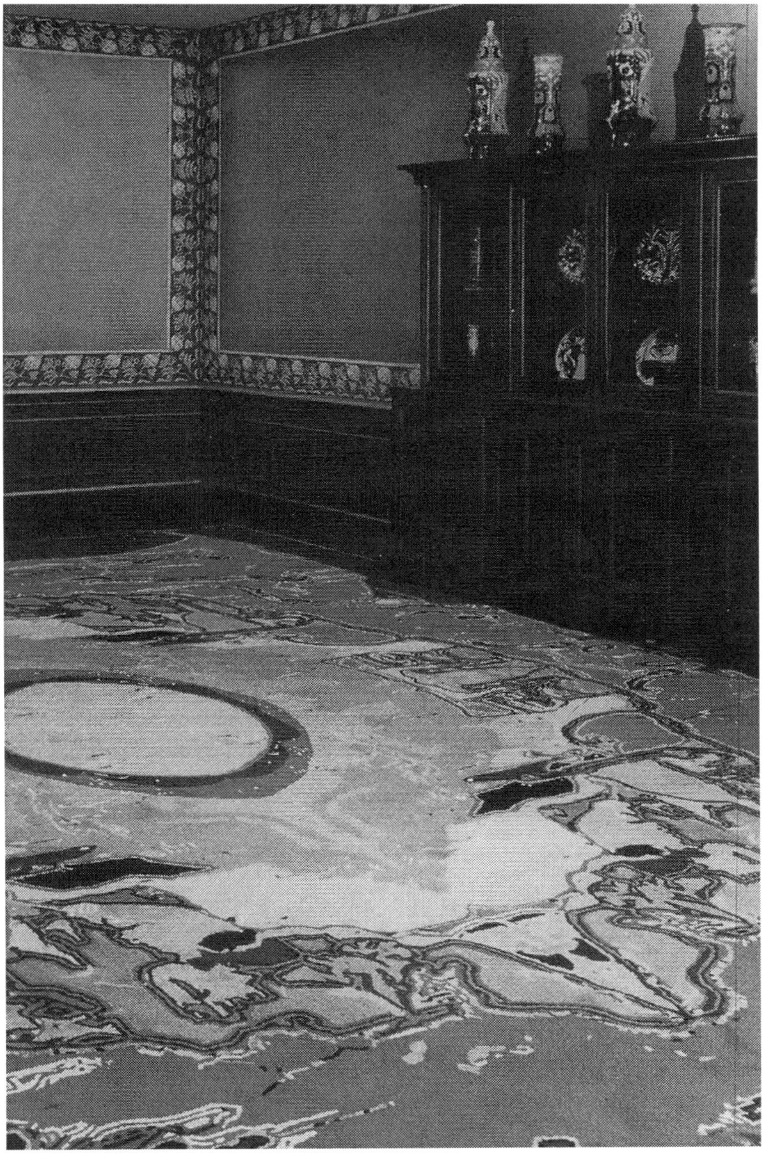

able to imitate the time-consuming traditional process by mechanical means and could thus compete with native batik makers. Artists such as Gerrit Dijsselhof, Carel Lion Cachet and Johan Thorn Prikker began to experiment with the technique after admiring the batiks in the Museum of Applied Arts in Haarlem and the Ethnographical Museum in Amsterdam.[31] This was limited to small-scale projects, apart from Thorn Prikker's designs made on a

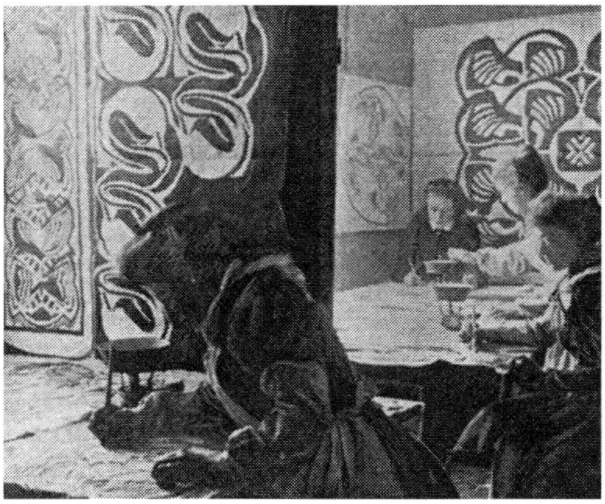

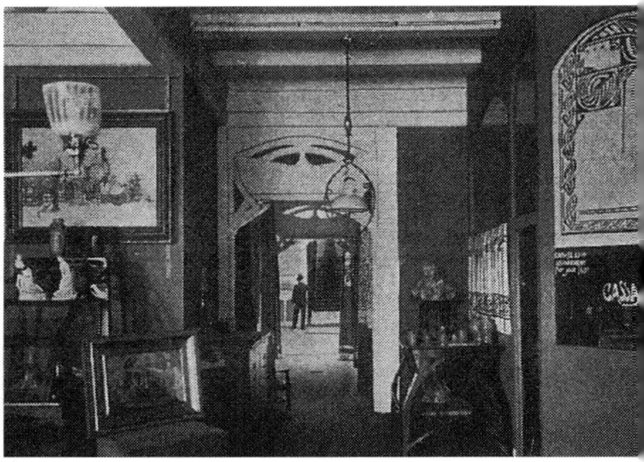

Women producing designs by Johan Thorn Prikker in the Arts and Crafts batik studio, Apeldoorn, c. 1901.

Interior of the firm Arts and Crafts in The Hague, 1898.

larger scale in the Apeldoorn-based batik workshops of Arts and Crafts in The Hague.[32] Examples of the company's batik fabrics were shown in Paris, while Sluyterman's decorative friezes adorning the exhibition (see above) had also been created in its workshops.

In 1900 Arts and Crafts was still the only firm of its type. It had been started two years earlier as the first workshop and gallery outlet for art and modern applied arts in the Netherlands. The gallery had been modelled along the lines of the Paris gallery Salon de l'Art Nouveau, run by Siegfried Bing, even though its name suggests a link with the English design movement. The painter and designer Johan Thorn Prikker was the leading artist for Arts and Crafts, but work by Jan Altorf and Theodoor Colenbrander, and by foreign artists like Henry van de Velde and George Minne, was sold there as well. The products of these two Belgian artists instantly provoked fierce criticism at the gallery opening in August 1898, while Thorn Prikker's batiks and furniture, clearly inspired by what the Belgians were doing, also came under fire from certain reviewers. Berlage, for instance, wrote in *De Kroniek*, a month after the official opening of Arts and Crafts, that Prikker made 'a step from the sublime to the ridiculous' with his arbitrary furniture designs constructed from all sorts of 'pieces of wood' and the 'most distasteful combinations of lines' and 'impossible forms'.

What Berlage himself stood for was shown in Paris by the Amsterdam-based company J. B. Hillen.[33] This was a medium-sized firm industrially producing furniture. In Paris, however, Hillen presented a unique and robust wall unit in oak, eight metres long, designed by Berlage and decorated in flat relief carving by A. C. Oosschot. Furniture was supplied in Paris by two other companies: the studio of Van Wisselingh in Amsterdam, with unique and extremely luxurious, handcrafted objects designed by the artists

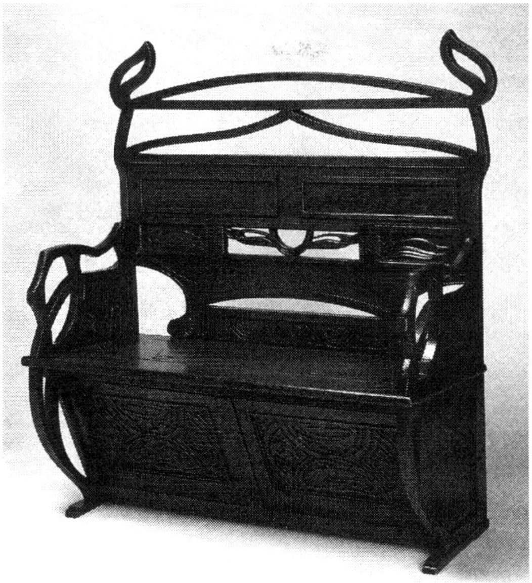

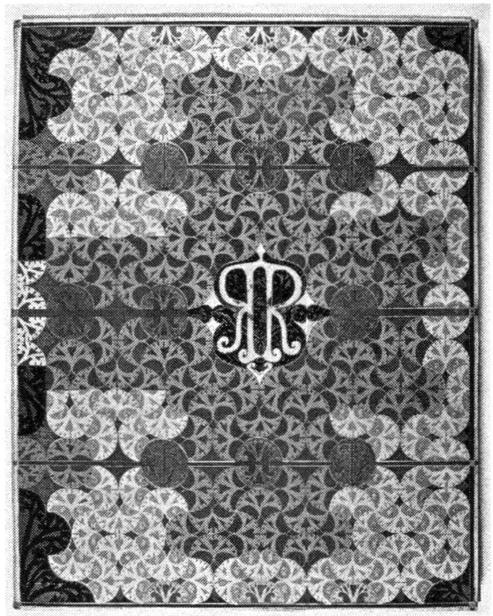

Johan Thorn Prikker (Workshop of Chris Wegerif, Apeldoorn), oak bench, 1898.

C. A. Lion Cachet (Scheltema & Holkema Amsterdam), Rembrandt portfolio, batik on linen and parchment, 1899.

Carel Lion Cachet, Theo Nieuwenhuis and Gerrit Dijsselhof, and the workshop of Pierre Cuypers in Roermond.

To complete this survey of the Dutch offerings in Paris, objects in precious metals chiefly came from the silver firms of Van Kempen and Sons in Voorschoten and Begeer and Brom in Utrecht. Van Kempen was the oldest and largest silver manufacturer in the Netherlands and had an imposing artistic and artisan tradition.[34] The firm had been one of the few Dutch participants at the Great Exhibition in London in 1851. Around 1900, this company, alongside its traditional designs in historic styles, made products

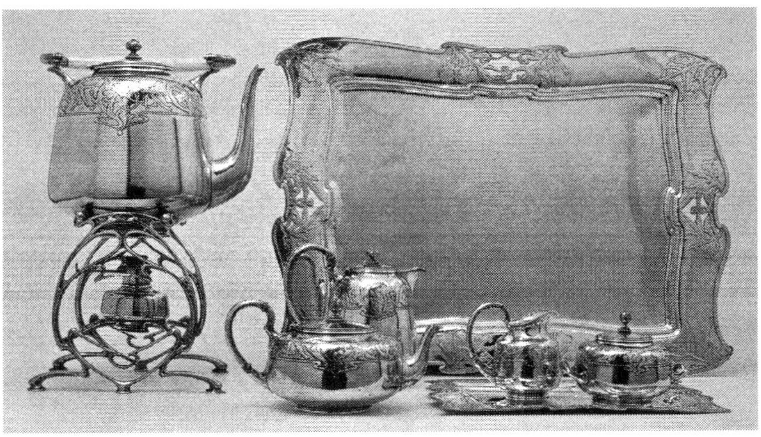

A. F. Gips (C. J. Begeer silver factory, Utrecht), silver coffee and tea service, 1900.

in the then modern, floral international Art Nouveau style. Among the items shown in Paris by C. J. Begeer, participating for the first time at an international exhibition, was a coffee and a tea service in a similar style designed by the architect A. F. Gips. Several smaller silversmith companies were also represented. The firm of Hoeker & Zoon, for example, displayed exceptional designs by the metalworker and craftsman Jan Eisenloeffel, who made strikingly simple tea services clearly influenced by Japanese applied art.[35] Frans Zwollo showed pitchers, vases and dishes in silver and copper.

Decorative Art in Turin, 1902

The trend for artistic handcrafted products was evident throughout Europe. Two years after Paris, Turin organized a major international exhibition devoted to this 'modern decorative art form'. International competition in this field was immense and the Netherlands decided that 'typically Dutch' should be the starting point for their entry. The nation had shown in Paris that it had great capacity 'on this new territory', but in Turin it could demonstrate that 'decent work of sober conception and good taste would in the end be more valuable and enjoy a better reputation then the more pompous and capricious work of our neighbours' – such, according to the organizing committee, were also the opinions one could read in magazines.[36]

The experienced Karel Sluyterman again designed the Netherlands stand, but this time, at the express request of the other committee members, it was in a completely different style. Now the design had to be plain and

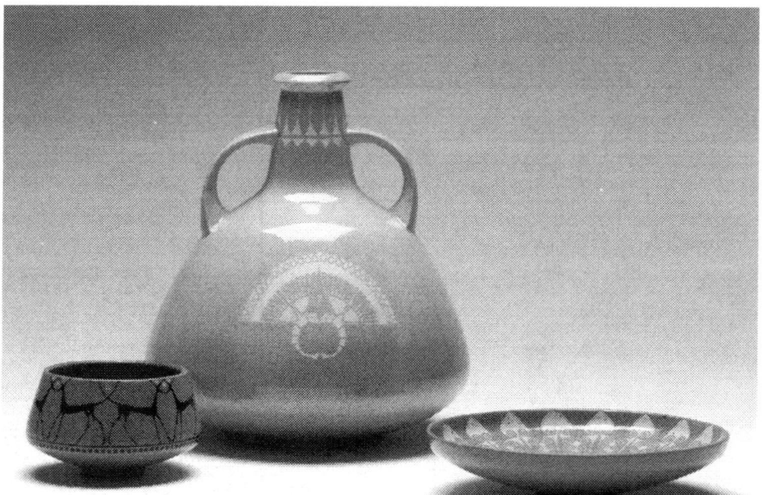

C. J. van der Hoef (Pottery Amstelhoek, Amsterdam), small bowl (1902), vase (1906) and saucer (1900–03).

M. Duco Crop (P. Fentener van Vlissingen & Co. Helmond), hand-printed cotton, c. 1896; photo from *Bouw-en Sierkunst*, 1901.

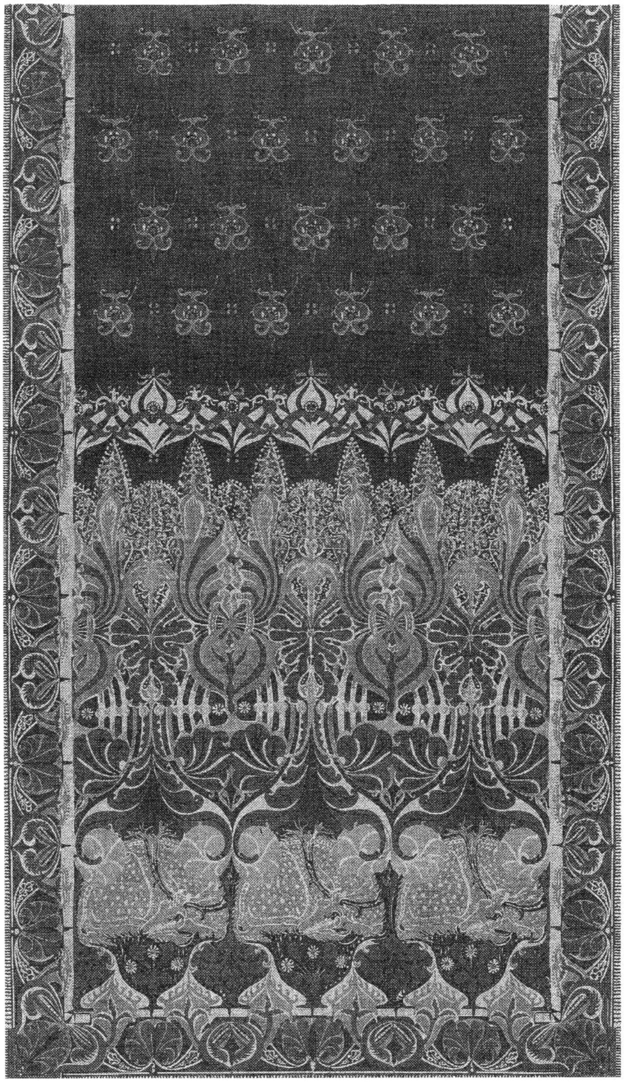

restrained, in keeping with the new image with which the Netherlands hoped to distinguish itself from all the anticipated excess of the other nations' pavilions. The truth is, however, that at the heart of this request there also lay an extremely tight budget. With the help of an exhibition stand made of wooden slats and canvas, an attempt was made to make a virtue out of financial necessity.

Despite the completely different objectives and approach, there were still striking parallels with the Paris exhibition. Most space in Turin was again reserved for the Rozenburg and De Porceleyne Fles ceramics. In view of the character of the exhibition, it comes as no surprise that the Maastricht factories were still unrepresented. Rozenburg once more pulled out all the stops with its eggshell porcelain, much sought after internationally. Whether this work fell within the organizers' objectives is a different matter, since just how 'plain and simple' were these refined decorative objects? Nonetheless, Rozenburg was one of the few manufacturers to do good business in Turin. Amstelhoek, however, a small pottery with its simple vases and cups by Chris van der Hoef and Lambert Zijl, did not go unnoticed, while work by the potter W. C. Brouwer also sold rather well.

The Netherlands' two largest silver manufacturers again participated, although it was obvious that Begeer had not taken the aim to exhibit 'honest' and 'simple' design too seriously. As well as two new, decorative Art Nouveau vases, the Utrecht company again displayed the successful floral decorated service by Gips. The metal workshop of Hoeker & Son submitted work by Jan Eisenloeffel once more, while the traditional working silversmith Frans Zwollo again participated with richly chased, silver decorative objects. Drawn towards Theosophy, Zwollo was now designing according to geometric systems. Additionally inspired by Japan, this led to a highly personal design idiom. The Delft firm of Braat was present with objects of base metal, while H. P. Berlage submitted a brass clock made by Becht and Dijserinck of Amsterdam.

Again, as in Paris, the textile industry was poorly represented, although considering the specific character of the Turin exhibition this is not so surprising. One notable entry in this context, however, was by the Helmond textile printers P. Fentener Van Vlissingen.[37] This firm showed modern cretonnes by the artist Michel Duco Crop, inspired by English Arts and Crafts fabrics. In 1894 when Crop made his first design, Veth had translated into Dutch Walter Crane's *Claims of Decorative Arts*, in which the mechanized printing of cotton was lauded as an inevitable modern development. Thus the Duco Crop-designed curtain fabrics for Van Vlissingen are probably the earliest examples of a fundamental and deliberate collaboration between artist and manufacturer in the Netherlands.

Four furniture companies were each invited to design a complete room for Turin: these were J. B. Hillen, which once again displayed H. P. Berlage's robust designs; Arts and Crafts from The Hague, also present in Paris; plus two recently founded companies, 't Binnenhuis from Amsterdam and Onder de Sint Maarten from Zaltbommel. The new firm of 't Binnenhuis, founded by Berlage in 1900 as a 'cooperative for the sale and design of furniture and other applied arts', was intended as a downright provocation

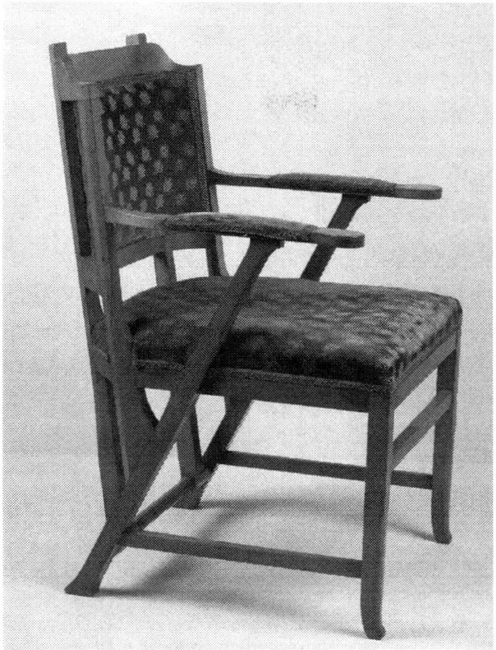

H. P. Berlage ('t Binnenhuis, Amsterdam), cherry-wood chair with moquette upholstery, c. 1900.

Jan Eisenloeffel (Amstelhoek, Amsterdam), copper enamelled tea service, 1900.

to Arts and Crafts.[38] As we have seen, Berlage was among those who had levelled unusually harsh and hostile comments at the Hague gallery. Not only did he loathe the 'affectation' of the products on sale but he also condemned the firm's international and purely commercial basis. In total contrast, 't Binnenhuis was to propagate in a non-commercial way the supposedly 'healthy rational' Dutch principles and thus challenge the 'falseness' of the un-Dutch products of Arts and Crafts.

In setting up his own retail outlet, Berlage had gained the backing of the Amsterdam jeweller W. Hoeker, the book dealer H. Gerlings and the Hague financier Carel Henny. Through their new company furniture by Berlage, as well as by such designers as Willem Penaat and Jac. van den Bosch, could be purchased or made to order. Berlage made it known he wanted to design everyday furniture for 't Binnenhuis, products that were affordable for ordinary people. His furniture makers were allowed to use only straight pieces of native Dutch oak and Berlage asked them to leave the joints clearly visible, even occasionally giving them decorative accents. Other decoration was applied sparsely in shallow relief or with contrasting wood inlay in the flat parts of the objects. Berlage believed fittings had to be sturdy and clearly emphasize their specific function. He looked to early seventeenth-century, Old Dutch (*oud-Hollands*) furniture design as his chief source of

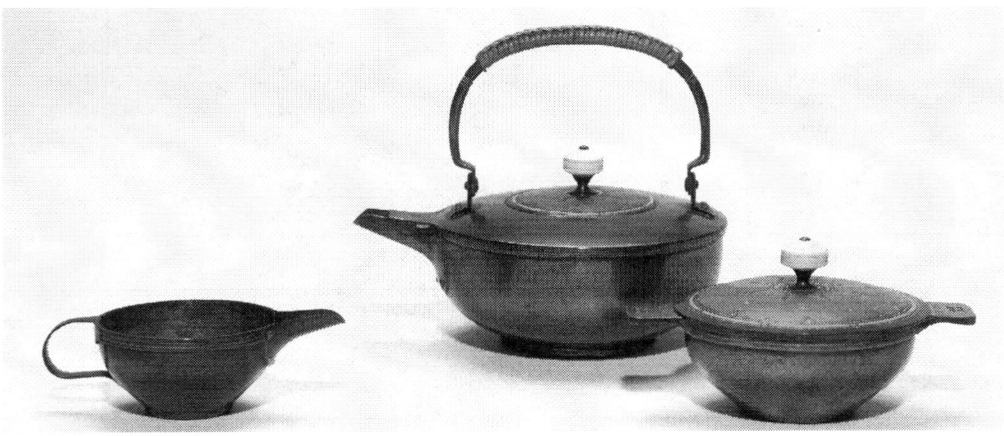

inspiration. Honesty of materials, simplicity and 'rationality' were no doubt the most important starting-points of the Golden Age.

The vases, pots and dishes by Van der Hoef sold by 't Binnenhuis were traditionally made at Amstelhoek from native types of clay. They were decorated with simple, flat, inlaid clay motifs in a contrasting colour or with traditional *ringeloor* or slip decoration. Jan Eisenloeffel's silver and copper services were created from plain, geometric shapes firmly secured to each other with rivets. The finished result was then decorated with unfussy lines, simple openwork patterns or with inlaid decorative motifs in enamel. Frans Zwollo's metal objects were traditionally embossed and chiselled with stylized natural motifs.

Within a few months it was already clear that these designs, mainly traditionally handcrafted to lofty principles, were in practice far too expensive for a wider public. Moreover, the cooperative principles on which 't Binnenhuis was based had proved unworkable. Most of the firm's affiliated artists turned away from Berlage, blaming him for putting his own commercial interests above those of the company. In fact, during the Turin exhibition, apart from Berlage himself, only the furniture designer Jac. van den Bosch was still officially attached to the company.[39]

Since the Paris exhibition much had also changed at Arts and Crafts. The designer Johan Thorn Prikker had left and in Turin the gallery displayed a striking interior by Chris Wegerif, who originally was responsible only for financing the company. This self-taught designer had combined elements of

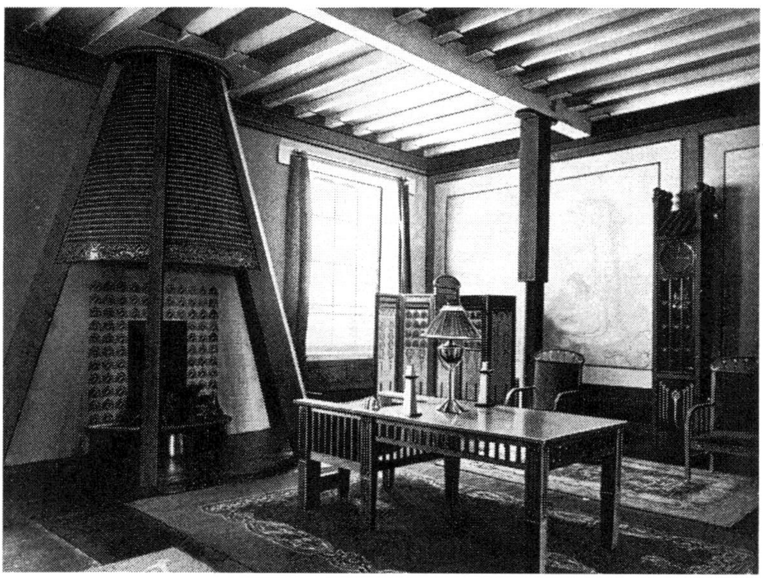

C. Wegerif, hall at the International Exhibition of Decorative Arts in Turin, 1902.

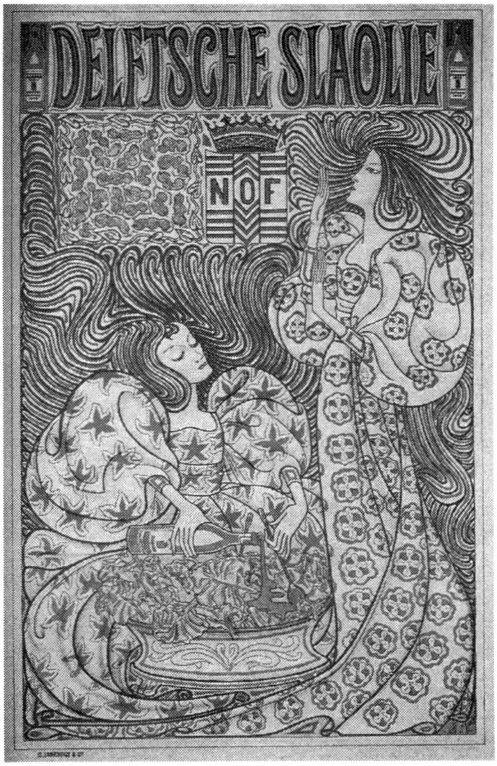

Jan Toorop, poster advertising *Delft Salad Oil*, 1895.

modern English and Austrian furniture design in a delightful and eclectic manner. These costly, very un-Dutch designs were decorated with linear motifs of contrasting and expensive inlaid woods. Wegerif completely ignored the dogmatic Amsterdam designers who feared commercialism – besides, the foreign press was quite taken with his designs.

The fourth complete interior was submitted by Onder de Sint Maarten, a workshop and selling outlet in Zaltbommel, set up just the previous year, that sold furniture and copper work in a style clearly inspired by Berlage and his Binnenhuis.[40] Remarkably, the simple furniture designs were by Karel Sluyterman, a designer who had recently forsaken his earlier, effusive Art Nouveau style.

Van Wisselingh did not have a complete interior but submitted individual furniture designs. Gerrit Dijsselhof was represented by, among others, a dividing screen that was prized by the reviewer of *The Studio* as one of the most noteworthy of the Dutch exhibits: 'the grand polyptych, with several panels . . . on which are represented various animals, such as roe deer, peacocks, cranes, storks, fish etc, admirably drawn by Mr Dysselhoff, printed by the Batik process, and finished off with remarkably clever silk embroideries by Mme Dysselhoff'.[41] Carel Lion Cachet had a very expensive chair and tea-table inlaid with ivory and ebony, with the seat and back of the chair covered in costly batik parchment. Karel de Bazel showed some individual items of furniture that had been designed for 't Binnenhuis. It is likely that the conflict over policy at 't Binnenhuis had led to De Bazel not showing his designs in Turin under that firm's name.

As well as entries from companies, much work by individual artists could be seen in Turin. Committee member Philip Zilcken had made sure that entries were also received from graphic designers. Thus Theo van Hoytema's exquisite lithographs for the children's book *Uilengeluk* (Owls' Fortune) could be admired, as well as a lithographed calendar by Theo Nieuwenhuis, posters by Jan Toorop and book covers by Johan Thorn Prikker, Antoon Derkinderen and Chris Lebeau. The last submitted a beautiful batik velvet copy of the novel *De Stille Kracht* (The Hidden Force) by

Louis Couperus. One of Toorop's posters submitted was an advertisement for the Calvé Oil Factory, Delft, executed in delicate colours in a curvilinear style and featuring two women with long hair and elaborate garments pouring oil into a large bowl of salad. After this poster appeared in 1895, Art Nouveau in the Netherlands was often mockingly dubbed 'the salad-oil style' and the entry itself is not exactly an example of the organizers' declared desire for 'honest' Dutch design.

Many of the artists participating in Turin had begun their careers as painters. Thorn Prikker and Toorop, for instance, turned to the decorative arts only during the 1890s. This was more than simply a shift in artistic direction: the artists had undergone a development that had consciously led them to want to use their artistic talent for the benefit of the community.[42] In this they had been inspired by the romantic ideals of the Gothic Revival and the social ideas of the English reformist movement. Specifically, they had come into contact with the social and political ideals of John Ruskin and William Morris via Henry van de Velde and other Belgian artists.

Gerrit Dijsselhof had taken a similar route. After some years at the Academy of Fine Art, The Hague, he enrolled in 1884 as a pupil at the Amsterdam Rijksnormaalschool, which, like the National School of Applied Arts, was located in the Rijksmuseum. Here he trained as an art instructor. It was quickly apparent that the decorative arts appealed more to Dijsselhof. The classes of Pierre Cuypers and Jacob de Kruyff inspired him to study medieval and Eastern art and ornament, and the romantic image of the Middle Ages he then acquired formed the basis for his high ideals regarding the artist's duty to society. This had been an era when artists and artisans still worked with great conviction on major joint projects, peacefully and in an environment untainted by commercialism. Thus, Dijsselhof saw his exciting watercolours of fish, with which he made his debut in 1891, more as decorative applications for a wall than as autonomous art works. Three years later, in 1894, Dijsselhof devised the illustrations and the exceptional cover for Veth's translation of Walter Crane's *Claims of Decorative Arts*. Shortly after its publication, Dijsselhof was commissioned to decorate a room in a doctor's house in Amsterdam. For this he produced batik and embroidered wall panels depicting stylized birds and deer, wood panelling and doors with highly original flat-relief carving, as well as the room's furniture. After various diversions and modifications, this unique interior was finally installed as the 'Dijsselhof Room' at the Gemeentemuseum in The Hague in 1935. It is the earliest Dutch example of an interior in the 'modern' style in which all the components are in keeping with one another.

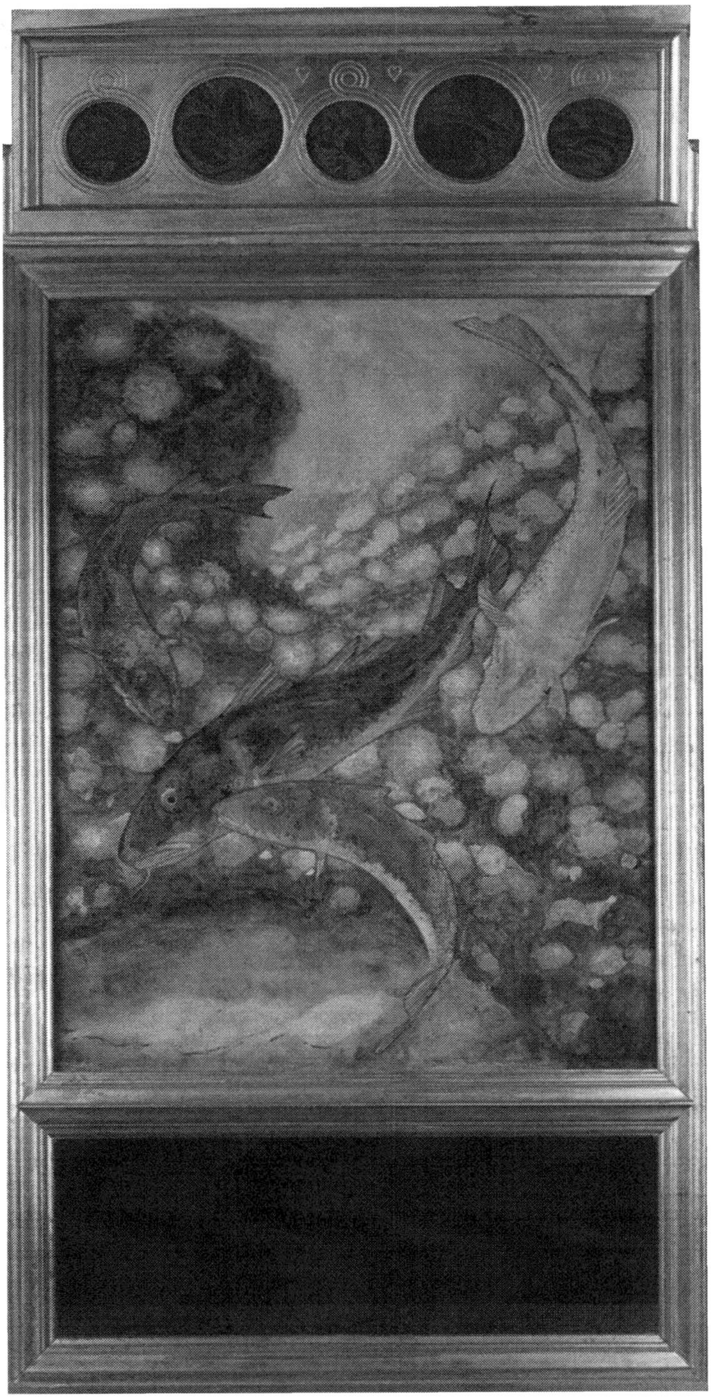

G. T. Dijsselhof, chimney piece with large watercolour, five cut and painted wooden panels and mirror, 1892.

Admired by many of his contemporaries as much for his immense literary and historic knowledge as for his serious and idealist attitude, Dijsselhof was responsible for a shift in thought among several fellow artists. His decision to make completely handcrafted products as a matter of principle also changed their manner of working. They too were now convinced that their products should not only be available for a rich elite. Everyone should be able to experience the 'purifying' influence of the 'art of everyday life'. Simple domestic objects did not really have to be decorated with expensive ornament. The artistic aspect and the – just as important – personality of the artist were also recognizable in the simple painted motifs of a vase, in the ordinary carved ornament of a wooden cupboard and even in the hammered surface of a copper dish. Thus, mass-produced industrial products, which had none of these attributes, were loathed by them as being cold, impersonal and 'art-less' objects.

Break or Continuity: Berlage and the Forming of an Image

Was the Dutch entry in Turin successful? Did it conform to what the organizers wanted and did the new Dutch image come across sufficiently? The reviewer of the Dutch section in *The Studio* (see above) observed that the Dutch were more hostile to the naturalistic decorations than any other people and stated: 'With very view exceptions, a pronounced tendency will everywhere be found for geometrical forms, combined with certain decorative elements culled from the barbaric art of the savage races of the remote East.'[43] The Netherlands press itself was very happy overall. In a detailed account of the exhibition in the monthly current affairs magazine *Elsevier's Maandschrift*, one critic wrote: 'when the department was finished it was exactly as it should have been, plain and understated. No screaming nonsense with shrill colours and whimsical lines, but unpretentious and uncompromising, with warm tones in calm rooms.'[44] In the foreign press, however, there was little mention of this particular Dutch quality – instead, the more opulent art objects were admired. The sales accounts showed that foreign visitors were just as interested in Rozenburg eggshell porcelain and Chris Wegerif's designs for Arts and Crafts as in Berlage's plain and robust furniture, Amstelhoek's and Willem Brouwer's 'archaic' pottery or Jan Eisenloeffel's simple tea and coffee services.

Long after the Turin exhibition, the Netherlands continued to cherish the image of a successful reversal in the applied arts in favour of a more simple and restrained Dutch New Art (*Nieuwe Kunst*). What is more, this has been carried over into design literature to this very day. This is probably related to an all too easy and barely critical analysis of the contemporary debate about design of that era. Only recently have people begun to realize how art historians have largely allowed themselves to be led by the prejudicial way in which Dutch critics discussed *Nieuwe Kunst* at the time. It is now becoming evident just how much the image forming during this period was manipulated by its major theorist and spokesperson, H. P. Berlage. Just how 'new' the supposed *Nieuwe Kunst* was is also open to debate. Is it not more appropriate to describe this movement, with its nostalgia for old crafts and medieval ideals, as 'old' or at least 'old fashioned'?

H. P. Berlage was trained according to the classic principles of Gottfried Semper.[45] He was even one of the few Dutch students to attend the Polytechnikum in Zürich (from 1875 to 1878), where Semper himself taught until just before Berlage's arrival. Consequently, his earliest designs in the 1880s are distinguished by an abundance of Renaissance motifs. In the 1890s Berlage's views and style evolved slowly, partly under the impact of Pierre Cuypers and the writings of Eugène-Emmanuel Viollet-le-Duc, and later the ideas of the English Arts and Crafts artists. Thus the designer became increasingly convinced of the importance of the constructive principles of Gothic architecture and furniture from the Dutch Golden Age.

H. P. Berlage, boudoir of the Villa Henny, The Hague, c. 1900.

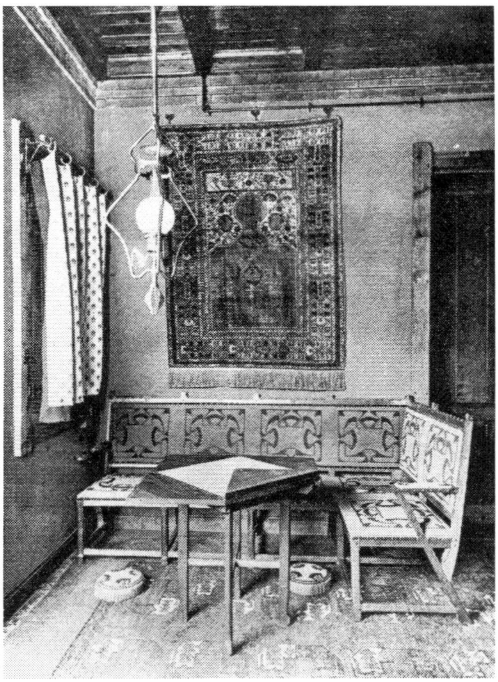

From 1898 it was possible to see the concrete result of Berlage's desire for greater simplicity in design, linked to his growing abhorrence of 'useless' detailing, rising skywards in the form of his uncompromising, linear and sparsely decorated Koopmansbeurs (Stock Exchange) in Amsterdam. Through his involvement in this acclaimed project of all manner of artists, who provided sculptures, ceramic panels, paintings and even inscriptions for the exterior and interior, Berlage promoted a new form of *Gemeenschapskunst* (community art).[46]

During the same period Berlage designed two residences in which his ideas on interior and furniture design were expressed in virtually the same demonstrative manner. The first house, the 'Henny Villa' in The Hague, was for the banker Carel Henny, who later financed 't Binnenhuis, and his family. The other house, 'Parkwijck', was built in Amsterdam for Leo Simons, the idealistic publisher of the *Werldbibliotheek* (World Library).[47] Both houses were spectacular due to their unconventional asymmetric design and the functional placing of the windows. The Henny Villa was remarkable for its yellow roof, while Parkwijck had a striking chimney of unusual design. Berlage's new and rational design principles were much more to the fore in the interiors of both houses. Thus the living rooms epitomized the requisite 'honesty of materials' with their exposed brick walls and furniture designed in an uncompromising and Spartan style. We do not know what Leo Simons and his wife thought of such an interior, but Carel Henny's wife and children did not find their new home that comfortable or 'cosy', even though Mrs Henny was 'allowed' by Berlage to plaster the walls in her own 'boudoir' and to furnish it with a settee, some small foot-cushions and a Persian carpet on the floor, and even to hang one on the wall.

Berlage regularly set out his views in great detail in print. They first appeared in book form in 1904 in *Over stijl in bouw- en meubelkunst* ('On Style in Architecture and Furniture Design') and some twenty other publications followed. All his published works have undoubtedly contributed to Berlage being considered the leading figure of the reformist movement in Dutch design. Time and again this version of design history has been affirmed: 'The guardian of new Dutch architecture and applied art; the founding father of new construction; the "synthetician" who unravelled and filtered the past in his own powerful spirit and brought it together in a new unity', was how, in 1926, the designer Harm Ellens linked the supposed revival of Dutch applied arts primarily to Berlage.[48] In 1929 the writer Jo de Jong gave Berlage a place of honour in her survey *De nieuwe richting in de kunstnijverheid in Nederland* (The New Direction of Applied Art in the Netherlands). Looking back at the turn of the century, she wrote:

> At this time, while all design outside the Netherlands, be it furniture or book covers, buildings or chandeliers, is overrun with the eternal coiling, whiplash lines of Art Nouveau or Van de Velde style, Berlage exposes bare materials and honest construction and puts forward functionality and simplicity as the first requirements of a domestic object.[49]

It was only in the late twentieth century that this 'personalized' historiography based on a deliberately constructed image was put into perspective. In particular, Berlage's image as the great Messiah who revealed and perfected the process of design reform in the Netherlands begun by Pierre Cuypers was gradually laid to rest.[50] Art historians have often hardly looked beyond the writings by Berlage himself, or those of his kindred spirits and devotees, just as the various biased discussions about the Turin exhibition were followed in a relatively uncritical manner. So again the question arises: how innovative was Dutch decorative design in Turin? Was there really a fundamental distinction between the entries in Paris and Turin, as contemporaries would like us to believe? Was it a question of a breakthrough around 1900, or rather one of continuity?

We have established that in a certain sense 1900 represents more the apex of a development that had already been under way for a few decades, in which a rethinking of artisanship and the democratization of the applied arts was pivotal. Thus 1900 chiefly marks the beginning of a period in which the design debate was led more by artists, architects and skilled artisans and less by industrialists, technicians and consumers. It was an age in which the artist-craftsman was central and in which, for the time being at least, there was absolutely no sign of a new, twentieth-century industrial design style. In other words, while industrial products were undoubtedly being produced, even at a steadily increasing rate, for the moment their design was hardly a theme for serious consideration in the worlds of art and architecture.

The Society for the Elevation of Craftsmanship

It was not just the romantics and applied artists drawn to medieval ideals who wanted a return to the values of the Dutch crafts tradition. The decline of small workshops and the disappearance of crafts people was also lamented by other groups in Dutch society, and the government itself had begun to see it as a major social and economic problem.

For these reasons the Society for the Elevation of Craftsmanship (Vereeniging tot Vereedeling van het Ambacht, VVA) was founded in 1897 on the initiative of Arti et Industriae.[51] The ever active Boersma had been appointed chairman of Arti et Industriae in 1890, the first Dutch society aimed at bringing art, industry and architecture closer together. Initially founded as a local Hague organization in 1884, it became a national society after Boersma joined the board. He saw several basic characteristics of the traditional Dutch work ethic united in the artisan, namely a sense of responsibility, versatility and autonomy. New, straightforward, mechanized devices were welcome as far as he was concerned, and even essential if work done by hand was not to degenerate into a mind-numbing competitor of industry. The Netherlands Society for Trade and Industry, at that time the major society for manufacturers, trade representatives, engineers, lawyers and politicians, supported the new society both financially and by being represented by certain of its leading members. Generally it was viewed as socially important that crafts people should continue to exist as a kind of buffer between the ruling class and the steadily growing proletariat. Thus, woodworkers, ornamental painters, potters, Delftware painters and silversmiths were increasingly regarded as a typical and indispensable group within Dutch society. On the basis of these social considerations alone, it was felt that crafts should be protected and cherished.

The VVA came up with a plan in which workers could take a master examination similar to the medieval guild system. This was fully supported by various established designers, including Karel de Bazel, Antoon Derkinderen and Pierre Cuypers. Between 1900 and 1907 De Bazel, as a member of the VVA technical committee, often drew up the designated assignments for the master examination for furniture makers, including 'a mirror frame from mahogany with inlay work', or 'an armchair with a curved back'.

The system was in place for only a few years, during which about one hundred craftsmen a year took the master examination. Not everyone was enamoured with the idea by any means. The artist Richard Roland Holst, for example, expressed his criticism in the socialist magazine *De Kroniek*, and its editor J. F. Ankersmit closed ranks behind him. Both felt that it ultimately made little difference whether you had machine or handcrafted production. What mattered most was improving the lot of workers. In the end the inevitable modernization of industrialized production meant that the VVA's idealistic plan never came to anything.

The Founding of the VANK

Developments in design education and the attendant emancipation of the artist-craftsman led shortly after 1900 to the establishing of their own professional body. Designers felt increasingly less at home in painters' societies or architectural associations, to which they had often belonged until then. So in 1904 the Association for Crafts and Industrial Art (Vereniging voor Ambachts- en Nijverheidskunst, VANK) was founded, the first professional body for designers in the Netherlands.[52] Most of the artists and designers who joined the association were those who carried out work in their own studio, or had their designs made up in small workshops where they usually had direct control of the production process. Pattern designers in carpet factories or cotton print-works, Delftware painters in pottery companies and cabinetmakers in furniture factories were not as yet members.

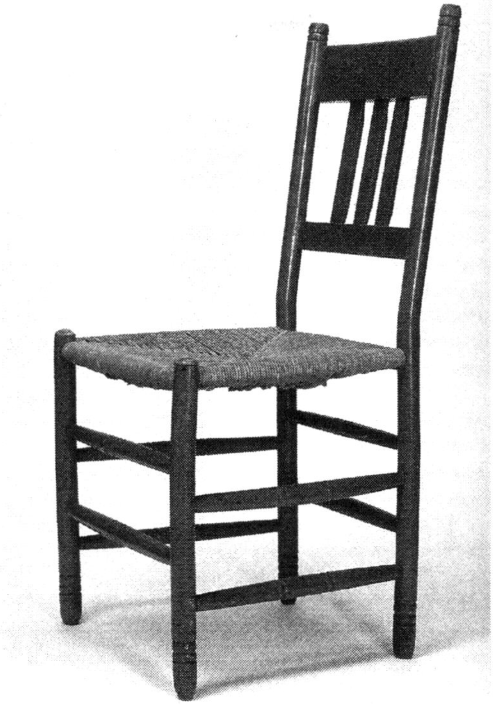

Chairman Klaas van Leeuwen had originally had the ambition to become a painter, but towards the end of the nineteenth century he met Mathieu Lauweriks and Karel de Bazel and was struck by their reformist views on design.[53] Along with De Bazel and Oosschot, Van Leeuwen set up the small-scale furniture workshop De Ploeg in 1904, while also teaching at various applied arts schools. In 1910, however, tired of the many conflicts with his colleagues and disillusioned by the scant return on all his efforts, Van Leeuwen turned his back on applied arts and began painting again. Other founding members included Jac. van den Bosch, who was also assistant manager of 't Binnenhuis, Amsterdam;[54] Chris Lebeau, who, with Jan Eisenloeffel, had founded De Woning, a production collective and selling outlet as well as an offshoot of 't Binnenhuis;[55] typographer Sjoerd de Roos, who would become the leading type designer of the first half of the twentieth century in the Netherlands and whose Hollandsche Mediaeval of 1912 was the first complete Dutch font;[56] and Herman Hana and R.W.P. de Vries, who were both decorative artists but ultimately wrote about and lent critical support to the ideals of the new applied arts.[57]

Willem Penaat (De Woning, Amsterdam), 'Farmer's Chair', 1899.

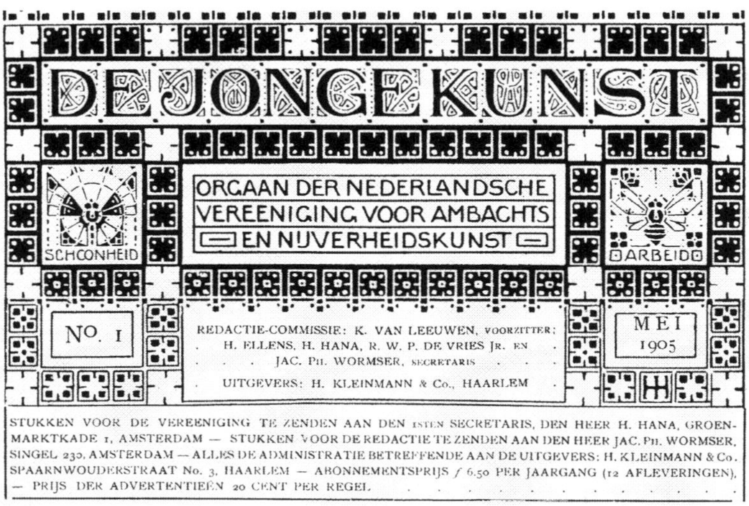

Herman Hana, frontispiece of *De Jonge Kunst*, magazine of the VANK, 1905.

Finally, there was the industrious Willem Penaat, who trained as a design teacher and became head of Amstelhoek's furniture workshop in 1900, becoming involved with 't Binnenhuis in this capacity, and then joined De Woning as a co-worker.[58] He was one of the few designers who, with his version of the traditional Culemburg peasant chair, took seriously the aim of making affordable, well-designed, machine-produced furniture. Within the VANK, after a few turbulent early years of internal friction, Penaat's level-headed and decisive chairmanship managed to bring the conflicting views and totally different personalities of its members into line. He also did useful work within VANK as a member of the Committee for Artistic Ownership and was involved in settling issues relating to plagiarism and design protection. His efforts in this led to the groundbreaking Copyright Act of 1912.

The ideals of the founders of the VANK were expressed in *De Kroniek* by the socialist journalist and politician P. L. Tak.[59] He described the group of designers as 'artists' and 'forerunners' who, in tandem and solidarity with the socialist movement, proclaimed the 'dawning of a new age'. With their striving for 'truth, honesty and realism' in their designs, they rejected the 'spiritless historic styles' that, according to Tak, no longer belonged to the modern age. He also mentioned the success of Berlage's Beurs and, allied to this, the 'clay pots, brass and simple furniture' he had no doubt seen in 't Binnenhuis. To him these were products with a logical construction, meaningful lineation and harmonious dimensions. He predicted that there would be many problems in putting across these new design principles since the wider public was not yet ready for them. He also admitted that it was impossible to make good, simple designs for people on a tight budget. Thus Tak gave a political dimension to the new movement in applied arts – he recognized a patently obvious resistance to capitalism. For these reasons, it was a matter of conscience for the artist-designer whether to use machines or not; by doing so one ran the risk of becoming a 'capitalist' manager of a factory.

Less politically charged, but just as idealistic and impassioned, were the words of Pierre Cuypers at the opening of the VANK's first national exhibition at the Stedelijk Museum, Amsterdam, in 1911. The elevation of craftsmanship and industry by bracketing them together with art should be the aim of the society, he believed. Every product could be an art work providing it was 'good, true and beautiful': 'good' if it could be used for the purpose for which it was made; 'true' if the design properly expressed this purpose, and 'beautiful' if pleasing to the eye.[60]

While most of the industrial artists who joined the VANK in 1904 still worked according to traditional methods, the organization's two-part name was not entirely misleading. The need to create better conditions for working with industry was certainly a major issue from the start. As early as March 1905 the board member Herman Hana had broached this important subject, feeling that his fellow board member Klaas van Leeuwen was far too negative about machines. Hana regarded the machine expressly as a tool of the modern age. He argued that 'machinalism', as he described the new machine-based aestheticism, should also have repercussions for decoration: an ornament machine, based on a kaleidoscope and a projector, was the result of this idea. In 1910 Hana invested all his money in the first Dutch house completely made from concrete, an experiment he carried out with a cousin from the United States and with aesthetic advice from Berlage. Although the project was a success, it was a financial disaster.[61]

The VANK's objectives included serving the interests of its members and the professionalization of the field. Repeatedly confronted with the cheap imitations of their designs by more mechanized firms, the struggle that eventually led to the establishment of the Copyright Act in 1912 was considered a success. Among its other duties the VANK also sought to draw up better regulations for competitions and improve design education through the publication of a trade journal, yearbooks, lectures and exhibitions, and by promoting its views on other social issues with one voice.

The backgrounds and ideals of its members, however, were to remain divergent for as long as the organization existed. Not all members were as politically aware and not all shared to the same extent the romantic, social idealism projected by Tak or Cuypers. Alongside members who regarded the VANK as part of the socialist democratic movement were others who saw it primarily as a modern trade union to serve their interests. Yet membership always remained low: in the first year this was 85 and never rose above 300. The first design exhibition in the Stedelijk Museum, which ran for six weeks in 1911, attracted only 3,500 visitors; a sharp contrast to the tens of thousands of people drawn to the major industrial exhibitions in the second half of the nineteenth century.

The possibly exaggerated image of the importance of the VANK and the radical changes it supposedly brought about is partly due to *De nieuwe richting in de kunstnijverheid*, written by the textile artist Jo de Jong to mark the association's twenty-fifth anniversary in 1929. As she explains in the Preface, the book was intended as 'a guide for teaching modern applied arts'. It was apparently felt necessary to provide future designers with a solid historical and ideological basis by giving them an overview of twenty-five years of 'crafts and industrial art'. In her zeal to give the developments more weight, she sketches a wide chasm between the nineteenth and twentieth centuries. In this respect the difference between this and the *Arti et*

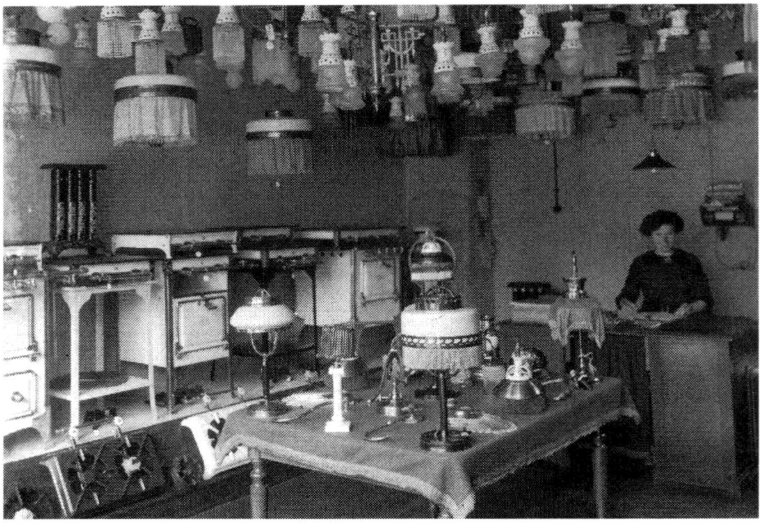

Interior of a shop for lighting and electric domestic articles, 1913.

Industriae jubilee book, written in 1909 by Karel Sluyterman to mark its twenty-fifth anniversary, is striking. Sluyterman more realistically recognized design in the early twentieth century as a logical continuation of developments that had already begun in the mid-nineteenth century: 'When, almost at the same time as in the other European nations, after the London Exhibition of 1851, one cheerfully looked for effective means to improve the industrial arts, even in our country the first signs of a revival were revealed.'[62]

The VANK certainly set the tone for the design debate in the early twentieth century, since it was its colourful members who made their presence felt, who taught at the schools of applied arts, who exhibited their work, were written about and often enthusiastically put pen to paper themselves. However, the hundreds of draughtsman in the burgeoning industry, the engineers with new ideas and the many foreign designers whose designs were purchased by Dutch companies formed a much bigger group. In one sense they were the actual precursors of the later industrial designers. Advertisements had become increasingly important since the turn of the century and determined the streetscape. The first Dutch cars were seen on the road and the number of bicycles increased exponentially. An increasing variety of kitchen and household goods could be bought that were partly from Dutch manufacturers and at some point had been 'conceived' by someone. Then the first electric ovens, vacuum cleaners, irons and heaters began to appear on the market. These too had been designed. The mushrooming chain stores were stuffed with tempting fashionable gadgets, which at a time

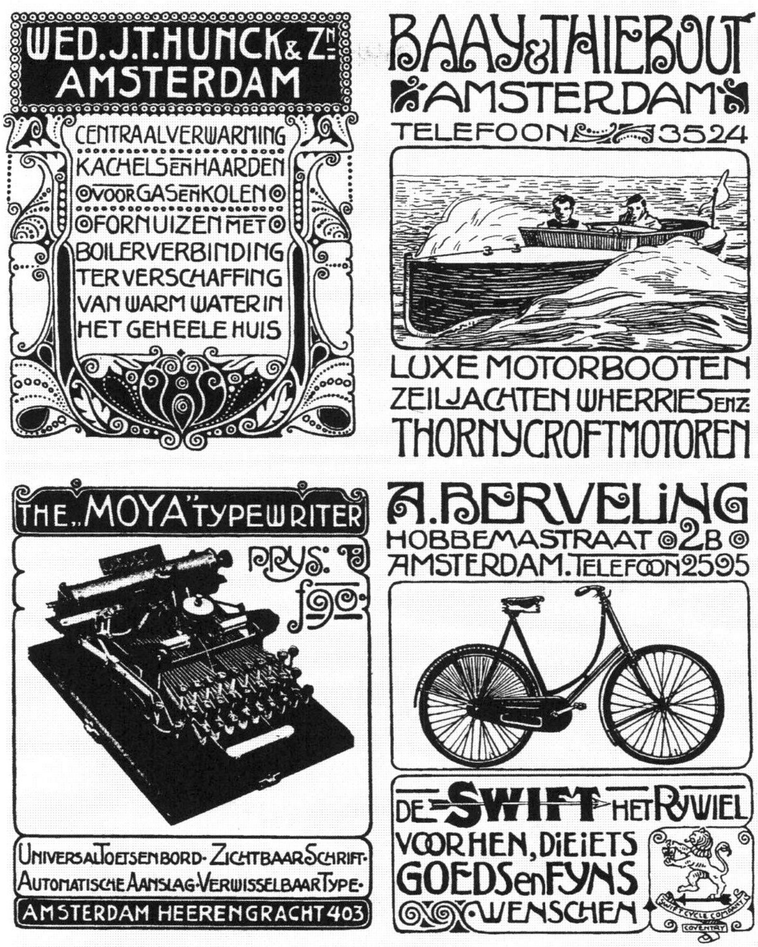

André Vlaanderen, page of advertisements in *Het Huis Oud en Nieuw*, 1905.

of increasing prosperity were eagerly snapped up by many shoppers.[63] All this determined the image of the Netherlands at the turn of the twentieth century far more than the hand-painted tea services, the beaten ashtrays and batik tea cosies of the VANK members.

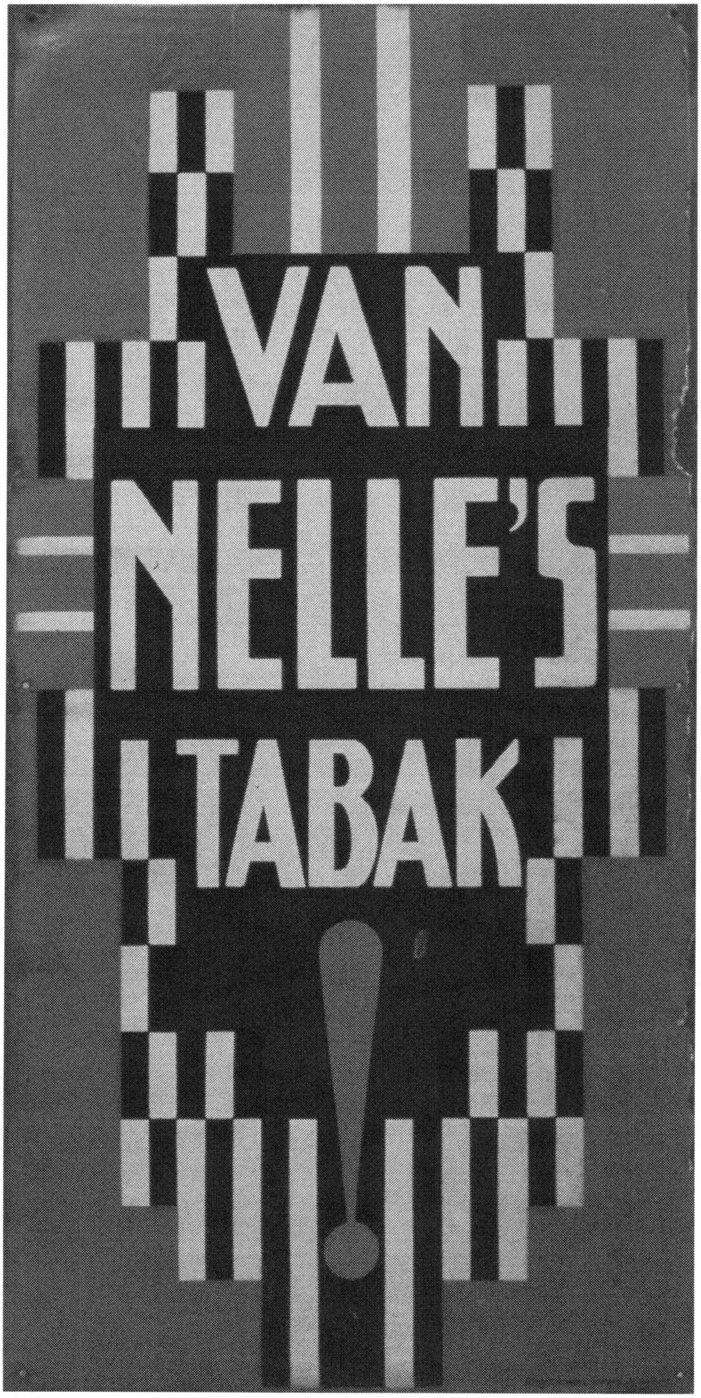

Jacob Jongert (NV De Vereenigde Blikfabrieken, Amsterdam), enamelled advertisement plate for Van Nelle's tobacco, 1925.

2

Design as Art, 1915–40

During the first half of the twentieth century most people in the Netherlands thought everyday objects, whether handcrafted or machine-made, were worth consideration only if they could be called 'art products' as well.[1] This idea was not specific to the Netherlands. When the German Work Federation (Deutscher Werkbund) was founded in Munich in 1907, people speculated on how far the artist's influence should be allowed to extend. Should the shape of an object be determined by the designer's individual artistic insight or should its appearance be dictated by factory manufacturing processes and economic considerations? These questions were also pivotal in the confrontation that took place in 1914 at the large *Werkbund* exhibition in Cologne, where Henry van de Velde, who championed the cause of the artist, was taken to task by Hermann Muthesius, who thought that technical and economic considerations should win the day. After this the members of the Deutscher Werkbund who argued for more industrial influence soon gained the upper hand.

In the Netherlands this ideological debate dragged on for much longer because the country had nothing like a fully developed consumer-goods industry, nor did the idealistic prospect of amalgamating art and design suffer the brutal disruption caused by the horrors of the First World War, in which the Netherlands remained neutral. Tens of thousands fled into the country from Belgium. The difficult situation challenged entrepreneurs to find new business partners and even new products.[2] Dutch designers could dream on in peace, which led to many continuing to focus on the 'arts' component as an essential factor in making and judging products until well into the 1930s.

Eisenloeffel versus Zwart

Just how far removed views on the role of art in design were from each other is evident from an historic public confrontation on the issue between two leading Dutch designers. In 1929, during the twenty-fifth jubilee celebrations of the VANK, Jan Eisenloeffel ardently defended the interests of art while Piet Zwart proved to be a hardened supporter of a more matter-of-fact and industrial approach.[3]

Jan Eisenloeffel, a metalwork artist, had a very unusual career before 1929. At the start of the century he had been one of the first designers to enlist the aid of machines to make simple domestic objects that everyone could afford. At the time, his uncompromisingly designed brass and silver tea sets, inspired by Japanese craftsmen, were already being manufactured on a small scale by Amstelhoek, using a spinning lathe. They were first sold by 't Binnenhuis, and later by De Woning.

The few months that Eisenloeffel spent in Munich in 1908, working for the Vereinigte Werkstätte as head of its metal department, made him change his ideas completely. There he was introduced to working methods in a large, rigidly organized firm, with a strict division of labour for each stage of the production process, and came to realize what cooperating with industry really entailed: handing in drawings and designs, complying with the limitations of existing machinery and an almost total absence of individual freedom to experiment. When he also faced a few cases of alleged plagiarism of his characteristic sober tea services in the Netherlands, Eisenloeffel decided to return and settled in the small village of Blaricum, where he devoted himself to making artistic decorative objects at his own studio: expensive, luxury items ordered exclusively by wealthy clients. The richly decorated lamps he made in the 1920s for the Rotterdam shipowner A. W. Goudriaan are outstanding examples of this type of work.[4]

Eisenloeffel considered these as nothing less than true works of art. At the jubilee gathering of the VANK in 1929 he spoke convincingly about 'artistic vision' and 'the enthusiasm and emotion' experienced by the artist designing products. He had captured this mental image as it came into his head by 'scribbling it down feverishly' onto a sheet of paper. He even went as far as denying the existence of industrial art. He was by then convinced that one had to choose between industry and art. A combination of the two was impossible: an 'industrial form', after all, is a machine-made product created in the same way as a pair of scissors, an oar or a racing car. The task of the designer working in industry, whom he thought should not be described as an artist, was 'to design forms from which ideally functioning domestic objects could be made in the shortest possible space of time, with a minimum of materials and with the lowest consumption of manpower and mechanized energy'. He felt that the true artist should have nothing to do with this type of manufacture, since it was 'corrupted by competition and cost-price considerations', but should turn his attention instead to the world of art, where he would have far more chance of finding his true self and where he could become 'a real person'; after this revelation he would be able to use his hands ('those miraculous tools') to testify to his new identity.

Piet Zwart represented the diametrically opposed viewpoint, undoubtedly influenced by the functionalist architects he knew so well, the ideas spread by the German Bauhaus (where he spent some weeks teaching in 1929) and by international avant-garde typographers.[5] He professed to pin his hopes on technology and industry and attached little importance to art. In his speech Zwart first wiped the floor with the supporters of the dated and naive Arts and Crafts ideal, many of whom were VANK members themselves, and then dismissed the semi-modern enthusiasts, whom he scathingly referred to as 'maiden aunts, and a few acquaintances living in the leafy suburbs', for still admiring a sort of 'machine romanticism'. For Zwart true technique was set to enrich the world with new materials and new methods of work that would force people to be inventive. This was a process that man could only direct, because 'the primal rhythm of human evolution' was inexorable. Zwart did not see design as a question of art or taste but rather as an expression of the designer's attitude to life. Above all else it was the artist's duty to identify with developments in modern society.

His own working methods are the best possible illustration of this standpoint. During this period Zwart produced his first revolutionary typographical products for the Dutch Cable Factory (Nederlandse Kabelfabriek) in Delft. These were commercial catalogues made by the most modern methods in which dynamic photography was combined with a totally new form of typography. The inspiration for his new idiom came partly from the Dada experiments of Theo van Doesburg, the German artist Kurt Schwitters, the Russian artist El Lissitzky and the German designer Jan Tschichold.[6] In 1924 Zwart and Berlage had together designed the now famous yellow, pressed-glass breakfast set for the Leerdam glassworks. While it was a standardized product that could be mass produced, its design was not very functional, because the handles on the heavy tea set had no opening for the finger. Zwart designed a set of glasses that were easier to use and even more suited to industrial production, called 'Anova'. These were made in 1928 for the Crystal Association Ltd (Kristalunie) in Maastricht. Less well known, but no

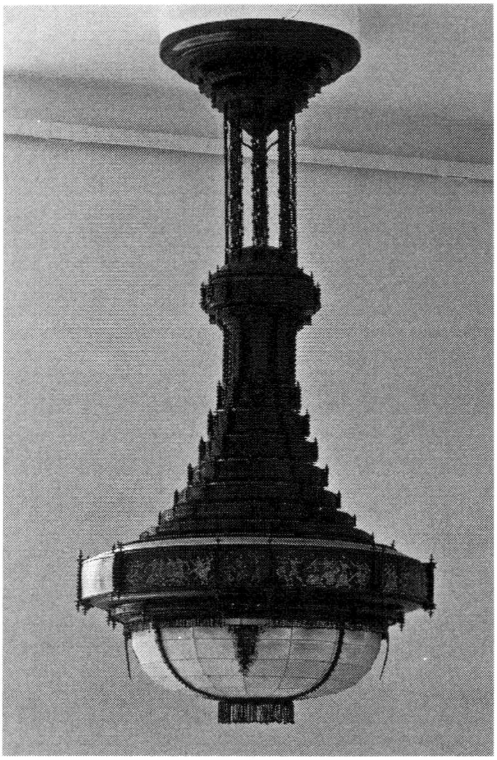

Jan Eisenloeffel, pendant lamp, tombac and translucent enamel, for A.J.M. Goudriaan, 1922–4.

Piet Zwart, 'tempo 2' cigarette packet, 1932–4.

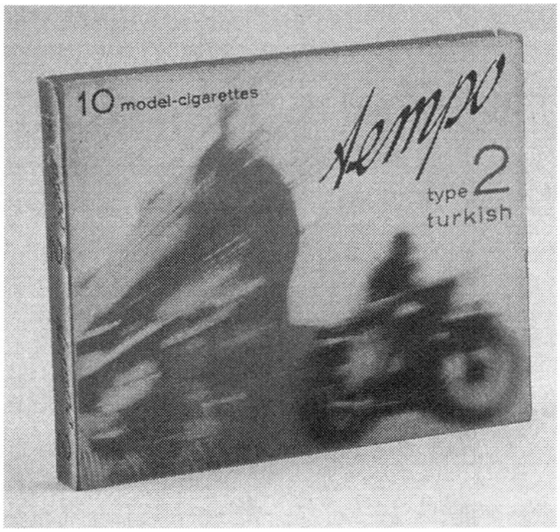

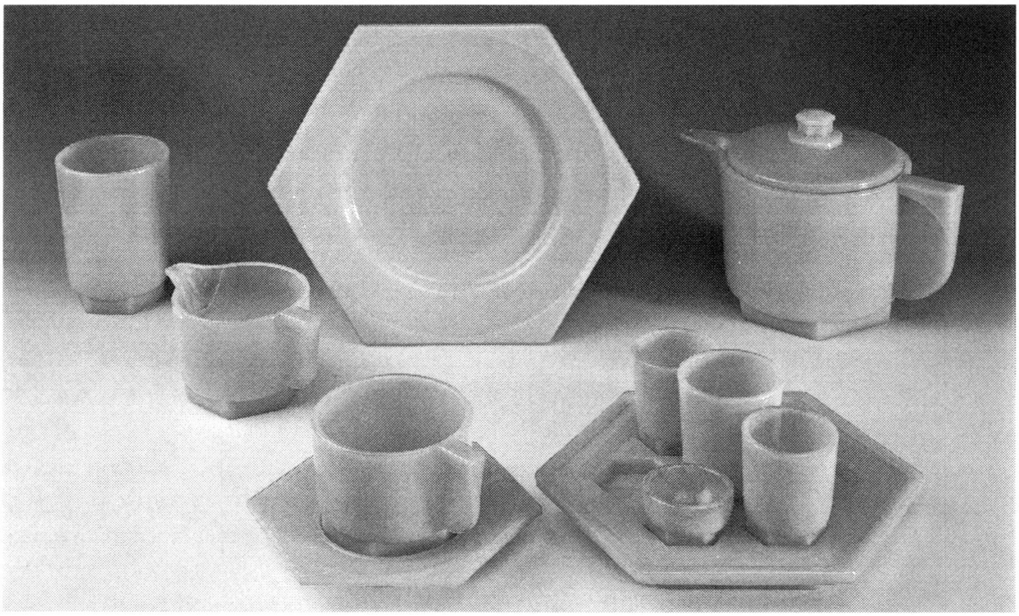

less illustrative of his ideas, are the simple domestic objects made from an entirely new synthetic material called Lignostone, composed of wood and asphalt, that he developed in the same year for a firm in Ter Apel.[7]

Jan Eisenloeffel and Piet Zwart represented the outer extremes of a wide range of views about the role of art in design during the first half of the twentieth century. The major difference between the German and Dutch public debate on the position of art in modern design was the timing: in Germany this took place in 1914; in the Netherlands it was fifteen years later.

The Arts in the VANK

Within the ranks of the VANK, discussion about the position of art and the role of the artist in industry had started earlier. This intensified in 1913 after the level-headed and determined furniture designer Willem Penaat took over the chairmanship from Jac. van den Bosch. He was supported by an equally dynamic colleague, Cornelis van der Sluys, who was secretary of the association. Both supported more close-knit cooperation between art and industry.[8]

Penaat and Van der Sluys were typical transitional figures, with their roots in the idealistic, artisan tradition, but sympathetic to modern production methods. While they were more inclined to design plain, oak furniture similar to Berlage's, using simple constructions with sparse decoration, they did not shrink from using machines to produce them: Penaat's machine-made 'peasant chair' was being produced as far back as 1899. In essence this was a very simple, traditional chair with a rush seat, like those seen in Dutch seventeenth-century paintings and that had been mass produced in the nineteenth century by chair manufacturers in Culemborg and the surrounding area. Penaat simplified the chair even more by curving the back of the chair outwards, which made it a great deal more comfortable. The only decoration consisted of a few 'beads' in the rungs and at the bottom of the chair legs.

Van der Sluys, who started his career around 1900 with Arts and Crafts in The Hague, soon after began designing for the German wallpaper and linoleum industry.[9] In so doing he was following the lead of Michel Duco Crop, one of the first Dutch artists to work as a designer and producer of industrially manufactured furnishing fabrics. From 1910 onwards Van der Sluys worked for a number of newspapers and weekly magazines, which gave him regular opportunities to air his modern ideas on affordable, machine-made, yet responsible design. In 1921 he published an informative

H. P. Berlage and Piet Zwart (Leerdam glassworks), yellow pressed-glass breakfast service, 1924.

survey entitled *Binnenhuiskunst* (Interior Art), followed in 1925 by *Machinale textielkunst* (Mechanized Textile Art) and in 1931 by *Onze Woning en haar inrichting* (Our Home and its Interior).

Exhibitions, both in terms of organizing and participating, were a good way for the VANK to generate publicity and achieve its aims (for the first presentation of the association members' work, at the Stedelijk Museum in 1911, see chapter One). The fact that a museum of art was chosen as the location says much about the ambitions and pretensions of the designers who were members of the association. On show in the Stedelijk Museum were complete interiors by Jac. van den Bosch, H. P. Berlage and H. J. Walenkamp, which had all been displayed at the Exposition Universelle et Internationale in Brussels the previous year.[10] These were complemented by work from designers such as Lambertus Zwiers and Cornelis van der Sluys. The poster that Walter van Diedenhoven made for the occasion is equally illustrative of the lofty ideals promoted by the association's members. It shows a woman with arms outstretched (an allegory for art?). In one hand she holds an orb, from which shafts of light radiate down to revitalize a flower. In the other hand is a bleeding heart, from which drops of blood trickle down to regenerate a flower. The message of its caption, 'Exhibition of Works of Art in the Stedelijk Museum', must presumably be that knowledge and passion are capable of regenerating the industrial arts.

Under Penaat and Van der Sluys, however, the association did not eschew commerce. They created a Publicity and Propaganda Committee and in 1921 set up the Institute for Decorative and Industrial Art (Instituut voor Sier en Nijverheidskunst, ISN), where the business world and government could find information about prospective designers. Moreover, at VANK's instigation, a National Exhibition Committee was established in 1922, in which the most important architectural associations were also represented. In forming this committee the association was also assured of a stronger position at large international presentations, the first of these being the Exposition Internationale des Arts Décoratifs Industriels et Modernes in Paris in 1925.

VANK was offered two galleries for its permanent use in the Stedelijk Museum in 1932. A specially created committee was appointed to develop a continuous programme of exhibitions, where work presented by association members was also for sale.[11] The VANK twice organized exhibitions on advertising in the Stedelijk Museum, in 1917 and 1935. On both occasions the title chosen for the exhibition was *Art and Advertising* and at both an attempt was made to define the position of the designer-artist within this new, modern field. To what extent could advertising actually be called art?

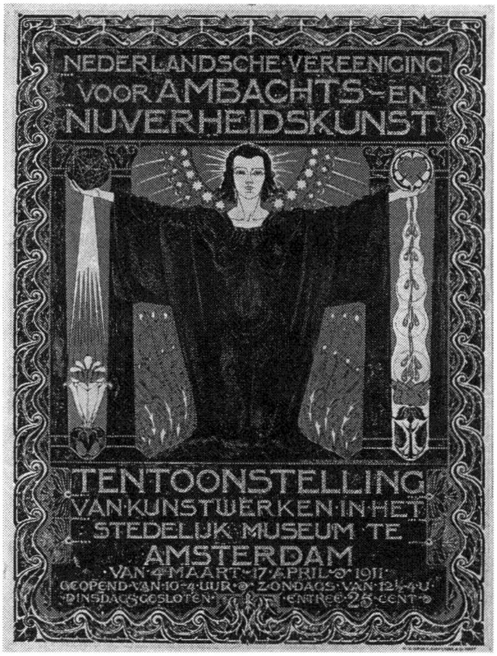

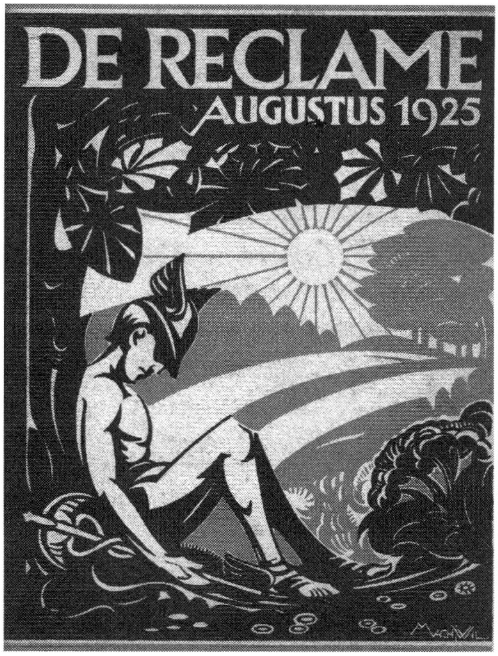

Walter van Diedenhoven, poster for an exhibition of the VANK in the Stedelijk Museum in Amsterdam, 1911.

Machiel Wilmink, cover of *De Reclame*, 1925.

And how could one produce advertisements 'meeting aesthetically appropriate standards' without selling out to industry?[12]

These questions had first been posed at the end of the nineteenth century in response to the posters made for the Delft oil factory (for Jan Toorop's poster for Turin in 1902, see chapter One). The factory's director, J. C. van Marken, was a man with a keen social conscience. The 'artistic engraved plates' commissioned from Richard Roland Holst, Jac. van Zon, Carel Lion Cachet, George Hendrik Breitner and Theo Nieuwenhuis should also be considered. Not only was Toorop's print criticized, but the painter Breitner's design came in for a great deal of censure too. His choice of subject was accused of being arbitrary (impressionistically painted horses) and of having been chosen to suit himself rather than the product. But in fact almost all the posters show that the artists were novices in this new field.

Richard Roland Holst, in his speech at the opening of the exhibition in 1917, addressed the question of whether an advertisement should 'scream out' at you or merely convey a 'quiet message'. At the opening in 1935, however, the then chairman of the association, J. R. van Royen, observed that these two viewpoints had now been integrated and that the aim of making advertisements that were both 'eye-catching and beautiful' had already been realized.[13]

A number of people commissioning publicity work thought it was important that good advertising met the required artistic standards. Those generous in handing out commissions included not only Van Marken at the Delft Oil Factory, but also the directors of the Van Nelle factories and Van Berkels's Patent in Rotterdam, the glassworks in Leerdam, the Bruynzeel Company (timber) in Zaandam, the lamp factory Philips in Eindhoven, various shipping companies, the railways, and the Dutch Trade Fair. According to the design critic W. F. Gouwe in 1930, however, the designers often complained about 'continued censuring and pressing of all sorts of requirements that got the grit out of the design long before the production itself was started'.[14] Nonetheless an advertising sector was established, and as early as 1921 the journal *De Reclame* replaced its more idealistic predecessor *De Bedrijfsreclame*. The new journal was much more realistic about advertising in a modern capitalistic society, the 'art of making a big noise', as Jo de Jong put it in 1929.[15] Even though industrial artists continued to think of it as art, they recognized that the commercial message had to be reduced to its essence, instantly legible as people whizzed by it in the tram.

Yearbooks and 'Applied Arts in the Netherlands'

An important platform for discussion about the role of art in design was the series of VANK yearbooks, published between 1919 and 1932 by W. L. and J. Brusse in Rotterdam with the financial support of the Dutch government. In addition to photos of new work by members (and non-members), the yearbooks were full of contemplative as well as critical articles written by the members themselves, as well as contributions from outsiders that were thought to be of interest. For the present-day reader the style is sometimes too florid and verbose, but usually a clear picture is given of how the designer's task was conceived and the place that art was meant to take in his work. When it came to these subjects the writers had very lofty ideals, but they did not always see eye to eye.

The architect and furniture designer Jo van der Mey, for example, in an article entitled 'Modern Furniture Art' from the first of the association's yearbooks (1919), describes the exact requirements a 'modern' artistic piece of furniture is expected to meet.[16] In bombastic style, he launches into a criticism of the designs produced by the 'rationalists', which in his view were too greatly influenced by the intellect. A piece of furniture, in Van der Mey's opinion, should primarily be 'an object of feeling', should invoke 'emotion' and should even offer 'support and comfort'. These elements were entirely absent in designs by Berlage, Dijsselhof, Lion Cachet and Nieuwenhuis,

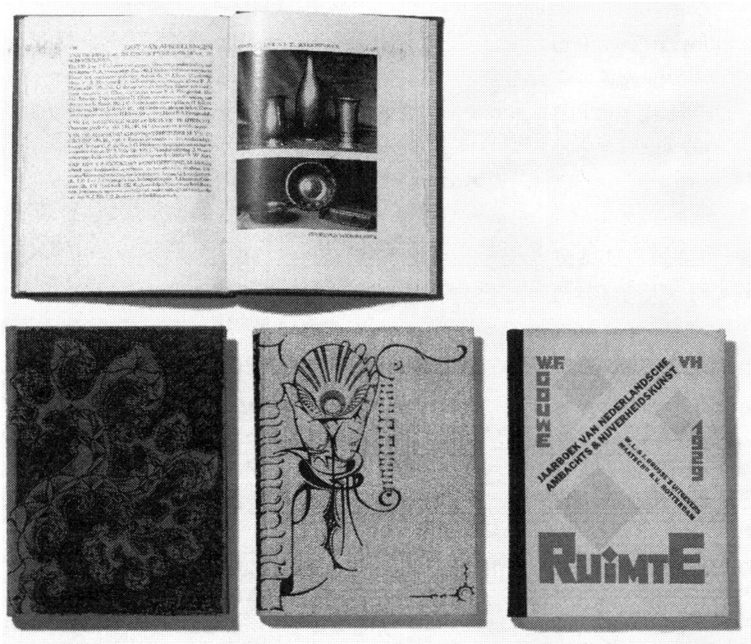

Four VANK yearbooks (W. L. en J. Brusse's publishers Rotterdam), covers designed by: J.L.M. Lauweriks (1922), N. J. van de Vecht (1923–4) and V. Huszar (1929).

whom Van der Mey considered to be among those who aspired to pure construction and true craftsmanship. They may indeed have made 'beautifully designed' furniture, but the emotion was missing. The main culprit in Van der Mey's opinion was socialism, 'purely materialistic' and governed by ideas aimed solely at improving the working and living conditions of the labouring classes. If this was the way these designers thought then they could never be inspired to design furniture that would trigger emotional feelings. Another unhelpful factor, however, according to Van der Mey, was the 'Philistine attitude towards art' prevalent in the Netherlands. The Dutch were interested only in trade and making a profit: 'Wouldn't it be wonderful if the ships roaming the Seven Seas could present a good image of "Dutch artistic ability"!' He must have had Lion Cachet in mind when he said this since he had already been involved in the interior design of some seven liners.

In that same yearbook a very different tale was told in 'Industry and Art' by the director of the the Leerdam glassworks, P. M. Cochius, in his assessment of the first results of the collaboration between his glassworks and the artists Karel de Bazel and Cornelis de Lorm.[17] The simple sets of glasses they had designed were admired by the critics and were selling well. In Cochius's view, bringing in artists was part of modern management, just as specialists were being brought in to other departments in increasing numbers. But even for him the specialist skills that the industrial artist contributed were still unmistakably 'art', the contribution of which would have a civilizing effect.

The loftiest ideals about the position of art in design can be read in the yearbook for 1923–4, which is filled with six 'philosophical dissertations' by H. C. Verkruysen, Prof. W. van der Pluym and Prof. R. N. Roland Holst.[18] In these articles, which are now difficult to digest, all stops are pulled out to place design on a pedestal by dint of its relation with art. The art historian Willem van der Pluym writes about the artist-designer in general:

> Even there, in the most unpretentious setting of day-to-day living where he creates products for domestic use, it will be his mission to raise the level of his work above the unimaginative practicality of everyday material objects, so that in using them humankind, to whom he imparts his talents, will enjoy a happier life.

These absurdly lofty pronouncements contrast sharply with the real position and significance of artists and designers in Dutch industry at that time, since in the 1920s it was still in its infancy.

Halfway through the 1920s more modern ideas began to make their way into the VANK yearbooks. There did indeed appear to be members who approached the position of art in design from a more progressive angle. One such was the manufacturer and designer Willem H. Gispen, who in his 1925 contribution entitled 'Art as a Necessity and as a Form of Playfulness' observed critically and ironically:

> The 'World of Arts and Crafts' is a remote village, far removed from heavy traffic. People keep themselves to themselves and a lot is still done the way it was twenty-five years ago. There are strict regulations to stop cars driving too fast, and people still hammer out artistic, brass ashtrays costing twenty guilders each. The hammered ashtray, entirely hand-made, has a symbolic meaning . . . it is a sweet little object of beauty, a work of art, but is still without style.

In this long epistle Gispen goes on to argue that something can only be said to have real style when its form is in harmony with its function: an ashtray is for collecting cigarette butts and for tapping out ash from your pipe: it is not a work of art.[19]

In the ten years preceding this article Gispen's beliefs and working methods had undergone an enormous change.[20] At his metalwork firm,

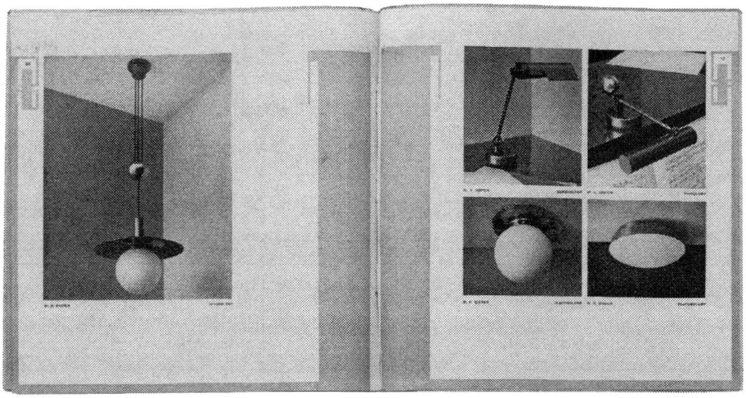

W. H. Gispen, lamps reproduced in *Wendingen (Techniek en Kunst)* (1928); with the piano lamp designed with J.J.P. Oud, above right.

established in 1916 in Rotterdam, he initially made handcrafted objects like railings, fireplaces and signboards, in a distinctly decorative style. His products were then still referred to as 'decorative metalwork'. In 1919 he changed the name of his firm to Gispens Metalwork factory, which hints at more contemporary ambitions. In that year Gispen became a member of the VANK and subscribed to *Wendingen*, the journal of the architectural association Architectura et Amicitia; the next year he subscribed to *De Stijl*, the avant-garde periodical established by the artists Theo van Doesburg and Piet Mondrian in 1917.

In the meantime his firm continued to grow, and for Gispen the design and production of simple modern lamps became increasingly important. It led him to become engrossed in the very latest ideas about lighting engineering, such as those developed in Germany, with a special interest in the advances being made in America, where conforming to uniform standards or 'normalization' was becoming increasingly important. Moreover, through his friend the architect J.J.P. 'Bob' Oud, he was introduced to the progressive views about design taught at the Bauhaus. This was how Gispen came upon his unique 'Giso glass', a process in which crystal glass was covered with a thin, precisely calculated coating. This resulted in a minimal loss of light from the lamp, while having the advantage of not blinding the onlooker.

The Giso lamps were presented almost simultaneously at two exhibitions in Germany in 1927. At *Europaïsches Kunstgewerbe* in Leipzig, which displayed the crafted decorative object, Gispen's lamps were a bit out of place because of their industrial character. They were far more suited to the Werkbund exhibition *Die Wohnung*, in Stuttgart, where Bob Oud and Mart Stam installed them in their show houses in the Weissenhofsiedlung. Gispen's enthusiasm for the products and ideas he was introduced to at this

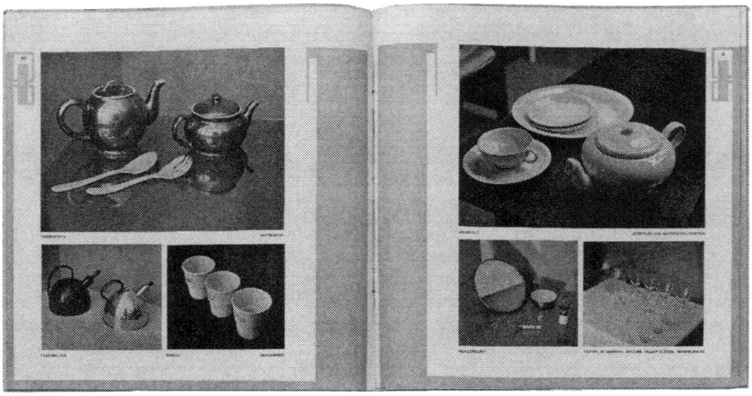

Paul Schuitema, photographs of the exhibition 'Artless Domestic Objects', reproduced in *Wendingen Techniek en Kunst* (1928).

much talked about breakthrough of the international New Building (*Nieuwe Bouwen*) in Stuttgart was clearly expressed in a lecture given to members of the Rotterdam Circle on his return. His continuing preoccupation with the place of art in these new developments is shown by the title of his lecture, 'Artless Domestic Objects not Anti-Art, but Anti-Industrial Art'. To add a little lustre to his speech, he assembled a modest selection of simple, undecorated, industrially manufactured domestic objects, including one of the first of his own tubular steel chairs. The collection was printed in a special issue of the periodical *Wendingen* entitled 'Technique and Art' (1928).[21]

Gispen's continued interest in the more artistic, formal aspects of design is shown in the striking piano lamp he developed together with Bob Oud. This object, which was later to become so famous, was more of a spatial composition, such as the furniture designer Gerrit Rietveld was making at the time, or more an experiment in terms of balance, as practised at the Bauhaus, than a serious industrial design. The first example of the lamp was developed in July 1927 as a wedding present for a couple with whom Oud was friendly and was made of brown burnished copper. Shortly afterwards Gispen started to produce the small lamp in series with a shiny nickel

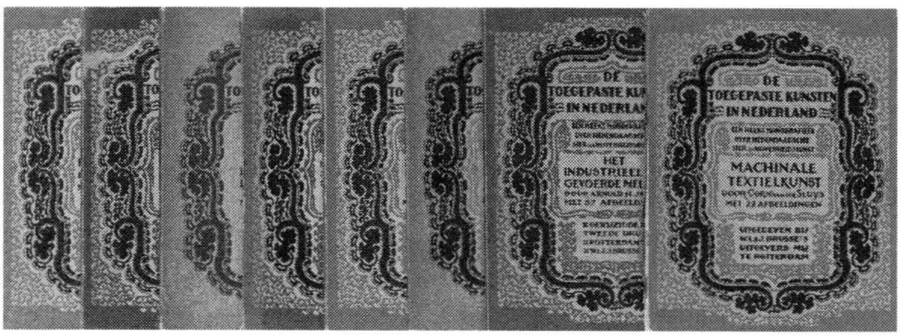

R. Gerbrands, covers of the series *De Toegepaste Kunsten in Nederland* ('The Decorative Arts in the Netherlands'), 1923–35 (W. L. en J. Brusse's Publishers, Rotterdam).

finish, under the commercial name of 'Giso 404'. In 1929 sales of chairs also increased, thanks to a large order from the Van Nelle coffee, tea and tobacco company in Rotterdam.

The tendency of many VANK members to stress the artistic and artistically minded aspect of design in the 1920s and '30s emerges from the contents of the twenty-four booklets of the 'Applied Arts in the Netherlands' (*Toegepaste Kunsten in Nederland*) series published between 1923 and 1935. Unlike the yearbooks, this 'series of monographs on contemporary decorative and industrial art' was not a VANK publication, but closely linked in terms of its contents, authors and advertisements. They were intended for craftsmen, designers, students, industrialists, art dealers and anyone else who showed an interest. To increase their usefulness they listed important applied art shops, schools and museums, and even the advertisements served as an important source of information for readers.

Most of the booklets in the series are dedicated to themes that have a direct relationship with house and home, like Gispen's *Sierend Metaal in de Bouwkunst* (Decorative Metal in Architecture, 1924), and *Machinale Textielkunst* (Mechanized Textile Art, 1924) by Cornelis Van der Sluys, mentioned earlier. Two booklets were published on furniture or, to use the association's terminology, 'furniture art': *Het Moderne Meubel* (Modern Furniture, 1924) by Just Havelaar and *Het Industrieel uitgevoerde meubel* (Industrially Produced Furniture, 1925) by A. H. Jansen. Contrary to what the title might suggest, Havelaar represented the traditional point of view in which the furniture designer, and thus the artist with his specific talents,

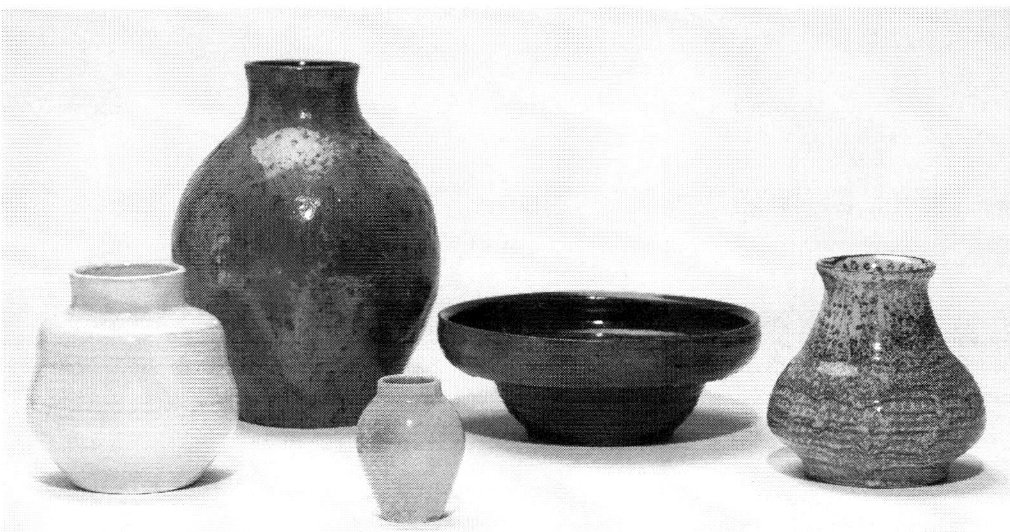

C. J. Lanooy, four hand-crafted vases and a bowl, 1910–20.

was expected to inject character and beauty into his designs. Of course, his products had to fit in with the times and it was for this reason that a modern piece of furniture had, in Havelaar's view, to be 'democratic': 'an expression of a socially adjusted, working population, that no longer stresses class distinctions'. The style of this 'modern' furniture was expected to evolve all by itself in a period in which sport, flying machines and big industries were to become everyday reality. The artist's task was to interpret 'aesthetically what society can create in a material way that is of value'. Those who may have assumed that Havelaar's line of reasoning would inevitably lead to industrially manufactured furniture, however, were quite wrong. In his view only a furniture artist working traditionally was capable of making a style of furniture appropriate for the modern age.

In his booklet the furniture designer Arnold Jansen turns his attention to cheaper furniture for the workers. More than Havelaar, he wonders if it is then possible to embark on large-scale, industrial mass production of cheap, yet well-designed furniture in the Netherlands. It seemed apparent to him that the initiative still lay with the artist-designer and that factory owners had shown little real interest in this type of project. The objection was that the intervention of the artist meant that the product, by definition, would be more expensive because not only did the artist have to be paid for his plans and suggestions but also furniture factory workers complained that this prevented them from working as quickly and efficiently as normally. Moreover, modern furniture did not often appeal to the masses and only a few 'pioneers' were interested in it. Since new models took time to become established, a more restricted form of serial production seemed to be a more realistic option for the present.

The title of the booklet on ceramics, *The Potter's Art*, is in itself revealing. The director of the Gemeentemuseum in The Hague, H. E. van Gelder, had good reason to choose this title in 1923. He was an admirer of T.A.C. Colenbrander's designs for the Rozenburg pottery factory produced from the 1880s, which he considered works of art. As early as 1916 he ensured that a great quantity of Colenbrander-designed Rozenburg earthenware was accepted into the collection of the Gemeentemuseum. At the same time, however, he increasingly came to appreciate ceramics by pure potters such as Chris Lanooy and Bert Nienhuis. In the 1920s these artists were no longer making small series of vases, but only unique objects that were compared with abstract works of art both by the artists themselves and by art lovers such as H. P. Bremmer, the leading art authority in The Hague, and his following.[22] The vases were even given titles and each piece was signed. *The Potter's Art* and another booklet, *Glass and Crystal* by

Karel Wasch, were so popular that they were reprinted in 1927, when the writers seized the opportunity to update their texts.

The series is a rich source for anyone wanting to study design (and art) in the Netherlands in the 1920s. Besides the VANK yearbooks, *Binnenhuiskunst* (Interior Design, 1921) by Cornelis van der Sluys and *De Nieuwe Richting in de Kunstnijverheid in Nederland* (The New Trend in Applied Art in the Netherlands, 1929) by Jo de Jong are the two most important sources. This has led to the products illustrated in them functioning as a sort of canon for good modern design in the Netherlands in the interwar period. The question is whether this does justice to the real significance of these pieces. The fact that most of the writers were VANK members means that the objectivity of these publications cannot be guaranteed. Many Dutch products and designers are not dealt with in these publications and thus have remained unknown. Notable omissions, for example, are the many modern coffee and tea services that were produced in the Maastricht potteries during these years. This is even more remarkable considering the enormous quantities that were made.[23] You will also look in vain for information on bicycles and cars, household appliances, kitchen aids, sports equipment or office supplies and garden furniture. If more insight is to be gained into the production of consumer goods in the

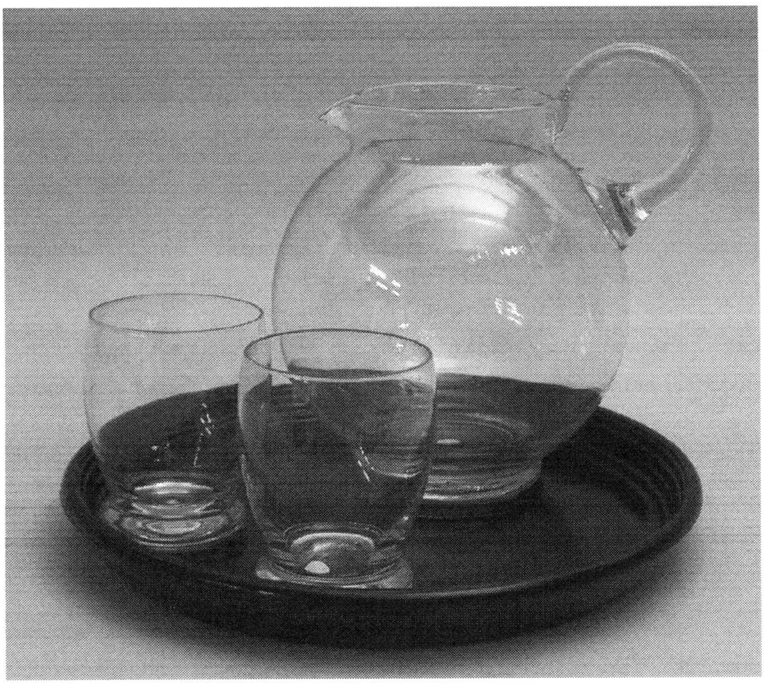

W. J. Rozendaal (Kristalunie, Maastricht), jug with beakers, 1932.

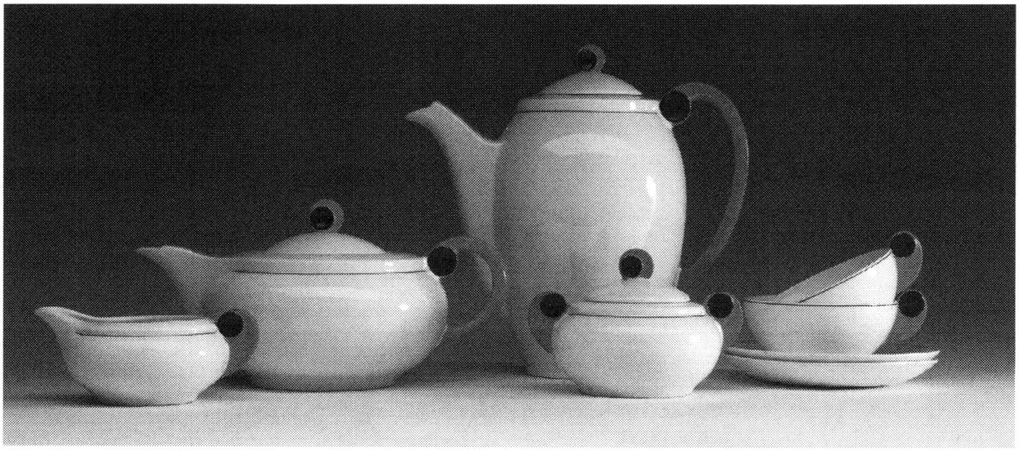

E. Bellefroid (Ceramic factory De Sphinx, Maastricht), 'Strand' service, c. 1933, exclusively produced for the HEMA in more than 100,000 copies.

Netherlands it will take a great deal of research, and more detailed studies will have to be carried out on subjects such as the merchandise available in large department stores.

The Amsterdam School

Amsterdam architects who were also active as product designers sometimes joined the VANK, but continued to feel most at home in their own association, Architectura et Amicitia, in which industrial art and interior design were also important. At the 1915 exhibition in the Stedelijk Museum to celebrate the sixtieth anniversary of Architectura et Amicitia, decorative and industrial artists were represented in outstandingly large numbers. To many people's surprise, it became clear at this exhibition that a completely new generation of architects had emerged who had been almost unnoticed before, most notably Michel de Klerk, Piet Kramer and Jo van der Mey (see above).[24] These three architects were also involved with aspects of interior design. One of the projects they were working on at the time was the interior of the revolutionary Scheepvaarthuis (Shipping Trade House), one of the icons of what was later to become known as the Amsterdam School.[25] This office building in Amsterdam, housing a number of shipping organizations, was designed by Jo van der Mey in collaboration with the large architectural firm of A. L. van Gendt, although, as far as his interior designs were concerned, from 1912 onwards he worked with a large group of like-minded architects and industrial artists.

These architects began to rebel against the otherwise undisputed authority of their older colleague Berlage. What they mainly complained about was the lack of an artistic element in his buildings and furniture. They even went as far as airing their grievances in the special issue of the *Bouwkundig Weekblad* (Architectural Weekly) published in honour of Berlage's sixtieth birthday in 1916. Michel de Klerk reproached the famous architect for not being an 'artistic builder' and said that, at best, all he had achieved was the introduction of a few technical innovations for the benefit of the building trade.[26] De Klerk and his sympathizers argued that buildings and their interiors should become more 'expressive' again. They soon had the opportunity to realize this in the most exemplary way in a series of public housing projects – 'workers palaces' – built in the early 1920s in the southern suburbs of Amsterdam. The fantastic and sometimes overwhelming sculptural Amsterdam School style, however, can be admired in other parts of the city and in several other towns and villages.

De Klerk, Van der Mey and Kramer knew one another from the time with Eduard Cuypers's firm of architects. Cuypers, after training with his famous uncle P.J.H. Cuypers, had resolutely put his Gothic Revival building legacy behind him and had opened his mind to all manner of contemporary reform movements both at home and abroad. At the end of the nineteenth century he had founded in Amsterdam the Studio of Decorative Art 'Het Huis' (The House) to complement his architectural firm. There they had no qualms about using the Dutch *Nieuwe Kunst* idiom on a grand scale to complement the international Art Nouveau style and the German and Austrian *Jugendstil*. Examples of this can be seen in the beautifully illustrated magazine *Het Huis* that Cuypers published from 1902. The firm's young designers were stimulated to read foreign periodicals and to travel, one of the destinations being Scandinavia. Eduard Cuypers was also involved with industrial art in the Dutch East Indies. On the island of Java he designed buildings for the Javasche Bank and he was also responsible for the design and furnishings for the pavilion representing the Dutch colonies at the Exposition Universelle et Internationale in Brussels in 1910. The tolerant and flexible atmosphere created in his firm undoubtedly acted as a strong stimulus to the growth of the Amsterdam School.

As building activity came to a standstill due to the war, about 1914 Michel de Klerk started to concentrate on designing furniture.[27] J. F. Zeeghers, director of the furniture makers and retail outlet 't Woonhuys, gave him every opportunity to do just that. The sumptuous interiors and expensive furniture that De Klerk drew for Zeeghers were only for extremely wealthy clients. His unique designs with their complicated symbolic design idiom could only be made by the most skilled craftsmen. The rich mahogany table, for instance, commissioned for the office of the director of

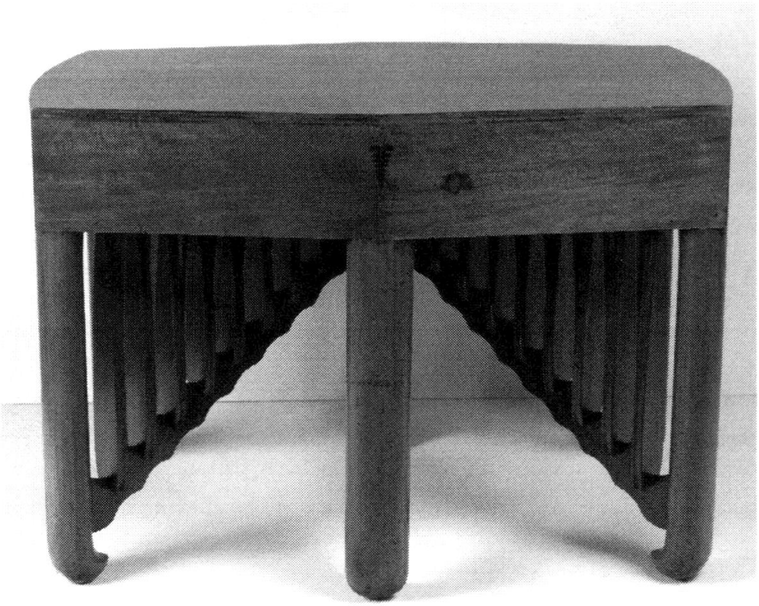

M. de Klerk, mahogany table for the boardroom of the Netherlands Steamship Company, Shipping Trade House, Amsterdam, 1913.

the Netherlands Steamship Company, has an elongated hexagon tabletop supported by six legs attached to the table in an almost inconceivably excessive constructional design. For the wedding of the rich Amsterdam lawyer J. H. Polenaar, De Klerk designed furniture for the salon and the bedroom made from mahogany, poplar and birch plywood, everything extravagantly decorated with black stained and sculptured ornaments. This early suite of furniture is in the collection of the Rijksmuseum in Amsterdam. In total De Klerk probably designed around a hundred pieces of furniture, of which only about twenty-five items have been preserved.

Piet Kramer had begun designing furniture as early as 1911.[28] His earliest models are still in a measured decorative-rational style. It was not until 1917 that his designs also began to take on the lavish and expressive sculptural character that was so characteristic of De Klerk's work. Kramer had his own firm of architects and was also an aesthetic adviser for the Amsterdam Municipal Council. In this capacity he was responsible for scores of sculptural bridge railings and metal bridgeheads of a characteristic type that can be admired in the city to this day.

Although the number of furniture designs produced by De Klerk, Kramer and Van der Mey was relatively small, their expressive sculptural style had an enormous impact on design in the Netherlands. Many of their contemporaries were directly inspired by them, especially the sculptor

Hildo Krop, in whose designs the sculptural element is even more apparent.[29] This is not so surprising in view of his main occupation. The furniture designs made by Piet Wormser, Piet Vorkink, Jan Antonie Snellebrand and Adolf Eibink, with their original design idiom worked in expensive varieties of woods, show a strong affinity with the work of De Klerk, Kramer and Van der Mey.

In the soft furnishings chosen for their interiors, the Amsterdam School designers often displayed a preference for mock-velvet.[30] This robust, shiny upholstery fabric was already familiar as it had been used in the Netherlands since the eighteenth century, when it was known as *velours d'Utrecht*, but between 1910 and 1920 the mock-velvet weaving mills modernized their classic designs. The new patterns were not printed by a mechanized process, but were applied using a 'new' method based on traditional block-printing. The daring and original designs of the Rotterdam decorative artist Jaap Gidding, with their strong and sometimes almost exotic black patterns on a shining purple, stone-red or gold-coloured background, were especially popular with the Amsterdam School architects.[31]

The attention paid to luxurious and unusual materials, decoration, craftsmanship and aesthetics was applauded in wider circles, not least because it left so much more scope for artistic experiment after the plain, austere furniture and other domestic products influenced by the 'rationalists'. The obvious influences and vague echoes of the Amsterdam School can again be seen in the work of pupils at applied art schools and in the illustrations of various VANK publications. Around 1920, for example, a strikingly large number of exuberantly designed clocks were made by many metal artists, both famous and less well known.

Thanks to Jaap Gidding, colourful carpets were once again fashionable in the mid-1920s, while his expressively painted glass for the Leerdam glassworks and his designs for the colourfully decorated ceramics made at the Zuid-Holland and Regina pottery factories can also be connected to developments in Amsterdam. Gidding owes his national fame mainly to his decorations in the lobby of the Tuschinski Theatre in Amsterdam, which still exists today as a cinema. Built in 1921, and sumptuously decorated both inside and outside, it illustrates the appeal that the Amsterdam School idiom had for many of those working in design at the time. Pieter den Besten, Charles Bartels and Willem Bogtman also worked on the interior of the Tuschinski Theatre, while Cornelis van der Sluys designed the coverings of the lounge chairs – even this confirmed Berlage devotee now allowed himself a little more luxury and decoration in his furniture designs than he had in the past.

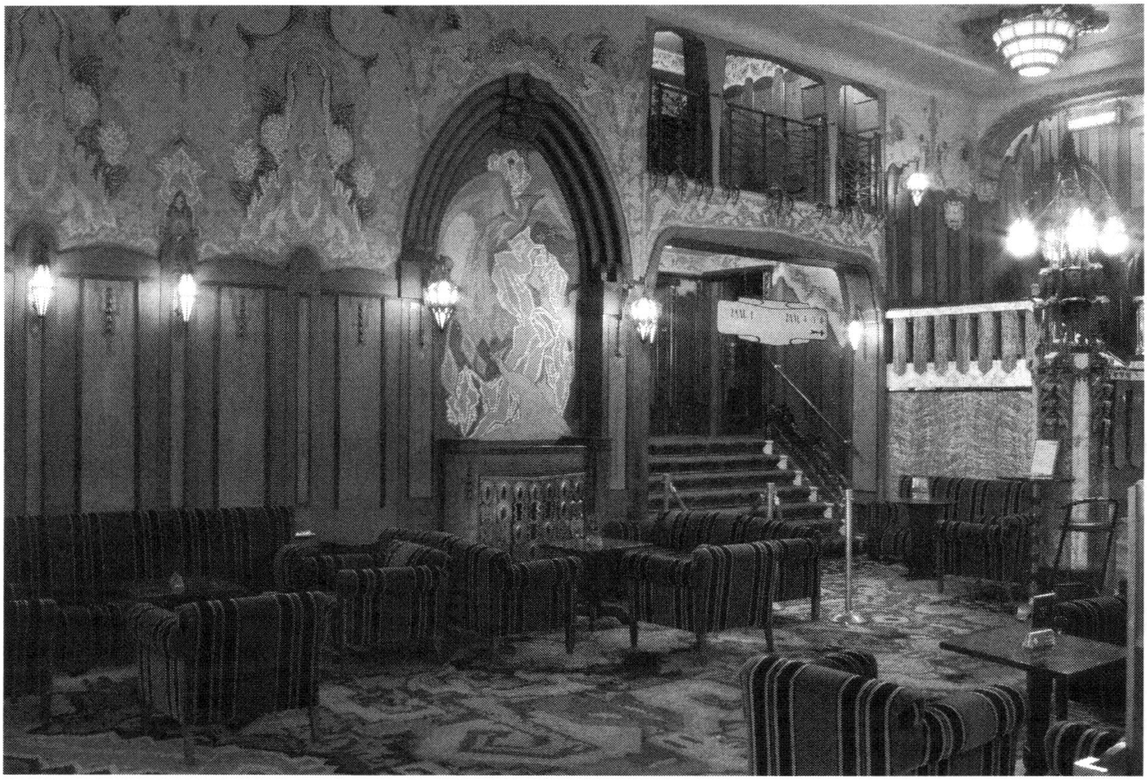

Jaap Gidding, lobby of the Tuschinski Theatre, Amsterdam, 1921.

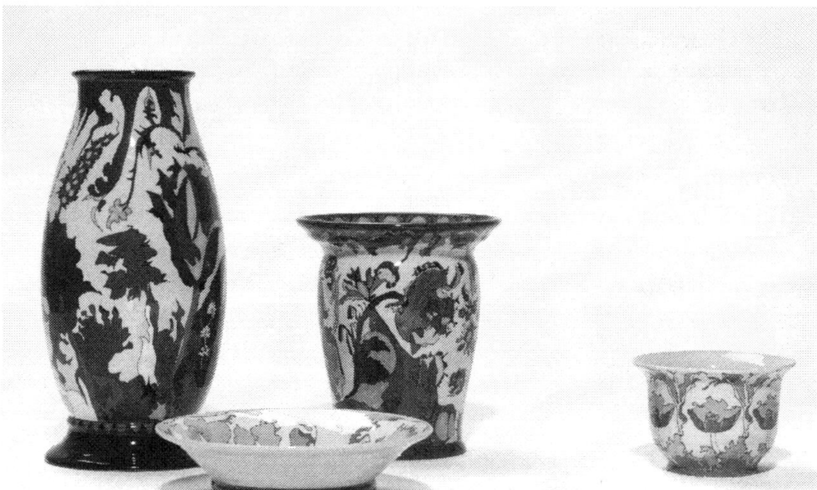

T.A.C. Colenbrander (Plateelbakkerij Ram, Arnhem), two vases, bowl and flower pot, 1922–5.

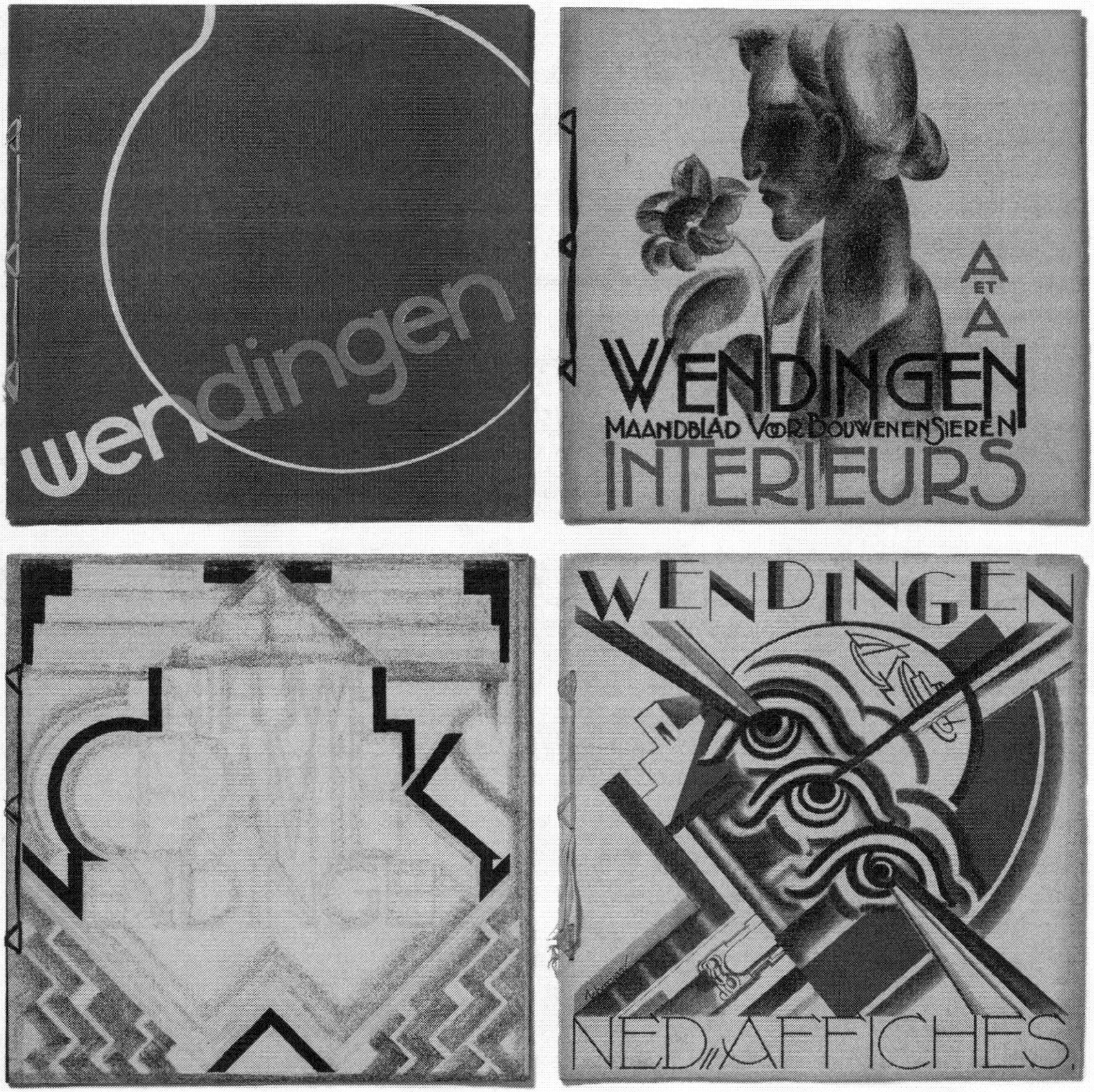

Four *Wendingen* covers by A. D. Copier (1930); S. L. Schwarz (1931); Otto B. de Kat (1927) and Tine Baanders (1927).

Another significant influence was the reassessment of Theo Colenbrander's work. A number of wealthy enthusiasts for the colourful Rozenburg earthenware made thirty years before, even went as far as to set up a separate factory in Arnhem dedicated to the execution of Colenbrander's new ceramic designs.[32] From 1921 the idiosyncratic designs of the octogenarian artist were produced by the Ram Delftware Factory. The artistic image of the vases and dishes, with their ostensibly abstract and colourful designs, was emphasized by their poetic titles.

Wendingen and *De Stijl*

Under the influence of the Amsterdam School, members of the architects' society Architectura et Amicitia increasingly shifted their focus to the aesthetic side of building. The architect H. Th. Wijdeveld played an important role in this change.[33] At his request the painter Richard Roland Holst joined the editors of *Architectura*, but this was not enough for Wijdeveld, who was as idealistic as he was determined. Believing that the new artistic zeal among the members of the society warranted an entirely new periodical, in 1918 he established the monthly journal *Wendingen* (Inversions), which became the new mouthpiece of this movement in Dutch architecture and design.[34]

Wijdeveld, who like so many others had been trained at the firm of P.J.H. Cuypers, subsequently worked for a time with L. M. Cordonnier, the architect of the Vredespaleis (Peace Palace) in The Hague. In 1913 he established himself as an independent architect in Amsterdam, where he became friends with F. M. Wibaut, Amsterdam's first social democratic alderman, and received his first commissions as part of the development of the city towards the south. By Dutch standards Wijdeveld was an exceptionally free, creative and internationally minded designer. During a long trip abroad in 1920 he became friends with such divergent figures as Erich Mendelsohn, Adolf Behne, Henry van de Velde, Gordon Craig and Amédé Ozenfant. His interests went beyond the bounds of architecture: he was also fascinated by urban development, theatre, music and literature, while even designing stage sets, theatrical costumes and posters. Influenced by the work of Mathieu Lauweriks, Wijdeveld developed an individual type of graphic design for *Wendingen*. For the headlines he used printing composing material, which he incorporated into his design in a surprisingly decorative manner. The result was not the most legible of designs but it was extremely expressive. A critic remarked that Wijdeveld did not design typography, but 'art', using typographic aids. *Wendingen* was subtitled *Maandblad voor bouwen en sieren* (Monthly Journal for Building and Decoration), and for each issue a designer or artist was chosen to make a new cover design.

Thus *Wendingen* was more than an ordinary architectural journal, rather a platform for art of the future in all its varieties. The print run for this exquisitely produced journal remained fairly modest and usually ran to just over 1,000 copies. In that respect the year 1921 was a high point: 2,400 copies per issue were printed, of which 600 were in German and 600 in English. In the early years the emphasis was strongly on the work of architects from expressionist Amsterdam School circles, but foreign designers were also featured. In 1924 special issues were devoted to the French designer Eileen Gray and the German Expressionist architect Hermann Finsterlin. Also in 1924, the year of Michel de Klerk's untimely death, three issues were filled with his work. In 1925 and 1926 seven entire issues were devoted to the American architect Frank Lloyd Wright.

After Wijdeveld had left the editorial board in 1926 and was replaced by H. C. Verkruysen, writings by the modern functionalist designers were allowed more space. Apart from architecture, there were contributions on related arts like sculpture and murals, but space was also regularly reserved for furniture design and furnishing fabrics. Entire issues were devoted to stage design, posters, bookplates and even shells. The last issues of *Wendingen* appeared in 1932.

However, despite its wide range of special interests, covering everything to do with art, architecture and design, *Wendingen* did not represent the entire range of cultural avant-garde activity in the Netherlands. The journal *De Stijl* was published at almost the same period (1917–31).[35] Begun by Theo van Doesburg and Piet Mondrian, it attracted a group of painters and architects around it: while the total number of subscribers fluctuated between one and two hundred, about a thousand copies were printed of each issue. Although the painters Mondrian and Van Doesburg, together with Bart van der Leck and Vilmos Huszar, are seen as *De Stijl*'s main figures, the architects Jan Wils, Robert van 't Hoff, Bob Oud and the furniture-maker Gerrit Rietveld were no less essential for the main aim of the 'movement': the integration of painting and architecture. Mondrian thought up a new term to cover this: 'Neo Plasticism' (*Nieuwe Beelding*). The merging of the two disciplines was in his view an inevitable process. This is why the definite article ('De') was used in the title of the periodical. It was not just a new style. What it revolved around was the development of *the* style. Mondrian was the most important ideologist behind these utopian aims, but until his death in 1931 Van Doesburg was the driving force behind the periodical and the only stable factor from start to finish.

Thus, implicitly, design played a prominent role in the theory of the *Nieuwe Beelding*: the distinctions between architecture, art and interior design would indeed eventually merge, so that in a certain sense it would all be design, that is to say *Nieuwe Beelding*. For De Stijl artists the explicit issue of the position of art in applied art or in industrial design was thus irrelevant.

Gerrit Rietveld, whose work was first illustrated in *De Stijl* in 1919, Vilmos Huszar, Bart van der Leck and Bob Oud were engaged in designing products. Theo van Doesburg designed stained-glass windows at the start of his career. Without a doubt Rietveld's furniture is the most interesting.[36] The Utrecht furniture designer initially worked in the Berlage idiom and was also influenced by his instructor P.J.C. Klaarhamer. From around 1916, however, he plotted his own pioneering and innovative route. Nevertheless, Rietveld was, and remained, an ordinary furniture-maker, an experimental pragmatist and a man of few words: his theoretical reflections on his avant-garde work were few and far between. No matter how modern, conceptual, spatial and expressive his pieces of furniture might have been, Rietveld saw them as domestic objects and not as works of art. His universally known armchair, built of strips and planks of wood, was conceived back in 1918 – thus well before Rietveld came into contact with Mondrian or Van Doesburg. It was not until 1923 that the chair was painted in the primary colours of red, blue and yellow.

In the 1920s Rietveld gradually began to concentrate on the opportunities opened up by mechanized furniture production and architecture. His work maintained its experimental character; not a single pre-war design of Rietveld's became a commercial success. His influence, however, on design culture in the Netherlands – and abroad – has undoubtedly been immense right up to the present day, due to its idiosyncratic, pioneering character. Rietveld's association with the Congrès Internationaux d'Architecture Moderne (CIAM) and the numerous orders he received from the renowned Amsterdam furniture store Metz & Co. strengthened his position throughout the 1930s both as an architect and as a designer. Even after the Second World War Rietveld continued to be an authority on design culture in the Netherlands and was to remain so right up to his death in 1964. We will return to him in later chapters.

Paris 1925

The relationship between art and design was also the central theme in the Dutch entry to the Exposition Internationale des Arts Décoratifs et Industriels Modernes in Paris in 1925. This was the express aim of the French organizers, but nobody could have foreseen that this would have led

The Dutch exhibition at the Exposition des Arts Décoratifs et Industriels Modernes, Paris, 1925.

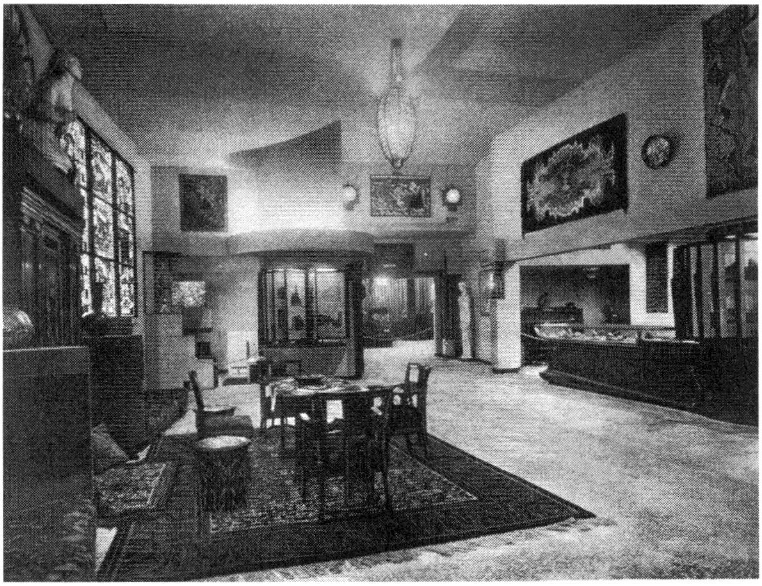

to such a heated debate back in the Netherlands, since it was while preparations were being made that the question arose as to which of the Dutch designers best represented the chosen theme: the *Wendingen* supporters or their colleagues from *De Stijl*?[37]

The De Stijl designers today have a worldwide reputation, whereas the names of many of the *Wendingen* designers are no longer well known; in the 1920s, however, the situation was the reverse, as was shown clearly by the Paris entry. As was usual in the Netherlands at that time, trade representatives had once again taken the initiative to ensure Dutch participation in the exhibition. It was only at a later stage that the Dutch government proved willing to make a financial contribution and it was not until 1924 that a Preparatory Committee could be appointed. The Dutch pavilion was designed by the Amsterdam architect Jan Frederick Staal. H. Th. Wijdeveld coordinated the interior design. All Dutch designers were asked to send in work, but an assessment committee set up by the National Exhibition Council, consisting of H. P. Berlage, J. Gratama, W. M. Dudok, J. Mendes da Costa, Hildo Krop, R. N. Roland Holst, J.L.M. Lauweriks and C. A. Lion Cachet, ultimately decided who would be allowed to participate. The result therefore was that the character of the entries going to Paris was determined to a large extent by Amsterdam School architects and a few other related designers.[38]

The Dutch pavilion was theatrical and furnished in somewhat dark colours, with the emphasis on hand-crafted, luxurious, 'artistic' furniture and decorative products. Jaap Gidding's stained-glass windows and carpets, together with Colenbrander's ceramics, provided the occasional colourful accent. Apart from their work, Berlage, De Bazel, Lauweriks, Lion Cachet, De Klerk and Kramer were well represented.

Almost immediately there were arguments about the way the entry had been organized, its interior design and the choice of objects. The backbiting was not restricted to representatives of De Stijl, who were furious because their joint-entry proposal under the leadership of Theo van Doesburg had been rejected; other, more moderate designers such as Cornelis van der Sluys, himself a participant, criticized the fact that the most modern and progressive forces in the Netherlands at that time were not represented at the Paris exhibition. It was still possible, however, to see a few pieces of furniture by Sybold van Ravesteyn and Bob Oud executed in a restrained De Stijl idiom. Furthermore, The Hague furniture factory Pander sent in a modern gentleman's room by Hendrik Wouda, including bright red furniture produced in a clear-cut 'cubist' style. Nowadays work by Wouda and a few of his fellow townsmen is referred to as 'The Hague School' and is considered to be the most representative example of Art Deco in the Netherlands.[39] Others working in the same style included Cor Alons, Jan Wils and Frits Spanjaard. Their designs are related to De Stijl, but there are also clear indications of the influence of Berlage's later work and that of the American designer Frank Lloyd Wright. Although the lines are more clean-cut and it looks more modern, The Hague School design, like that of the Amsterdam School, was in general fairly luxurious and calculated to produce a specific effect.

The Bond voor Kunst in Industrie and its Precursors

All the commotion caused by the choice of the Paris entries in 1925 clearly showed yet again the importance many that Dutch people still attached to art in the design of the everyday environment. But it cannot be emphasized too often that this did not mean the automatic rejection of mechanized products with an artistic component. The conservative voices of Jan Eisenloeffel and others working in the traditional crafts continued to be audible in those years, although most designers recognized that more intensive collaboration with industry was inevitable, and perhaps ought even to be welcomed.

A decade earlier Willem Penaat had performed important preparatory work leading to the gradual change of attitude and the various initiatives taken in the 1920s and '30s. The great success of the Deutscher Werkbund –

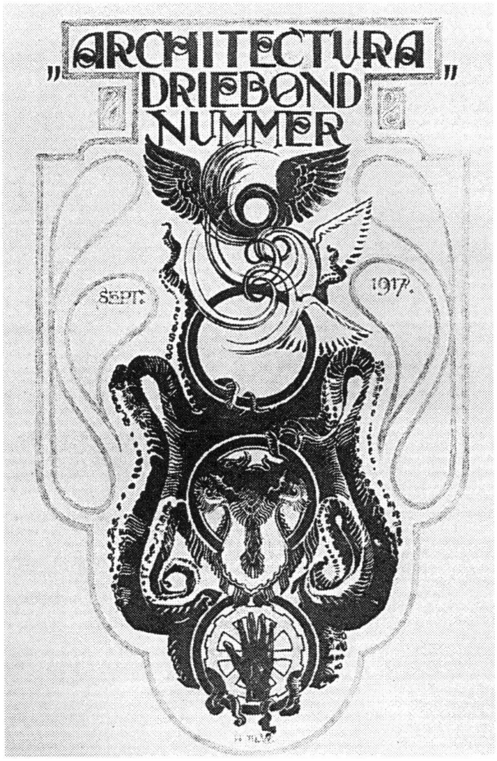

H. Th. Wijdeveld, cover of *Driebondnummer Architectura* (1917).

with the inbuilt danger, he felt, of German industry gaining too much supremacy – inspired Penaat as early as 1914 to make a proposal to establish a Dutch counterpart, the 'Nederlandse Werkbond'.[40] A few Dutch designers had then been members of the Deutscher Werkbund for several years, including Johan Thorn Prikker, who had been working in Germany since 1904, and Hendrik Petrus Berlage. Penaat unfolded his plans in VANK's publication *Orgaan* (Mouthpiece) and suggested they call it the 'Driebond' (Triple Alliance). This was a reference to the intended collaboration between art, labour and society, also referred to as beauty, technique and culture (or, if you prefer, artist, manufacturer and others who were interested). In Penaat's proposed Driebond the only artists eligible for membership would be those working in a 'modern' idiom – and only then by invitation. He wanted to exclude manufacturers who only wanted to work with artists for commercial gain. The social group that represented these 'interested parties' was supposed to consist mainly of art critics, museum directors and other art lovers. What seemed at first glance to be a revolutionary proposal proved on closer examination to be a fairly defensive move, and furthermore it was chiefly planned with the artists' interests at heart. One of Penaat's ideas was that the new organization would be supervising the work to ensure that the fast pace of industry did not undermine the quality of the designs, and that new materials and modern techniques did not lead to inferior products.

The architect Jan Gratama, from Architectura et Amicitia, also put down his thoughts on the Dutch version of the Werkbund. His proposal was more aimed at the modernization of Dutch industry than Penaat's. The two proposals were discussed by a small group of interested parties in Amsterdam. By pure chance, at precisely the same moment Berlage was giving a lecture in Cologne on the influence of the Deutscher Werkbund in the Netherlands.

Penaat's and Gratema's proposals struck a chord with the members of the preparatory committee, but its enthusiastic and energetic plans were thwarted by the outbreak of war. They did not really get down to work until

1917, and by then they were even more motivated to use all available manpower as efficiently as possible for the benefit of national industry. The disappointing artistic quality of the Dutch products on show at the First Dutch Trade Fair in Utrecht in that same year gave them an extra impetus to pick up the plans where they had left off.

The members of the reconstituted preparatory committee, in addition to the initiators Penaat and Gratema, were the architect Karel de Bazel, the graphic designer Sjoerd de Roos, the manufacturers Pelt, Van Dissel and Braat, and the Rotterdam publisher Brusse. Their first action was to place a lengthy article revealing details of their plans in the daily *Algemeen Handelsblad*, with an attached questionnaire to gauge opinion. Eight artist-designers, seven manufacturers and three private individuals with an interest responded, and their answers were then discussed in a special issue of *Architectura*. The artists thought that there was only a slim chance that such cooperation would soon lead to good, fine-quality products.

The firms that responded – De Porceleyne Fles ceramic factory, Leerdam glassworks, Philips lamp factory, Braat metalwork factory, Dieperink printers, Van Dissel weaving mill and the furniture factory Labor Omnia Vincit – were distinctly in favour of the proposal. Of course, this came as no surprise: these were all factories that had previous positive experiences of involving artists in the production process. Finally, the three 'interested private individuals' who filled in questionnaires, W. Martin, the Leiden professor of art history, Karel Sluyterman, now professor of decorative art and ornamental drawing at Delft Polytechnic, and H. E. van Gelder, the director of the Gemeentemuseum in The Hague, were all positive about the idea.

The issue of *Architectura* in which the responses to the questionnaire were discussed was far more carefully designed and more richly illustrated than usual. This so-called Driebond-nummer (Triple Alliance Issue) was in some ways a kind of trial run for the periodical *Wendingen*, which was to appear a few months later.

The decorative artist André Vlaanderen provided a somewhat more detailed argument in his own contribution to the Driebond issue.[41] He was almost more resolute in his arguments than his colleagues: it was high time artists stepped down from their ivory towers and started to design everyday industrial objects. He thought that the packaging of simple workaday products like matchsticks, biscuits and toothpaste deserved to look every bit as good as other products and that the designs should conform to contemporary artistic standards. Indeed, everyday products of this sort were perhaps even more deserving of special attention, since the civilizing influence of such a simple mass-produced article was, after all, far greater than that of expensive luxury products. André Vlaanderen was an authority on the subject, since he was familiar with the more exclusive, elitist side of the designer's world. He had been the manager of Eduard Cuypers's interior design studio during the years when De Klerk, Kramer and Van der Mey were employed there. Moreover, he had been the owner of one of the first graphic advertising agencies in the Netherlands since 1904. The very earliest advertisements he drew were published at the back of the journal *Het Huis*, and some of them also had an accompanying photograph. From 1915 Vlaanderen designed the advertising brochures for the Gazelle bicycle factory in Dieren, near Arnhem. By the time he ceased working for them in 1953 he had made a total of more than five hundred advertisements and scores of publicity folders for the factory.

The Triple Alliance in its projected form never got off the ground. Wijdeveld had foreseen this in his cover for the Driebond issue: the three parties concerned had too many conflicting interests at the time. Willem Penaat therefore decided to change his strategy and suggested establishing an advice and information office (see above), where manufacturers and other interested parties could find information about Dutch designers. This Institute for Decorative and Industrial Art (Instituut voor Sier- en Nijverheidskunst, ISN) was founded on 5 March 1921 with financial support from the Ministry of Education, Arts and Science and worked in association with VANK's Advisory Board and its Publicity and Propaganda Committee. As much documentation as possible on artist-designers, including many photographs, was gathered at the Institute's office in The Hague.[42] Willem Penaat was responsible for daily management of the ISN. The lawyer Jean François van Royen, in daily life secretary of the Dutch Post Office Board and a great typography enthusiast, took over from him as VANK's chairman, an office he held until 1940.

Penaat remained in office for only a very short time, because in 1923 he was appointed director of the Museum of Applied Arts in Haarlem. W. F. Gouwe succeeded him as director of the ISN. Dynamic as ever, Penaat used his new position to put forward some important enterprises, starting to build up a collection of modern applied arts, organizing temporary exhibitions and trying once again to stimulate cooperation between industrialists and artists. The result was the establishment on 26 March 1924 of the Dutch Federation for Art in Industry (Bond voor Kunst in Industrie, BKI). Unlike the contemplated Triple Alliance, this was primarily an employers' organization, but nevertheless its name made it immediately clear that promoting the interests of 'art' was its principal motive. As the name implies, it was not just about art and industry going hand in hand but most definitely about art in industry.

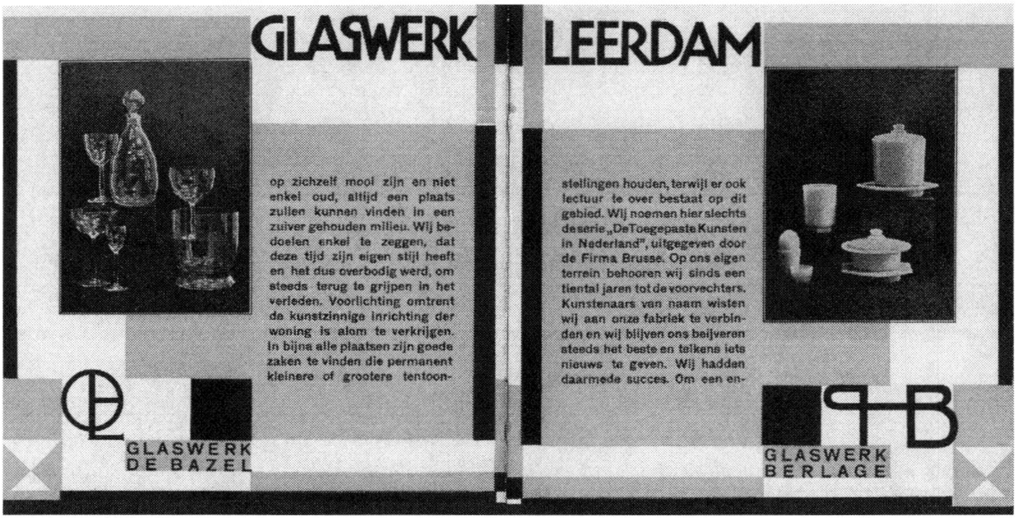

Machiel Wilmink, *Neemt het Schoone* ('Take the Beautiful') brochure (Leerdam glassworks), 1927.

The three manufacturers that Penaat found willing to work with him on the setting up of the BKI, the directors of the Leerdam glassworks, Van Dissel damask weaving mill and the Labor Omnia Vincit furniture factory, already had a tradition of collaboration with artists. Furthermore, all three had previously been enthusiastic about his Triple Alliance plans: they were no ordinary directors of three traditional firms.

Members of the BKI

The progressive, idealistic director of the Leerdam glassworks, P. M. Cochius, was not only interested in the well-being of his employees, but he was also absolutely convinced that a good, artistic product contributed to the happiness of the human race. He was active in the Theosophical movement in the Netherlands and there met the architect Karel de Bazel. Around 1915 Cochius had asked him to devise an alternative for what he felt were unsatisfactory pressed imitations of polished crystal, which Leerdam had brought onto the market in large quantities.[43]

Although De Bazel had never designed anything in glass, he accepted the invitation on condition that he would first be allowed to concentrate on blown glass. As he was accustomed to doing with his architecture and furniture, these glass products too were designed on a geometric basis. Beauty, according to his Theosophical beliefs, was after all in essence a specific correlation between size and number. In practice this meant that the golden section and squared paper were the starting point for the nine complete drinking sets that De Bazel was eventually to design for Leerdam between 1915 and his death in 1923. Aesthetically and commercially the experiment was a success. Whether due to the Theosophical principles or not, the critics and the public appreciated the restrained and unfussy design of De Bazel's sets, which for some even conjured up memories of 'old-Dutch' crafts.

After four years of experimentation De Bazel designed his first pressed-glass breakfast sets, which could be made in larger batches using moulds than his designs for blown glass ever could. Geometric principles are far more obvious here than in earlier objects. The success of De Bazel's work stimulated Cochius to commission other artists, starting with Cornelis de Lorm, who was followed by Chris Lanooy, Hendrik Petrus Berlage, Chris Lebeau, Jaap Gidding and many others, including the American architect Frank Lloyd Wright. Their designs were marketed as a special Design Collection by Leerdam. Furthermore, Andries Copier, Jacob Jongert and Machiel Wilmink developed special advertising material for this exceptional part of the firm's collection. The glassware itself was distributed through new channels: it could be seen at exhibitions in museums and art galleries and was sold in exclusive applied art shops.

Between 1923 and 1927 H. P. Berlage and Piet Zwart, who was then working at Berlage's firm, designed the sturdy, canary-yellow pressed-glass breakfast set mentioned above. Like De Bazel's set, the basic design principle behind this set was mathematical, being based on a regular hexagon and circle. Decorative vases for Leerdam were mainly designed by Chris Lanooy, Chris Lebeau and Jaap Gidding. Between 1925 and 1928 Gidding's original and colourfully painted examples were extremely popular. Although the vases in the Design Collection were not unique objects, neither could they be regarded as examples of industrial design. Artistry or special artistic merit continued to be linked to traditional methods of production.

Andries Copier, who at first was engaged to design advertisements for the glassworks, was the only designer Cochius employed on a permanent basis from 1923.[44] Copier was a jack-of-all-trades. Both technically and artistically he made an important contribution to the quality of the Design Collection and the good name of the factory. One of the ways in which he demonstrated his expertise in glass techniques and chemistry was his development of the totally new glass material graniver. This was used for floors and decorative mosaics, but Copier also designed a number of brightly coloured graniver cactus pots.

In contrast to De Bazel, Copier usually based his designs on shapes to be found in nature, as the names of his sets 'Comfrey' and 'Pear' illustrate. His

'Gildeglas' from 1929 is still produced today. The industrial character of this set of glasses was initially highly debatable because it was blown by mouth from crystal glass, which also made it very expensive. It was only after the Dutch government presented a Gilde service to Princess Elizabeth of Great Britain as a wedding gift in 1949, and after it won an award in 1954 at the Triennale in Milan, that it came to be manufactured industrially in 1958 and finally became a well-known mass-produced product. Although the glass has a natural appearance, the form was very consciously thought out after seeking advice from members of a professional wine-tasters' guild. All this led to it becoming one of the icons of Functionalism in the Netherlands.

While designing glass for everyday use, however, Copier was ultimately still more interested in traditional methods of glass-blowing and in the artistic effects that he could achieve in close collaboration with the glassblowers working at the factory, using colours, special raw materials and different techniques. His first 'free' forms were already being sold in 1924 under the product name of Unica. The most successful of these were subsequently produced in small series, and brought onto the market under the name Serica. Chris Lebeau also designed a great number of free artistic vases, dishes and bowls. All these objects for Leerdam were clearly signed, thus stressing their unique, artistic character. Together with the special Design Collection this amounted to only 10 per cent of total production at the most. The greater part of the range in Leerdam consisted of container glass. It is true that this was often designed with help from Andries Copier and Jacob Jongert, but unlike the prestigious Design Collection these everyday mass-produced products were certainly not proudly displayed at exhibitions in museums at home and abroad.

By 1924, the year when the BKI was founded, W.P.J. van Dissel, the director of a linen-weaving mill by that name in Eindhoven, also had plenty of experience of involving artists in production.[45] His firm had started calling on their services in 1905. Until then the only products being woven in Eindhoven were traditional, simple, chequered kitchen cloths made with a variety of household uses in mind. Inspired by the innovative work of a few Dutch artists in other fields, Van Dissel purchased a special loom at the beginning of the century, intending to start weaving modern, artistic 'Dutch' patterns. The artist Chris Lebeau, who had just created a furore with his refined batiks, was approached to see if he would be willing to make similar designs for Van Dissel. By 'Dutch' patterns the director of the weaving mill did not mean traditional seventeenth-century damask patterns, but Dutch *Nieuwe Kunst*, that is flat-decorative Dutch Art Nouveau motifs. The expensive jacquard loom Van Dissel had purchased for the job was a hand-loom, which in view of the tendency towards mechanization in linen-weaving factories at the time was unusual. His preference for such a loom clearly shows what objectives he must have had in mind when taking this initiative: not modernizing the production process but giving it a contemporary artistic injection, combined with harking back to the fine-quality craftsmanship to be found in the famous Dutch damasks of the Golden Age.

The beautiful tablecloths and napkins, with their rigorously stylized plant and animal motifs, that Chris Lebeau designed for Van Dissel in the following years were produced in the finest quality linen. The artist immersed himself so thoroughly in the complicated damask technique that it was possible to realize even his most subtle designs. Like the glassware for Leerdam, these tablecloths and napkins were presented in art galleries, museums and art magazines as 'artistic products'. The artistic table linen was marketed separately from the rest of Van Dissel's collection, which continued to be kitchen cloths, while with the aid of new looms large quantities of traditional, unpretentious ecclesiastical fabrics were also woven.

G. Pelt, the third manufacturer involved in the setting up of the BKI, was the director of Labor Omnia Vincit (LOV), an outstanding example of an idealistic furniture factory in Oosterbeek, near Arnhem.[46] Since the foundation of the factory in 1910 it had been operated on a system that gave the employees a say in company affairs and a share in the profits, although unfortunately there was seldom much profit to share. This factory was also one of the first in the Netherlands to introduce an eight-hour working day and provide a few days of annual holiday for the employees.

As well as supporting these social ideals and promoting the well-being of the workers, however, Pelt wanted to make good, distinctive, reliable products that were reasonably priced for a broad group of consumers. This was the reason why Pelt was absolutely not opposed to mechanized production and rejected the elitist, romantic views on handcrafted furniture of the English Arts and Crafts artists.

The architects H. F. Mertens, F. Spanjaard and, especially, J. A. Muntendam are the best-known designers who worked for this firm. It is patently obvious that their sober designs were influenced by Berlage's work. Generally speaking, LOV produced small series of these designers' works. A few Amsterdam School architects, including J. B. van Loghem, had their unique, much more expensive and elaborate designs made by the factory as well. Pelt's desire to be artistically innovative was without doubt less pronounced than it was for Cochius and Van Dissel: his idealism was more directed towards social reform and the well-being of his workers than towards a level of artistry.

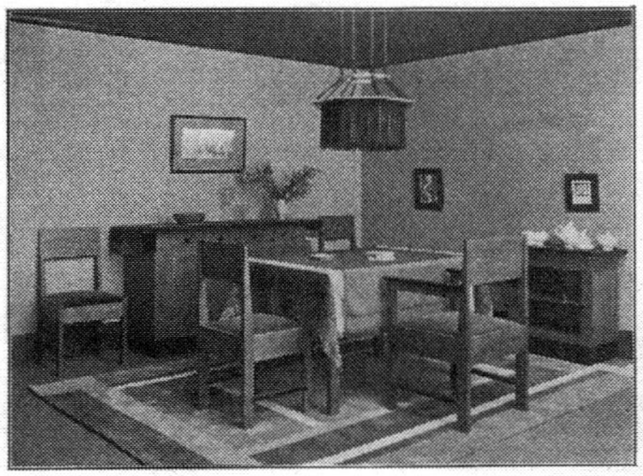

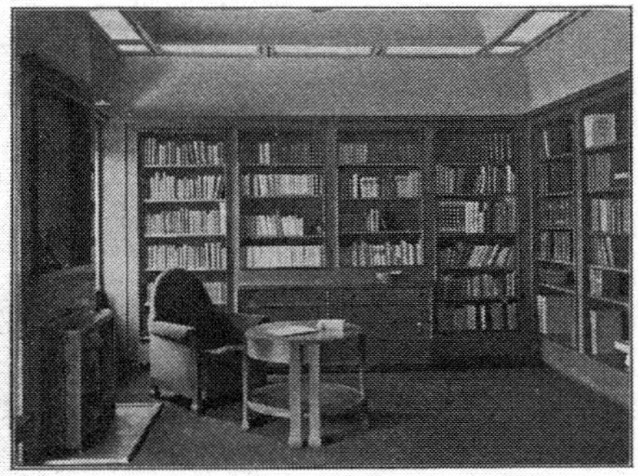

Photographs in the LOV brochure, *Wat overwint*, 1930, showing interiors by G. van Buuren and J. A. Muntendam.

The publishing firm W. L. & J. Brusse in Rotterdam was another of the early BKI members.[47] The firm's director, J. Brusse, the brother of the founder W. L. Brusse, was a member of the federation's influential board from the outset. This board determined the daily running of affairs and also decided which firms were allowed to become members. Ever since its foundation in 1903, Brusse's publishing house had pursued an idealistic policy. W. L. Brusse was an active member of the Rotterdam branch of the Social Democratic Workers' Party and his company's lists contained a great amount of socialist literature. His idealism, however, also extended into the artistic side of operational management. A great deal of care was taken with book covers and illustrations. In 1913 Brusse was the first publisher to use the Dutch Medieval font designed by S. J. de Roos. The most outstanding characteristic of this simple, harmonious font, which was made by the Amsterdam Type Foundry, was its legibility. It was also used for the VANK yearbooks, for the 'Applied Arts in the Netherlands' series and for almost all of Berlage's publications. Apart from Sjoerd de Roos, Brusse commissioned works from many artists and designers, including Johan Briedé, Jan van Krimpen, Richard Roland Holst, Piet Zwart and Paul Schuitema.

The members of the first board of the Dutch Federation for Art in Industry, in addition to J. Brusse, were P. M. Cochius, J. de Leeuw, director of Metz & Co., W. F. Gouwe, director of the ISN, A. E. von Saher, curator of the Museum of Applied Arts in Haarlem, and the designers Berlage, Penaat and N. P. de Koo. Thus the number of arts-related members far outnumbered those who were entrepreneurs or factory managers. The number of members increased only gradually because the entry requirements were strictly adhered to and it was necessary to be invited to join by an existing member. One of the conditions that had to be met before joining was that manufacturers had to mention their designers by name in all their publicity

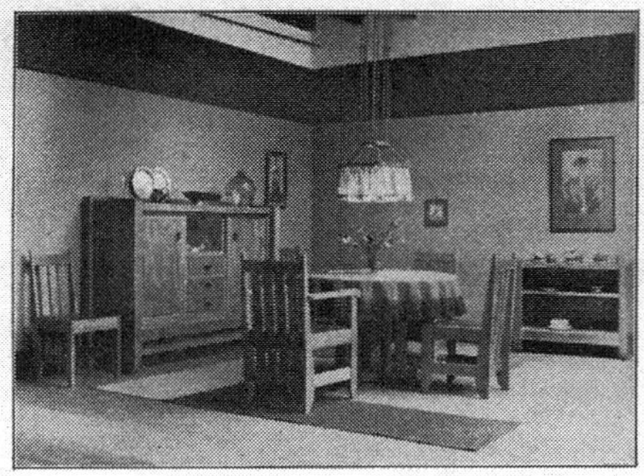

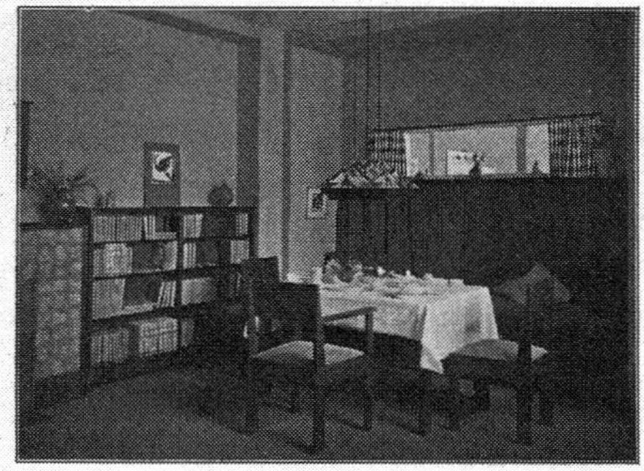

material. The Joh. Enschedé Type Foundry in Haarlem, the Pander furniture factory in The Hague, the Ram pottery factory in Arnhem, the Gispen metal factory, the Amsterdam Type Foundry and the Zuid-Holland pottery factory in Gouda were among the first firms to meet all the entry requirements. The number of members was never to be really large: in 1941 there were still only about thirty.

The BKI and Art

Until it amalgamated with the Institute for Industrial Design in 1950 (see below), the 'art factor' continued to be of importance to the BKI and its individual members. Almost all the manufacturers who met the requirements and were interested in joining produced products that were traditionally considered to be applied art or interior design, such as furniture, carpets, curtain fabrics, tableware and vases, with an emphasis on the more decorative and representational aspects of the field. However, popular partiality for decorated – meaning ornamented – products was gradually waning. In fact, the trend was going in the opposite direction; from the 1930s 'decorative' was no longer synonymous with 'decorated'. The preferred style that the federation implicitly propagated from then on was more that of modernism, with its uncompromising functionalism and abstract design idiom. The spherical vases and the Gilde glasses designed by Copier for Leerdam, and Gispen's modern lamps, could almost be termed prototypical for the 'artistically minded' industrial design promoted by the federation in the 1930s.

Joseph de Leeuw, the director of the prestigious furniture store Metz & Co., played an important role in the organization of the federation and in spreading its ideals.[48] This store had originally sold fabrics, having obtained

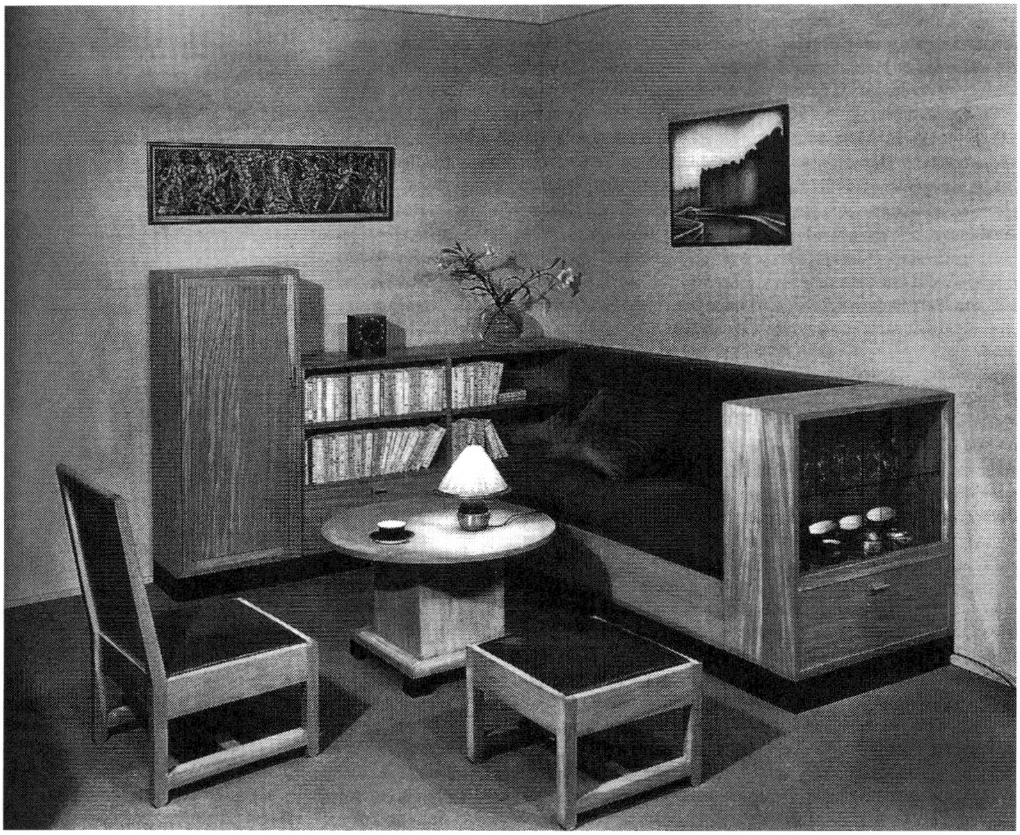

Willem Penaat (Metz & Co., Amsterdam,) joinable furniture, 1929.

exclusive rights in the Netherlands to sell those by Liberty of London. In 1918 De Leeuw decided that his firm was ripe for a rigorous form of modernization. By employing the interior and furniture designer Paul Bromberg, he made it possible for the firm to produce its own contemporary furniture. In 1924 Bromberg was succeeded by Willem Penaat, who had recently resigned his post as director of the Haarlem Museum of Applied Arts. Evidently Penaat had found his true vocation at last, for he was to stay with Metz & Co. for 25 years. In total more than 1,300 of his designs were brought onto the market by this Amsterdam furniture store. His 'joinable furniture', for example, proved to be a great success: the cubical designs could be combined in various ways, had a modern look and were space saving.

Undoubtedly it was through Penaat that Joseph de Leeuw became involved with the federation. De Leeuw linked idealism in modern design with a great talent for commerce, which led to the great importance of Metz & Co. to the development of design in the Netherlands in the 1920s and

’30s. As well as Penaat, designers whom De Leeuw managed to engage for the company included Gerrit Rietveld, Bob Oud, Bart van der Leck, Hendrik Wouda and Mart Stam. The company also sold designs by foreign celebrities such as Marcel Breuer, Erich Dieckmann, Le Corbusier and Alvar Aalto. The relationship between Joseph de Leeuw and the French-Russian artist Sonia Delaunay was even more exclusive, since for a few years she designed fabrics solely for Metz & Co.[49]

Not all the firms that were members of the Dutch Federation for Art in Industry were completely mechanized, and this was definitely not a precondition for membership. Therefore alongside weaving mills like Van Dissel and De Ploeg, calico printworks like Van Vlissingen and carpet factories like the Royal Carpet Factory Deventer, there were a few small-scale fabric studios run by craftsmen who were allowed to become members as well. Hand-weaving studios, such as De Knipscheer in Laren and Edmondt de Cneudt in Soest, together with the carpet-knotting and textile printing firm of Het Paapje, were also on the membership list. The latter company was particularly well respected. Under the capable management of Hans Polak, who had trained at the Rotterdam Academy, Het Paapje wove, printed and knotted artistic curtain fabrics and carpets for Metz & Co., De Bijenkorf and Pander, as well as producing special assignments for architects including W. M. Dudok and J.J.P. Oud.[50]

As well as the firms already mentioned, during the 1930s a few potteries working with craftsmen also became members, such as the one run by Pieter Groeneveldt in Voorschoten and Potterij De Rijn, where Meindert Zaalberg was in charge. The Zuid-Holland pottery factory was only allowed to become a member owing to the artistic quality of its special Modern Design Department, where designs by, among others, Jaap Gidding, Louis Bogtman, Jan Schonk and Erich Wichman were still painted by hand.[51]

A company that did not produce any products for living rooms, but still became a member, was De IJssel enamel factory in Dieren, which made simple kitchen utensils designed by Cor Alons during the late 1920s. Also on the membership list for a number of years were two factories, Frisia in Amsterdam and E. M. Jaarsma in Hilversum, which produced fire surrounds. The odd one out among the members was the chocolate factory of Van Houten in Weesp, which was allowed to join because of its special advertisements and packaging by Stefan Schlesinger, who was also responsible for a part of the graphic work for Metz & Co. and the Trio printing firm. This Austrian designer’s output attracted much attention in the Netherlands: it was elegant and refined, and had a more luxurious appearance than most Dutch packaging materials and advertisements at the time.[52]

National Post, Telephone and Telegraph Company

The National Post, Telephone and Telegraph Company (PTT) was not to be found on the members' list of the BKI. Although the company had not been a government institution since 1915, but a nationalized enterprise with its own budgetary accountability, because of its non-commercial status it did not qualify for federation membership. Yet it would have been an utterly 'model member' if it had been admitted. The major stimulus behind the company's progressive artistic policy was Jean François van Royen, whom we have already come across as the chairman of VANK and board member of the ISN.[53]

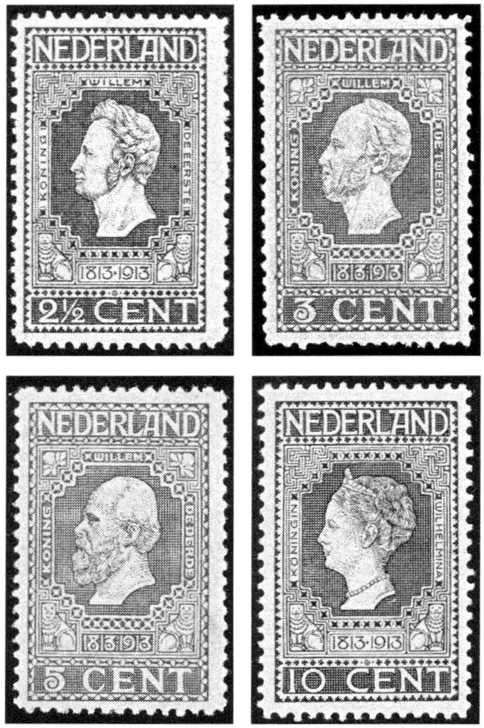

K.P.C. de Bazel, Jubilee stamps of 2, 3, 5 and 10 cents showing King Willem I, King Willem II, King Willem III and Queen Wilhelmina, 1913.

Van Royen took up his duties on the executive board of PTT directors in 1904. He showed a great interest in art and particularly in printing. Inspired by William Morris, in 1913 he became involved in a small firm of literary publishers, De Zilverdistel, which had been established a few years earlier by the poet P. N. van Eyk. As an amateur typographer Van Royen designed a few editions for this private press.

By drawing attention to the need to improve the postal company's graphic design, Van Royen was able to combine his work and his hobby in a useful way. By 1906 he had already secured a commission for Cornelis de Lorm to design new cast-iron signs to be placed outside the post offices. This was soon followed by a series of counter and wall plates. The text on the plates was executed in taut, sans serif lettering to which some minor geometric decorations had been added. The fact that these plates attracted attention, and that the special care taken over something so seemingly trivial was still extraordinary, is evident in a review written by Cornelis van der Sluys in the daily *De Hofstad*. Incidentally, it was this article that launched Van der Sluys's career as a 'design reviewer' in 1910.[54]

The first artist hired to design a new postage stamp was Antoon Derkinderen, who created a stamp drawing attention to tuberculosis in 1906. However, the jubilee stamps designed by Karel de Bazel from 1913 are better known, owing to their innovative, two-dimensional interpretation of traditional portraits of the four monarchs who had reigned in the Netherlands since 1813.

Van Royen thought that these initiatives were merely a 'drop in the ocean', because a great deal of work still needed to be done to improve the quality of Dutch government printing. In 1912 he denounced the standard of print as 'Ugly, ugly, ugly . . . the letter type, the typesetting, and the paper'.[55] His proposal to adopt the new Dutch Medieval font by Sjoerd de Roos for all national printed matter, however, was rejected.

When Van Royen became general secretary of the postal company in 1918, he took every opportunity, wherever and whenever, to improve the design of as many divisions as possible of this continually growing, multifaceted organization. Hundreds of commissions were issued by the postal company and the ISN over the years thanks to his mediation. The combination of his positions in VANK, the ISN and at the PTT meant that Van Royen almost became the embodiment of the views held on the relationship between art and design in the Netherlands during the interwar years.

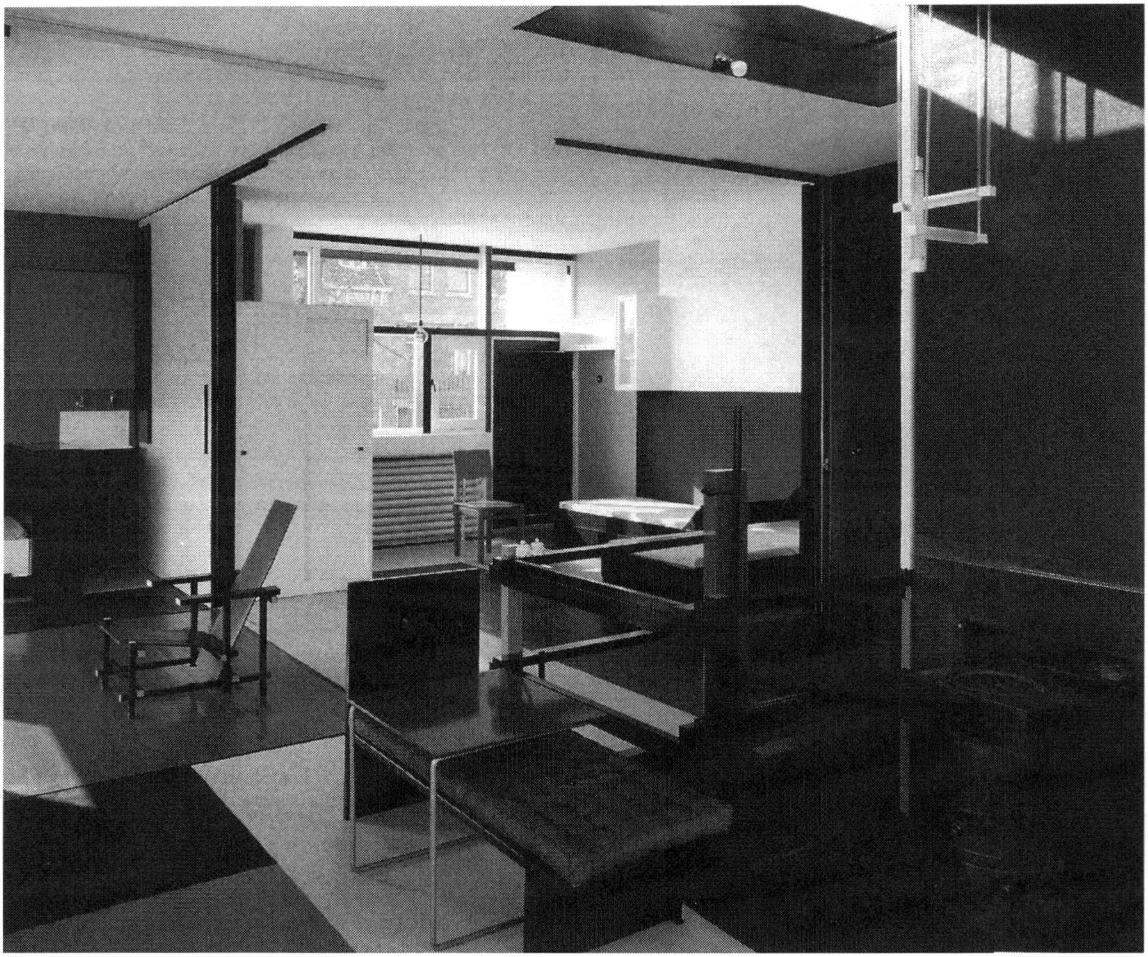

3

Good Design, 1925–65

The 1920s marked the start of a more radical, social and political form of idealism in Dutch design that came to be firmly linked to observations on modern functionalist architecture and progressive views on industrial mass production. In this context it is referred to as Dutch Moral Modernism.[1] This ethical aspect of design was only a relatively new phenomenon (see chapters One and Two). In the nineteenth century 'honesty' and 'character', for example, were central concepts in design, as were 'rationality' and 'sobriety' around 1900. From the 1890s designers were, on the whole, firmly convinced that those who used their products would be happier, in the belief that a handsomely designed object is bound to appeal to a person's better nature. Those who surround themselves with beautiful – and in the Netherlands this usually means unpretentious, sensible, honest – products will foster the same qualities in themselves.

This moralism was not restricted to the products but, as we have already seen, it also affected the perception of the artist's or designer's task. In the case of Pierre Cuypers and Gerrit Dijsselhof, ethical-religious motives determined not only their preference for Gothic style, but also their idea that the architect or designer should take the lead and become a binding element in the community.

At the start of the twentieth century, the founders and the earliest members of the VANK experienced all these social ideals in different ways, and in changing alignments and associations. For some it was socialist ideals that were of vital importance, for others Theosophy or Freemasonry. Usually the designers were actually inspired by a vague mixture of religious, ethical and utopian socialist-communal ideals. In an epoch in which a class-ridden society slowly but surely evolved into a modern community, these

Gerrit Rietveld, living room, Rietveld-Schröderhuis, Utrecht.

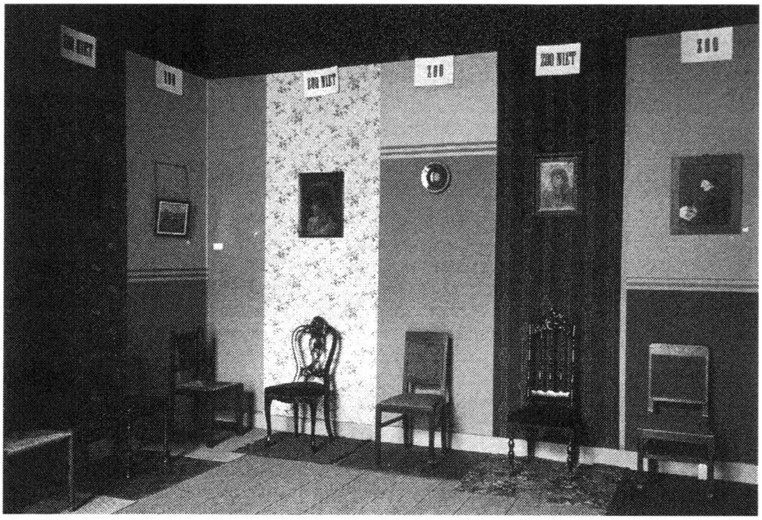

A view of the exhibition installed by Cornelis van der Sluys for *Kunst aan Allen* ('Art for Everyone') in The Hague, 1909.

industrial artists thought they could make their contribution by producing sensibly designed products that would have a civilizing influence. Designers like H. P. Berlage, Jan Eisenloeffel, Willem Penaat and Cornelis van der Sluys even went one step further and tried to ensure that beauty in general filtered down to 'the ordinary man'. They were involved in the founding of the Amsterdam association Art for the People (Kunst aan het Volk) and Art for All (Kunst aan Allen) in The Hague. The aim of these associations was to educate people on art, good taste and beauty. Exhibitions seemed a good way of promoting these aims in a tangible way. Shows with unambiguous titles like *Exhibition against Deceitful Taste* (*Tentoonstelling tegen Smaakmisleiding*, 1910) were set up where people could compare acceptable and unacceptable interiors and products, the differences being shown in clear letters next to the good and bad items on display.

The Rotterdam Opbouw (Advancement) Association: From Social Idealism to Moral Modernism

Until the 1920s the effects of this cultural offensive were marginal. The working classes had not the slightest need for the sober, sensible design that the well-intentioned cultural reformers were eager to foist upon them. Moreover, idealistic studios like 't Binnenhuis, De Ploeg, De Woning and Amstelhoek continued to make extremely expensive products that only the wealthy elite could afford. Most people were pleased with the growing supply of cheap mass-produced articles from abroad. And they were not

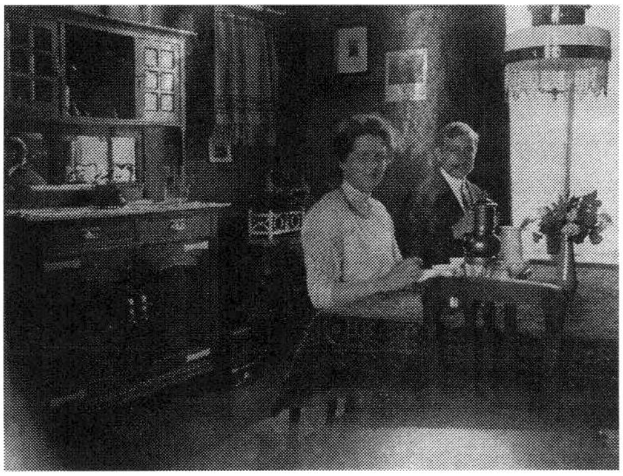

Newly wed maths teacher Joop Simon Thomas and his young wife in their new interior, 1912.

bothered by the fact that these products had been machine-made. The underlying principles were cherished by only a small select group of artists, the people who commissioned their work, and at most merely a few sympathetic teachers.

It was not until after the First World War that a number of designers became aware of their isolated position and the limited opportunities open to them for achieving their ideal of a better world through better design. The way in which the rather ineffective social-ethical design principles were gradually adjusted in the 1920s to form a decisive, but in actual fact equally moral form of modernism is well illustrated by the artists' and architects' association Opbouw (Advancement) in Rotterdam, and the changing views of a few of its most prominent members. Its existence also illustrates the increasing importance of Rotterdam in the design culture of the period.[2]

Opbouw was founded in 1920 by the architects Willem Kromhout and Michiel Brinkman. Among its earliest members were the decorative artists Jacob Jongert and Jaap Gidding, the manufacturer Willem Gispen, the architects Mart Stam, Bob Oud and Leen van der Vlugt, and the furniture designer N. P. de Koo; even the traditionalist architect M. J. Granpré Molière joined their ranks. They were a heterogeneous group with broad cultural objectives, although the name the members chose for their organization indicated their shared progressive mentality.

Jacob Jongert had moved to Rotterdam in 1918 on his appointment as Head of Decorative and Industrial Arts at the Rotterdam Academy.[3] After training at the National School of Applied Arts in Amsterdam, he was originally influenced by the socialist artist Richard Roland Holst and assisted him in 1911–12 in creating symbolically charged figurative murals in Berlage's Union of Diamond Workers building in Amsterdam. When Jongert saw the work of German industrial designers at the *Werkbund* exhibition in Cologne in 1914, however, his eyes were opened to the more contemporary ways open to artists willing to dedicate themselves to the service of society. Looking back at his life in his handwritten memoirs in the 1940s, he remembered being fascinated by the shopping streets in Cologne, where everything – from biscuits to electrical appliances – was packaged and exhibited in the shop windows in a modern way: 'This gave us a shock . . . here was a totally new terrain for us to work for, until then nobody else in Holland had been active there.'[4] From that moment he welcomed collaboration with industry and intensified his advertising work for the Purmerend drinks factory De Wed. G. Oud & Co. and the Leerdam glassworks. In 1919 Jongert started to work for the Rotterdam tobacco, coffee and tea factory of De Erven de Wed. J. Van Nelle. Influenced by the artists of *De Stijl* and the design of *Wendingen*, he incorporated geometric shapes and prime colours into his designs. His teaching at the Academy introduced others to his modern insights. In 1924 the newspaper *Nieuwe Rotterdamsche Courant* commented that 'Modernism' has now also penetrated through to the inner walls of 'our revered Academy'.[5]

Much of the packaging material and a large number of the advertisements produced between 1923 and 1940 for Van Nelle were designed by Jongert, who emphatically stressed the firm's modern, progressive image in his work. This allowed Jongert to fulfil his mission of spreading good design to as many layers of the population as possible. Nevertheless, no matter how progressive his packaging may have looked, and no matter how applied graphic art shifted in the direction of functional advertising through his labours, Jongert never took the definitive step that would have enabled him to create a commercial, purely industrial form of graphic design. The alphabet he used for Van Nelle, using modern sans serif script, was made by cutting out each letter piece by piece and sticking it onto paper. Unlike his contemporaries Piet Zwart and Paul Schuitema, he very rarely used the new medium of photography in his designs.[6] Indeed, Jongert recoiled from using the most modern equipment in pursuing his social ideal.

By contrast, Willem Gispen embraced the new age and its new technical prospects unreservedly and his lamps were industrially manufactured on a grand scale in the 1920s (see chapter Two).[7] He was also motivated by the firm moral conviction that the new 'pure and clean interiors' that could be created with the aid of these industrial articles would have a 'strong and stimulating influence' that would be of great benefit to the community. He loved the big city of Rotterdam with its cars and gigantic ships, the docks, the neon signs, its modern Lifting Bridge, its airfield and proper aircraft factory. But Gispen was no socialist. He was, and continued to be, a factory manager, and despite his social idealism his main interest lay in the economic performance of his company.

Mart Stam was probably one of Opbouw's youngest members when it was founded, since at the time he was barely 21 years old.[8] After completing his art teacher training course and a brief internship with the Amsterdam School architect J. M. van der Mey, he was able to secure an appointment in the office of the architect M. J. Granpré Molière. It was to be some time, however, before Stam would be in a position to take advantage of his

Opbouw membership, since that same year he spent several months in jail for refusing to do military service as a conscientious objector. This lonely period in his cell contributed to the radicalization of his views, so much so that he became a fully fledged communist. In 1922 and 1923 Stam worked in Germany at Max Taut and Hans Poelzig's office and got to know avant-garde artists such as El Lissitzky and the graphic designer Karel Teige. Straight after that he went to Switzerland and stayed until 1925 working, among others, for Karl Moser. On his return to Rotterdam in 1926 Stam became involved with the building of the Van Nelle factories. In that same year he invented the principle of the suspended, tubular, steel chair. Evidently he felt very much at home in Rotterdam during this period, because he twice turned down Walter Gropius's offers to take up the position of head of the architecture department at the Bauhaus. In the early 1930s Stam went to the Soviet Union for a few years accompanied by Johan Niegeman and Gerda Marx. There they joined Ernst May's group in Magnitogorsk, Siberia, where Stam met the former Bauhaus student Lotte Beese, whom he was to marry in 1934.[9]

They were not the first Dutch architects to take this principled step with such far-reaching consequences, since in 1925 J. B. (Han) van Loghem had set off for Kemerovo in Siberia.[10] Van Loghem was an architect and a furniture-maker who could be considered to belong to the Amsterdam School (see chapter Two). However, in addition to making expensive and luxurious furniture for the elite, he contributed to a few idealistic public housing projects. His views gradually became so radical that in 1919 Van Loghem joined the Union of Revolutionary Socialist Intellectuals and he reorganized his design office into a collective cooperative. His move to Siberia in 1925 was not altogether unexpected.

Han van Loghem stayed in the Soviet Union for two years and settled in Rotterdam on his return in 1927. In that year he also became a member of Opbouw and it was due to him that the aims of this organization were rigorously reformulated. From 1927 onwards membership was open only to architects and designers who openly and consciously supported the politically inspired ideals of functionalism and New Objectivity. From then on Van Loghem no longer designed 'artistic' wooden furniture, but turned to less striking, tubular steel chairs, which were produced by the Rotterdam furniture factory d3 (later known as Fana). His pioneering book *bouwen bauen bâtir building* (1932) proved to be one of the most fundamental Dutch publications on modern architecture in relation to politics, ethics and aesthetics. Covering everything from interior design to urban development, it was an important manifesto for functionalism. An architect was no longer supposed to restrict himself to drawing new houses but was to organize the whole 'process of living' from start to finish; the designer was expected to reinvent himself and develop into a 'producer'.[11]

The Problem of Public Housing

As in the rest of Europe, a growing number of Dutch designers were becoming increasingly conscious of the huge social problems facing society. This led many designers to consider affordable housing and everyday domestic objects as their principal mission. The Housing Act (1901) had already created a framework in which projects of this type could be carried through. From then it became compulsory for local authorities to become more active in controlling levels of hygiene, habitability and minimum rents. Initially it was the private building firms that mostly profited from these measures, but from 1914 the government began to take public housing into its own hands. One of the first sensational results was the Rotterdam housing complex Justus van Effen (1921), commissioned by Rotterdam City Council and designed by Opbouw member Michiel Brinkman.[12] Although trained in the classical tradition, he had presented them with a workable alternative to the compact and monotonous blocks of houses developed in the nineteenth century. The apartments were arranged in an ingenious and complex way, grouped around an inner garden; raised inner streets then ran up to the higher storeys. With their shared communal facilities, such as central heating, washrooms and children's playgrounds, Brinkman showed how an architect could make a real contribution to the well-being of his less fortunate compatriots.

The architect J.J.P. Oud was not only associated with *De Stijl* but was also a prominent member of Opbouw. During this period he began to take an interest in council housing and the problems of public housing provision.[13] In 1918 he was appointed to the position of architect at the Rotterdam Department of Housing, where he worked with Theo van Doesburg on the design of a few rows of houses in the Spangen district. Such a form of collaboration with an artist was entirely in keeping with ideas formulated by *De Stijl* on 'Neo Plasticism' (*Nieuwe Beelding*), but in practice working together still proved to be difficult. Van Doesburg designed stained-glass windows for the houses and created a special colour plan for both the exterior and the interior. In the living rooms the walls were yellow and the doors blue; the fireplaces were painted grey and blue and tiled with black, red and yellow tiles. In 1920 a show house was furnished in one of the blocks using Gerrit Rietveld's slatted furniture. After completion of their first joint project, however, Oud

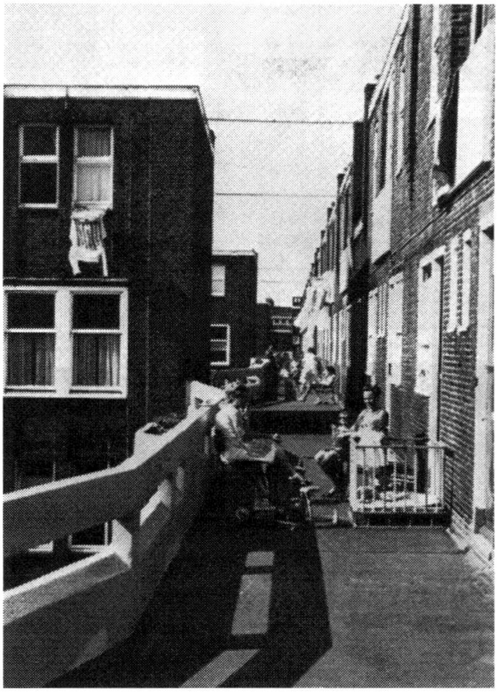

Justus van Effen, housing complex in Rotterdam by Michiel Brinkman, 1921. Public gallery on the third floor.

noticed that the new residents were very quick to paint or wallpaper over Van Doesburg's colours. Objecting to the even wilder colour scheme Van Doesburg planned for the next block, he thought that the colour suggestions for the front of the block flatly contradicted the static character of the architecture and the materials that were to be used. Oud's own, more 'decorative', ideas about the use of colour in architecture emerged a few years later in his design for the café De Unie. Here he made a striking statement with what was intended to be a temporary façade on the Coolsingel, right in the centre of Rotterdam, designed in complete contrast to its surroundings.

While working on his 1924 design for the Witte Dorp (White Village), a residential area in Rotterdam, Oud had to abandon most of his aesthetic aspirations for financial reasons. The only elements that referred to *De Stijl* idiom were the yellow and blue painted doors, window frames and guttering, together with the red roof tiles and the white plastered walls. Here, the simple, but well thought-out architecture and the planned urban development scheme were primarily intended to discipline and stimulate a group of troublesome residents into behaving like 'decent families'. In his 1925 plan for working-class housing in the Kiefhoek district of Rotterdam, Oud was given his first opportunity to carry out a number of more innovative ideas regarding the floor plans, standardization and industrial building. In this project his main aim was to design a standard house with a maximum amount of living space on a minimum budget. The houses were planned for the large group of economic migrants attracted to Rotterdam as the docks expanded, as well as for those who had been evicted from their dwellings in the old town centre due to a slum clearance programme. With these projects Oud earned an international reputation for applying new building techniques to public housing.

Nieuwe Bouwen (New Building)

The years 1927 and 1928 were crucial for developments in modern architecture and design in the Netherlands. It is not that everyone suddenly started to design and build in a more modern way – this still remained the exception rather than the rule – but from then debate on design and architecture did take a new turn. These years marked the start of a period in which discussion was influenced by international contacts, a positive attitude towards industrial building techniques and mass production, and greater political consciousness. Take 1927 for instance, a year in which the association De 8 (The 8) was founded by a group of then unknown Amsterdam architects. Their distinctly progressive views were published in a manifesto in *i10*, an international art and architecture journal.[14] In eighteen short 'business-like' statements the young architects state that De 8 aims to give a 'critical reaction to the architecture of this era', that they intend to be 'realistic and factual', that they are seeking opportunities for 'international collaboration' and are willing to make themselves absolutely 'subservient to the assignment'. De 8 does not aim 'to make affluent architecture sprouting from a sensuality of form created by talented individuals' and for this reason the authors of the manifesto claim that 'it would be better for the present moment to build in an ugly, purpose-designed manner than to erect showpiece architecture from poor plans'. De 8 worked more for the realization of 'building-science' than 'building-art'.

That was plain speaking, and for those who read between the lines it was clear that De 8 wanted to oppose the 'wrong' sort of Amsterdam School buildings. What was perhaps less clear was that directly under the surface of this ostensibly technocratic language lurked a moral standpoint. In their analyses of the needs of future residents in their houses there was usually an undercurrent of strong views about the way people should live and relax in their spare time. Similarly, these architects' beliefs on the role of women in society were also fairly conservative.[15]

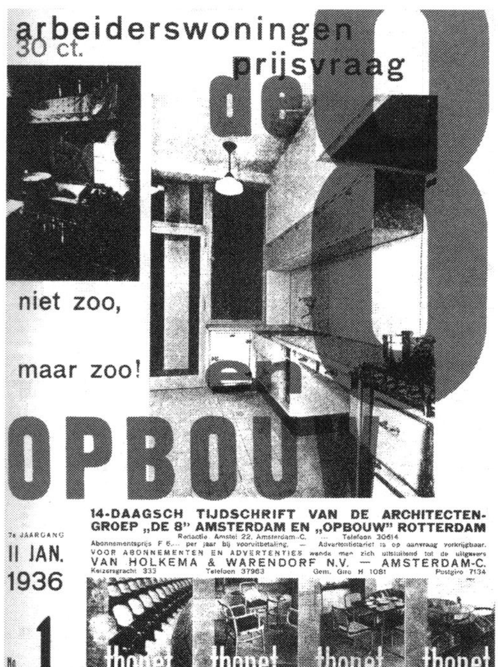

Cover of *De 8 en Opbouw*, 1936.

Members of De 8 took their inspiration from the Rotterdam Van Nelle factory, which was in the process of being built by the architects Brinkman and Van der Vlugt, and the Hilversum TBC Sanatorium, Zonnestraal, by Jan Duiker and Bernard Bijvoet. Both buildings were later to become icons of the international *Nieuwe Bouwen*, notably for their rational floor plans and structure, and for their disproportionate use of glass, metal and white-distempered walls. Jan Duiker and the more experienced A. Boeken,

Model housing by J.J.P. Oud at the Weissenhofsiedlung in Stuttgart, illustrated in J. B. Van Loghem, *Bouwen, Bauen, Batir, Building*, 1932.

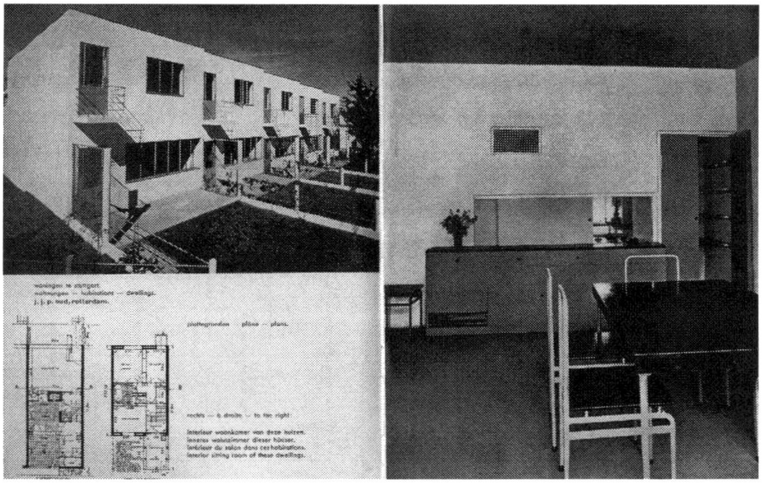

W. G. Wiebenga and Cornelis van Eesteren, who had recently been appointed chief architect to the Amsterdam Department of Urban Development, joined De 8 in the late 1920s and their membership was to win the society enormous prestige. In 1932 De 8 and Opbouw amalgamated and jointly published the periodical *de 8 en Opbouw*.

In 1927 Bob Oud and Mart Stam took part in the large international exhibition *Die Wohnung* in Stuttgart.[16] Part of this event, initiated by the Deutscher Werkbund, was a complete district full of show houses, the Weissenhofsiedlung, designed by sixteen modern architects from the whole of Europe, including Le Corbusier, Hans Poelzig, Peter Behrens, Bruno Taut and Ludwig Mies van der Rohe, all of whom at that moment considered it their mission to design a good house that everyone could afford to live in. With their suggestions for new, efficient, industrial methods of building and with practical solutions for the lay-out of the houses, the Weissenhofsiedlung was to contribute to solving the huge housing shortage in Germany and elsewhere in Europe. Oud and Stam both designed a row of semi-detached middle-class houses, just meeting minimum measurement requirements but providing a maximum of usable space. The lay-out of the houses was based on an analysis of the activities that should be carried out within its walls.

For the *Nieuwe Bouwen* architects the interior and everything related to it was every bit as important as the building itself. Moreover, not only did the house have to be arranged in as practical a way as possible, it also had to be built in the most favourable position for catching the sun and be easy to ventilate and clean. Stam also designed a large part of the furniture for their houses. The lamps were supplied by his friend and fellow townsman Willem Gispen.

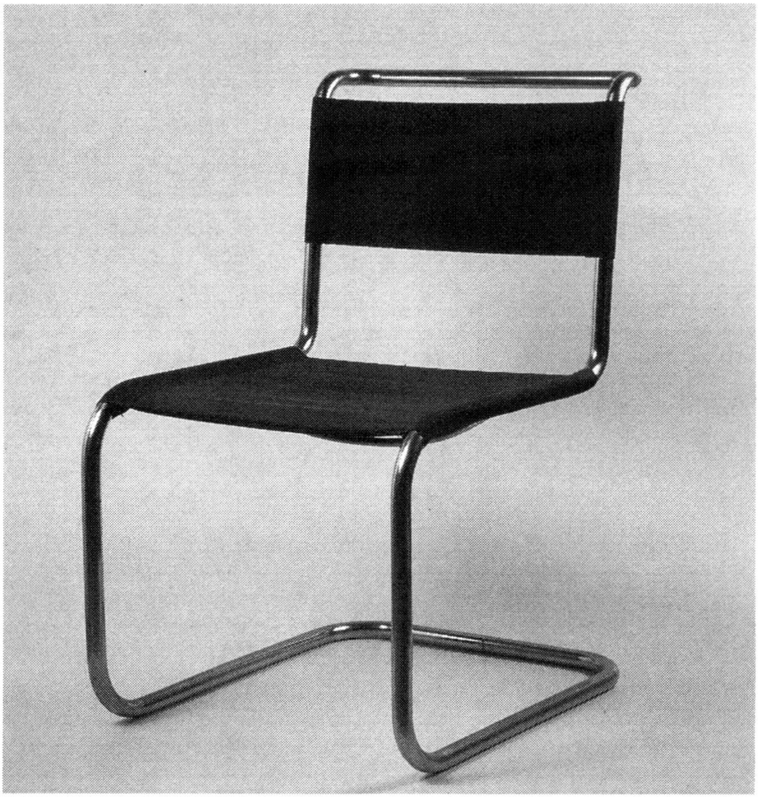

Mart Stam (Thonet), tubular steel chair, 1929 (designed in 1926).

Mart Stam first presented his free floating chair (without back legs) in Stuttgart. Much has been written about the background to this revolutionary idea and the true deviser of this new principle has often been disputed.[17] Whatever the truth, in 1926 Stam had completely worked out the archetype of the idea in Rotterdam, using gas pipes and fittings and two pieces of cloth – one to sit on and one to lean against. In November 1926 he talked about this experiment during a dinner party in Berlin (or, according to other sources, in Stuttgart) and made a drawing of it on the back of a menu for his colleague Ludwig Mies van der Rohe. A year later the German architect was able to show a splendid example of the *Freischwinger* made from shining nickel-plated tubes, whereas Stam still only had a makeshift example on show in his house. It is worth noting that in exactly the same period Marcel Breuer also designed his first chair made from tubular steel. Everything points to the idea being 'in the air'.

Oud also designed a few pieces of metal furniture for the Weissenhofsiedlung, including a dinner table with accompanying austere, minimalist

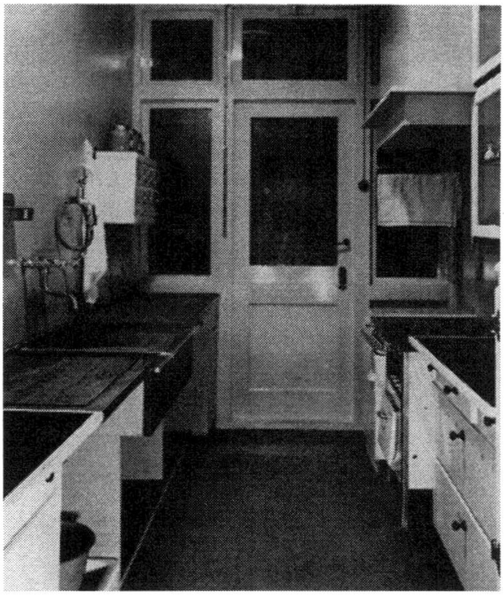

J. W. Janzen, 'Holland-kitchen', designed for the Dutch Housewives Society, 1931.

chairs, painted bright blue. The kitchen in his show house complied with the newest insights in the field of efficiency and hygiene as laid down by the German economist Erna Meyer in her book *Die Neue Haushalt* (The New Household Management, 1926). The Netherlands Union of Housewives took the initiative to have this book translated in 1928 and it too was to create a furore when published in the Netherlands under the title *De Nieuwe Huishouding*. The book spoke highly of the 'Frankfurt Kitchen', which had been developed by the Viennese architect Margarethe Schütte-Lihotzky for Ernst May's large social housing projects. Another title to appear in the Netherlands in 1928 was the Dutch translation of *Household Engineering: Scientific Management in the Home* (1915) by the American Christine Frederick under the title *De denkende huisvrouw* (The Thinking Housewife). Frederick was the first writer to base her views for the most practical kitchen design on F. W. Taylor's principles of operational management and labour division. It is a lesser-known fact that around 1930 the Dutch architect J. W. Janzen was commissioned by The Hague branch of the Netherlands Union of Housewives to create a design in imitation of the 'Frankfurt Kitchen' and, like the original, to base it on time and motion studies. This 'Holland kitchen' housed 'a housekeeping factory' and, as an added bonus, it incorporated the most modern labour-saving devices, including a refrigerator, a swivel tap and a small shower-head tap for rinsing the dishes.[18]

Gerrit Rietveld and *Nieuwe Wonen* (New Living)

The furniture designer Gerrit Rietveld also started to take an interest in low-cost housing in the 1920s.[19] His departure point was the interior. His first design from 1924, now known as the Rietveld-Schröderhuis, was not exactly what you might call a shining example of a house for someone living on a minimum income.[20] However, it was indeed an exercise in the 'liberation of superfluous objects' and that was to come in useful in later projects. Nothing in this house in Utrecht built for Truus Schröder-Schräder, the widow of the lawyer Frits Schröder, could be called traditional or 'ordinary', because almost everything – from floor plan to doorknob – had been specially designed to fit her and her three children's specific needs. The assignment was in principle to create a house in which living had been upgraded to an 'activity'. Rietveld had been granted permission to omit everything that was not strictly necessary, but only if this did not lead to a loss of comfort. However, the fact that Rietveld's idea of comfort was something quite different from encouraging idleness is shown in the final result. The unconventional, open structure of the house, the sliding walls that allowed the interiors to be used in different ways, the specially designed furniture and the conspicuous use of colour are the most important style characteristics. The fact that it was a brick building, and as such still traditional, scarcely detracts from the innovative character of the house.

Later on in the decade Rietveld became interested in public housing and designed a few small interiors and several model houses. He was one of the three Dutchmen present in 1928 at the foundation in La Sarraz, Switzerland, of CIAM, the international architects' organization.[21] The group claimed that they could make a realistic contribution to tackling the housing problem by opting for advanced solutions in the fields of technology, economy, hygiene, aesthetics and ideology. Apart from Gerrit Rietveld, the two Dutch members to sign the declaration were H. P. Berlage and Mart Stam. Berlage was invited by the organizers because he was seen as a pioneer of their innovatory ideas. J.J.P. Oud, then the Netherlands' best-known modern architect abroad, was also invited but was unable to attend due to the pressure of work.

Rietveld started to experiment about 1927 with new chair structures and different materials, making chairs by sawing sheets of fibre and triplex into a specific shape and then bending them in all directions. The results were revolutionary, 'a one-piece chair' that was none the less difficult to produce in large series. His 'bow chair' was more successful, combining fibre, and later bent triplex, with a metal frame. This was Rietveld's first, more or less successful, attempt at making an attractively priced chair for the masses. From about 1930 the chair, and variations on its design, was sold by Metz & Co. in Amsterdam, but at the time it was not cheap enough for everyone's pocket.[22]

In 1933 Rietveld designed a spectacular transparent glass pavilion on the roof of the Amsterdam Metz store in the Leidsestraat, where progressive functionalist furniture designs were displayed. Jan Duiker praised it as 'Metz & Co.'s Nieuwe Bouwen house'. Another Metz commission, however, a 1934 model house furnished by Rietveld in the new Bergpolderflat in Rotterdam, came closer to fitting this description. The architects Brinkman, Van der Vlugt and Van Tijen created living accommodation for a family

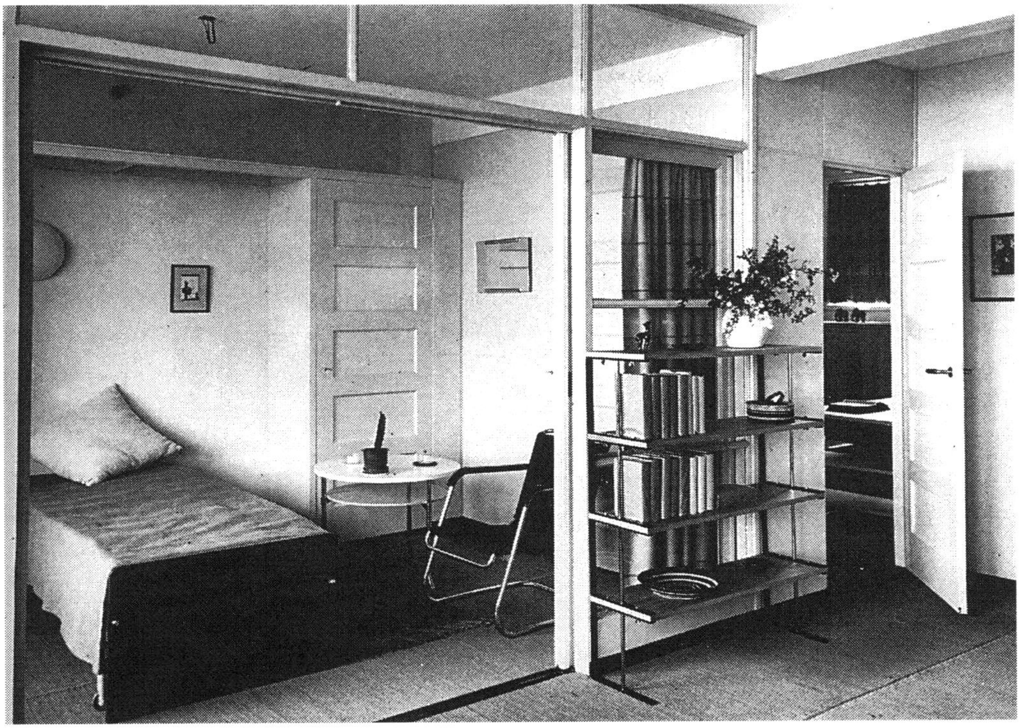

Gerrit Rietveld, interior of the Bergpolderflat in Rotterdam, realized by Metz & Co., 1934.

with two children taking up a surface area of only 45 square metres. An extremely efficient lay-out and large windows provided sufficient light, air and space, and encouraged hygienic living. The parents' bedroom could be added to the living space during the day by pushing back a sliding wall. Rietveld aimed at using just a few lightweight, moveable pieces of furniture and ensured that the space-devouring beds could all be folded up. This furniture was made of tubular stainless-steel or they were Thonet chairs, which were at least light and took up as little space as possible. Moreover, they were placed in such a way that they left as much open space as possible for walking and playing. The colours chosen for the interior were light and the soft furnishings subdued.

The White Villas

Although *Nieuwe Bouwen* architects considered the provision of houses for the people and their interior furnishing and decoration to be their main social task, there was little evidence of this in practice. The economic crisis in the late 1920s meant that very few public housing projects were actually realized. If something could be built, the high-principled functionalist designers seldom received the commission because very few authorities were enchanted by their work. The drawback with these architects was that, although they based their work on an analysis of the needs of the future tenants, giving their plans a scientific ring, or at least the pretence of inevitability, they did not really take the future tenants' wishes into account. The houses and blocks of flats built by modernist architects were thought to be much too bleak and functional and the show houses reminded people of offices and hospitals. Therefore, in practice, they seldom succeeded in convincing the tenants of their merits.

Still there were families other than Mrs Schröder and her children who were attracted by the promise of the whole New Living (*Nieuwe Wonen*) concept. Not surprisingly, they were to be found mainly among friends and relations of those in the modern architects' circles. Their new houses could not exactly be counted as homes for those living at subsistence level. In Rotterdam, for instance, two of the three Van Nelle factory directors had themselves measured up for a spanking new, ultra-modern design by the architect Leen van der Vlugt.[23] Kees van der Leeuw's house was built between 1927 and 1929 along the edge of a Rotterdam lake called the Kralingse Plas. It contained all the principal features of the functionalist house, from the sliding partitions, the smooth plain walls, the built-in furniture and the tubular steel chairs to the sports area at the top of the house, where the sliding glass roof could be partly opened to let in the sun and fresh air. The house was also equipped with the latest electrical and heating devices. The other director, A. H. Sonneveld, had a house built close to the centre of Rotterdam. It has recently been restored to its original state and is now a public museum. Before the Sonneveld family moved into their new white villa they had been living in an attractive, but dark and impractical, nineteenth-century building on a shadowy boulevard in Rotterdam. When they moved in 1933 the family left everything behind them in their old house in order to be able to get the most out of living in their newly equipped and newly furnished house.

The Sonnevelds were enthusiastic about modern urban life – Mrs Sonneveld was the first female car owner in Rotterdam. Van der Vlugt, their architect, assembled the whole family in advance to discuss every detail of their new house, including the interior decorations. Most of the furniture and the lamps came from the Gispen factory. The beds were supplied by Auping, a firm based in Deventer that specialized in comfortable, medically approved, hygienic, sprung beds as well as healthy mattresses. Metz & Co. took care of the soft furnishings, including the knotted carpet in the

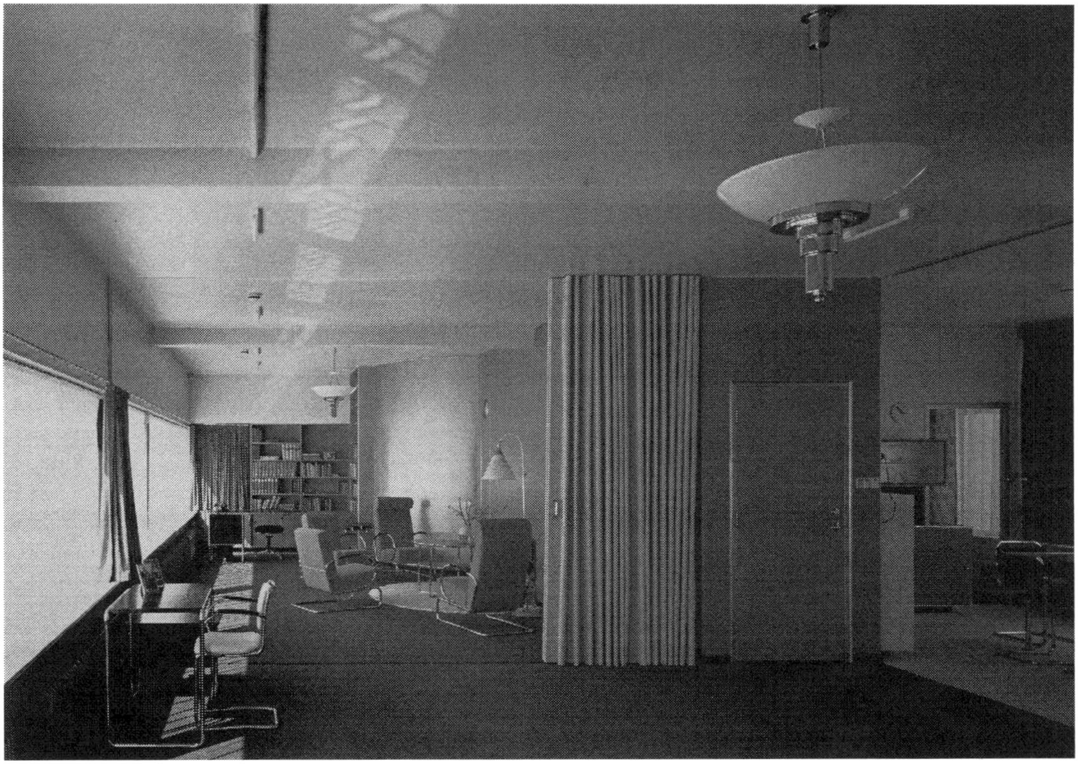

L. C. van der Vlugt, living area of the Sonneveld House, Rotterdam, with furniture by W. H. Gispen, 1935. Photograph taken following the building's restoration in 2001.

living room and the printed curtains in the kitchen, both after a design by Elise Djo Bourgeois. It almost went without saying that a great deal of linoleum was used in the house, bought from the firm Krommenie. Linoleum was waterproof, insulating, easy to clean and readily available for delivery in bright colours and was therefore the floor covering most recommended by all *Nieuwe Bouwen* architects.

Huize Sonneveld was a prototype for the *Nieuwe Wonen* concept, but a very luxurious version, for an excessive amount of attention had been paid to luxuries, comfort and aesthetics. It must have been wonderful to live in a house that was equipped with the most modern devices: from an internal telephone system to a goods lift, as well as radios and clocks built in and connected to a central network. Everything was in apple-pie order right down to the last detail, and had been chosen in accordance with the style. Even the flowers in the house were arranged in charming, undecorated vases in basic geometric shapes designed by Andries Copier for the Leerdam glassworks. The sets of glasses, just visible in photos from 1933 behind the sliding glass doors in the dining room, were also supplied by this factory.

The most striking thing in the house is the attention to colour. Whereas one would expect white and silver-grey, and at most a few primary colours, the atmosphere in the house was set by a multi-coloured palette ranging from copper and beige to tomato red, apple green, grey-green, yellow, greyish-blue and turquoise. Bart van der Leck, a former De Stijl artist, was colour adviser to Metz & Co. during this period and played a crucial role in the choice of colour schemes. Even today the effect of the turquoise bathroom, containing a bath, a shower with six shower-heads, two washstands, a toilet, a bidet and heated towel racks, is overwhelming. The family's two daughters and the two live-in maids each had their own separate bath.

Steel or Wood?

During the 1930s the design of modern but low-price pieces of serially produced furniture remained a priority for many social furniture makers, architects and other designers as they experimented further on the design and production of tubular steel furniture. The firm Auping no longer concentrated solely on beds but had moved on to other types of metal bedroom furniture. In Rotterdam, where Gispen was already established, a new firm was set up called d3.[24] Paul Schuitema was its main designer and joint director. In addition to his own work, a group of both little-known and well-known artists and architects, including Han van Loghem, Ben Merkelbach and Arie Verbeek, supplied new designs.

Meanwhile, foreign tubular furniture came onto the market in the Netherlands. Some of these items were well made, but others showed the early stages of a form of mannerism. The principled functionalist designers from the early days condemned these designs as pretentious, because they only 'appeared' to be modern but in reality were taking advantage of the snobbish, conventional requirements of a small group of *nouveaux riches* pandering to capitalist trade and industry.[25]

But even the Dutch functionalists found it difficult to hold on to their original ideals. The chairs Gispen designed and produced, with their comfortable seats and armrests, could also be classed as traditional armchairs; even Oud designed a few representative easy chairs with soft cushions and comfortable armrests. In the journal *de 8 en Opbouw* all new products were regularly reviewed and harshly critiqued: 'What can be worse for spiritual steel furniture than a combination of elegant thin metal with ungainly, heavy, thick cushions, filched from the club-chair design. This deer-elephant production is a grave threat to our new interior design', sneered Han van

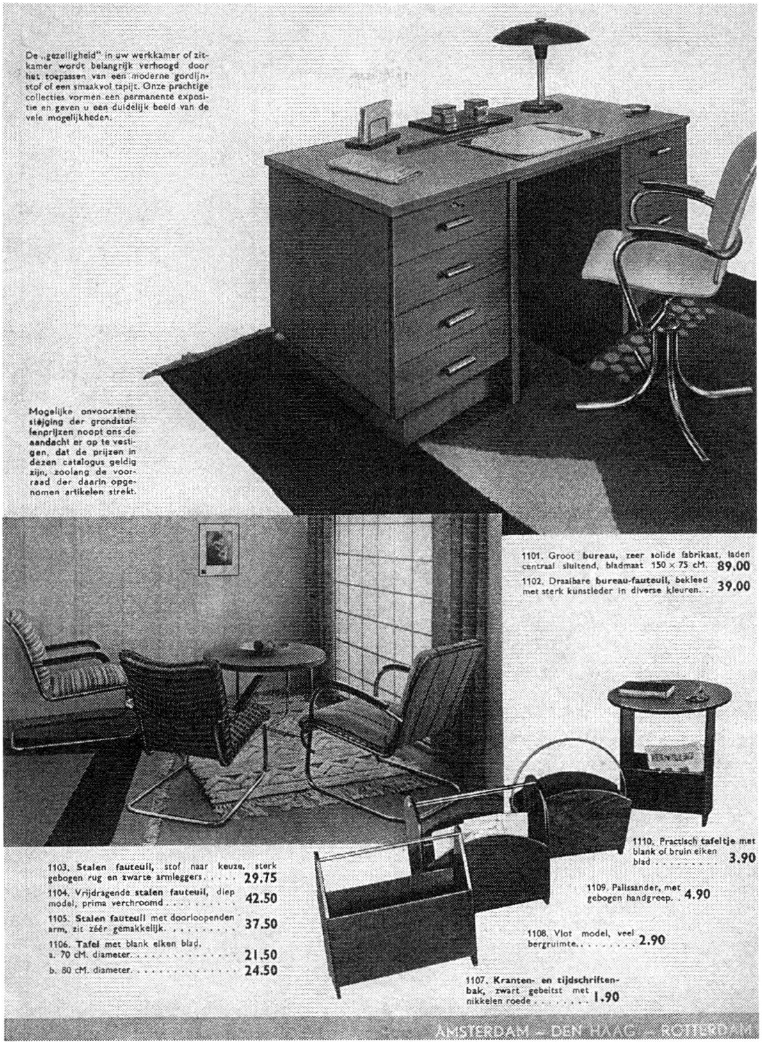

Paul Schuitema, brochure for *De Bijenkorf*, 1937.

Loghem in 1935.[26] As the 1930s progressed people also began to wonder whether tubular metal furniture was better than similar wooden models. After all, in practice wooden furniture could also be styled in such a way that it would be lightweight and easy to move around, and it was still cheaper to produce than stainless-steel. Meanwhile, designers and producers of wooden furniture had further rationalized their production lines. Cornelis van der Sluys, for example, who at the time had been active as a furniture and interior designer for more than 30 years, designed a completely updated collection in 1932 that he called his 'Normal Series'

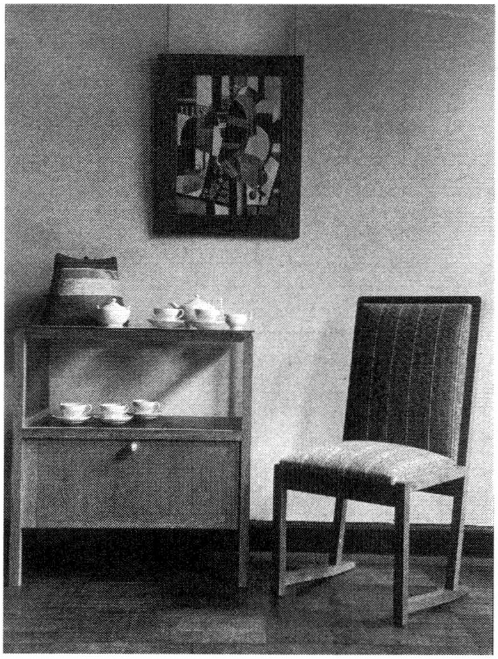

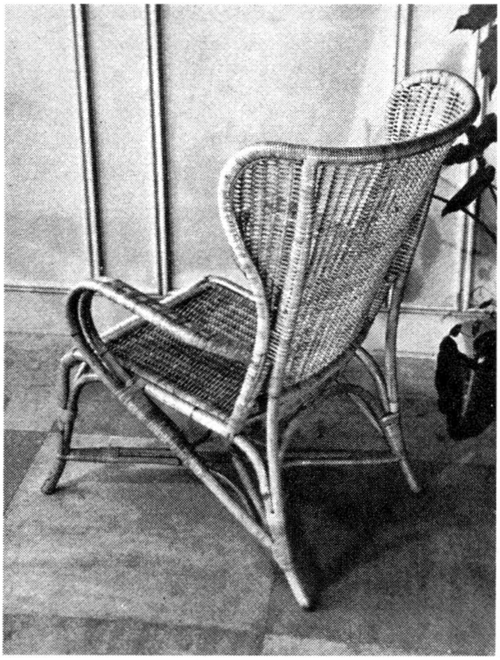

(*Normaal-Serie*). By standardizing measurements and construction practices they could be produced more efficiently and cheaply.

In the same period Willem Penaat devised his programme of 'modular' furniture for Metz & Co., a system that could provide a maximum of possible uses for a minimum amount of money. The larger furniture factories, like Pander in The Hague and the Utrechtse Machinale Stoel- en Meubelfabriek (UMS), also adopted elements from this innovatory movement.[27] By such means commercial modernism penetrated through to the wooden furniture market.

As far as interiors were concerned, the purest and most doctrinaire form of New Objectivity was now over. Even Mart Stam was to design a few pieces of wooden furniture in the mid-1930s. In the same period the deeply socially committed interior designer Ida Falkenberg-Liefrinck came up with an original and practicable alternative for both wood and metal.[28] She proposed a return to the use of traditional cane and designed a series of comfortable cane chairs, in which the advantages of both metal and wood were combined in a surprising way.

Cornelis van der Sluys, furniture from his 'Normal Series', 1932.

Ida Falkenberg-Liefrinck, rattan chair, 1936.

Domestic Objects

Although before the Second World War the most confirmed and idealistic modernists were chiefly active as architects and furniture designers, those socially and politically motivated to following the path of modernism could also be found working in other fields of design. Indeed, a strong socially and ethically motivated body of thought also inspired the entrepreneurs who joined the Federation for Art in Industry (Bond voor Kunst in Industrie) in 1924. Fuelled by social objectives, Cochius, the director of the Leerdam glassworks, had already started to collaborate with many artists and architects such as K.P.C. de Bazel, Cornelis de Lorm and H. P. Berlage as early as 1915 (see chapter Two).[29] The thinking behind this was that good, sensible products would bring beauty to the home and contribute to people's happiness. However, the pressure of the socio-economic situation and the hard reality of a large factory forced him to adjust his idealistic attitude over the years and to pursue a more pragmatic policy. He had to produce more efficiently and cheaply. Cochius had found an almost perfect interpreter of his modern, social-industrial design ideals in Andries Copier, who by 1924 was acting as the permanent designer at the glass factory and showed a great interest in the technical facets of glass-making.

Social idealism was also at the root of the De Ploeg weaving mill, which was established in 1923 in Bergeyk as a co-operative factory and commercial business enterprise.[30] Here in the late 1920s Frits Wichard and Jo Köhler designed the simple, striped and checked curtain materials that were sold under the trade name 'Colora Series'. These simple cotton fabrics, woven in bright primary colours, were in great demand among the architects of *Nieuwe Bouwen*. The plain 'Dobby fabrics' by De Ploeg, craftily uniting

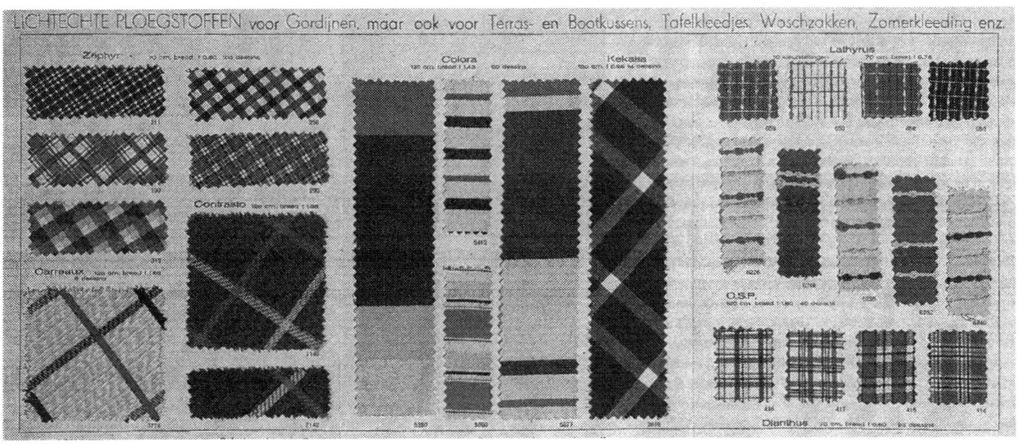

Cotton textile samples, Weverij De Ploeg, Bergeyk, 1930.

industrial and traditional workmanship in their design, continued to be popular until long after the war. In 1933 the German textile designer Otti Berger was involved in the development of a new collection. She had trained at the Bauhaus, where she had been one of the weaver Gunta Stölz's most talented pupils. During the four years she worked for De Ploeg she designed sixteen new patterns.

During the 1920s there was even evidence of socially motivated modernization in the traditional production of luxurious silverware.[31] After the firms Van Kempen and Begeer had amalgamated in Voorschoten in 1919, Carel J. A. Begeer took over the management in 1925. During this period he became increasingly convinced of the social importance of industrially produced silver consumer goods since more people would be able to afford them than the exclusive handmade wares. This is why he brought the Austrian designer Christa Ehrlich to the Netherlands in 1927. She designed four tea services for Van Kempen en Begeer in 'shapes telling they were made by machines'[32] – products with taut lines, basic geometric shapes and very subdued decoration.

Moral Modernism was even discernible in the early radios put on the market, under the trade name Erres, by the Rotterdam firm Stokvis. After the designer and journalist Otto van Tussenbroek had fiercely criticized Stokvis for improperly historicizing the design of contemporary products, he was commissioned by the very same firm in the late 1920s to design a new, more functional series.[33] It seems, however, that the entrepreneur's wish to promote good taste did not run very deep. When customers did not show much interest in Van Tussenbroek's austere designs, Stokvis soon reverted to the more luxurious and popular Art Deco design, combining

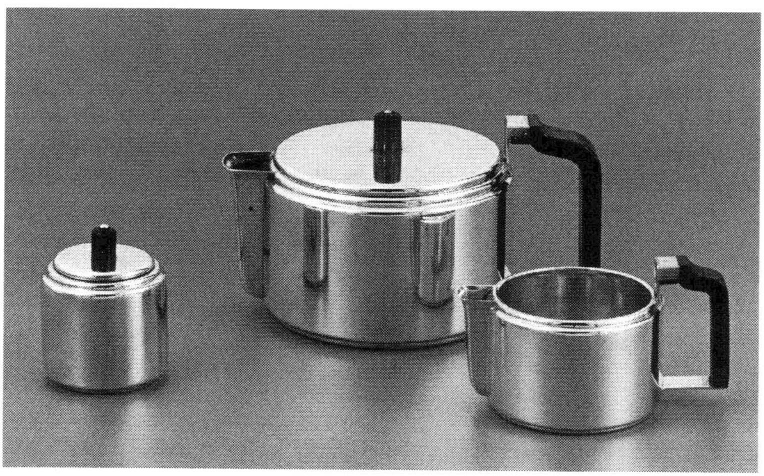

Christa Ehrlich (for the silver factory Van Kempen and Begeer, Voorschoten), silver tea service, 1930–31.

gleaming varnished wood and white Bakelite with gold piping and gold knobs. This was not the end of Van Tussenbroek's influence at Stokvis, however, since during the mid-1930s he was involved in the design of the first modern streamlined cylinder vacuum cleaner, also sold under the Erres name. At first the firm emphatically denied that this streamlined style, then so popular, had partly been selected for commercial motives; the torpedo-like shape of the appliance was said to have been chosen solely for its constructive, functional properties, not on fashionable grounds. Research into the patent history of this model has revealed that Stokvis was being somewhat economic with the truth.

Idealism in Modern Typography and Advertising

The most committed and convinced 'moral modernists' could be found in the field of applied graphics, typography and advertising. These included Jacob Jongert and, to a lesser extent, the Amsterdam designer Fré Cohen, but the most important names in this area are Piet Zwart, Paul Schuitema and Gerard Kiljan.[34] There were clear parallels between their strong views on the socio-political significance of graphic design and the role the designer played in this process, and those of the architects and furniture designers. That was not surprising, because Piet Zwart moved in architectural circles and had himself designed furniture in the past. Paul Schuitema, indeed, became a director of a furniture factory. In their graphic work Zwart, Schuitema and Kiljan fought with conviction against the extravagance and the 'unwholesome' decorations to be found on the old and, in their eyes, elitist decorated products. They waged war against 'dishonesty' in design, and against dated, inefficient, traditional, methods of work. They were entirely convinced that their new applied graphic art could contribute to a better world.

Zwart had been trained at the start of the century at the National School of Applied Arts in Amsterdam. His earliest pieces of furniture were designed in the style of Berlage and De Bazel. He and his wife made cushions in a decorative style associated with the Wiener Werkstätte. After the First World War Zwart converted to socialism and radically changed his ideas on design and the role of the designer. In 1919 he was among those, together with Berlage, Van Doesburg and Wijdeveld, who became members of the Union of Revolutionary Socialist Intellectuals (Bond van Revolutionair Socialistische Intellectuelen). Despite friendships with Vilmos Huszar, Jan Wils and Bob Oud, Zwart did not join De Stijl because the theories peddled in their journal were far too theoretical for his taste. His political convictions led him to make a conscious choice in preference of mass production and modern technology.

Zwart's first advertisement for the Netherlands Cable Factories (Nederlandse Kabelfabrieken, NKF) dates from 1923. His unconventional designs were notable both for their dynamism and their simplicity. It was obvious that the modern cable industry, which after all was about new and contemporary phenomena such as electricity, increases in scale and introducing conformity to a standard, was bound to appeal to his imagination. The designs also show that he knew the work produced by De Stijl. But Zwart was also well informed about the ideas of the Russian Constructivist El Lissitzky.[35] The influences of Dadaist design language and Kurt Schwitter's design solutions were visible in the original way in which he used type size, composition and highly imaginative visual and typographical jokes. Schwitters was a familiar figure in the Netherlands during this period through his contacts with Van Doesburg. Zwart began to incorporate photos in his designs for the NKF in 1926. A photographer would be engaged to take a series of close-ups of cables; Zwart would then process the images in an innovative way, selecting the colours red and blue for use in advertisements and brochures. He lavishly praised the use of photography in advertising as the most objective, realistic and honest way of supplying information.

By then Zwart had built up a large international network. In 1928 he taught for a short time at the Bauhaus and was asked by Kurt Schwitters to become a member of the Ring Neue Werbegestalter, a group of influential and progressive advertisement designers in Berlin. In the eyes of these left-wing designers advertising had not yet become an insidious way of increasing profits but was more a form of public relations and, as such, art for the masses. Zwart's already considerable international reputation in that year is also shown in Jan Tschichold's *Die neue Typographie*, in which the new principles involving international avant-garde typography are in some cases explained on the basis of examples taken from Piet Zwart's work.[36]

The lessons that Zwart gave from 1919 at the Rotterdam Academy of Art, where Jacob Jongert was in charge of Decorative and Industrial Arts, were equally steeped in these progressive views. His criticism of all other forms of, in his eyes, conservative Dutch design education was therefore harsh: 'bizarre purposefulness and individual wilfulness' was how Zwart described the typical characteristics of the students' work.[37] They seemed to be completely uninformed about modern techniques. On the basis of this Zwart concluded that the teaching syllabus was in no way able to meet the demands of the new age.

In 1928 Zwart even put forward the revolutionary view that it would be better to do away with lessons focused on autonomous art in favour of design. He developed a curriculum for a new 'design course' aimed at dispelling the

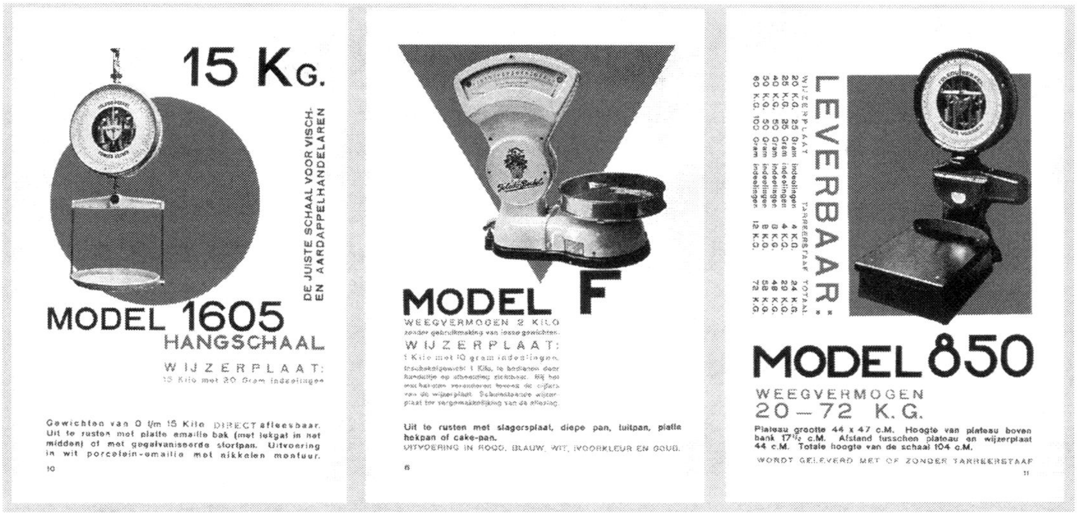

Paul Schuitema, advertisement for scales, patented by Van Berkel's, 1927.

'unimaginative, historicizing, passive intellect that stems from a conventional attitude towards life'. In order to accomplish this, Zwart tactlessly observed, it would be necessary to replace a number of the older teachers or retrain them. He insisted that photography and film should become basic subjects, but in his plan Zwart also mentions radio, sound movies, advertising, town planning and even television.[38] The Board of the Rotterdam Academy showed absolutely no enthusiasm for Zwart's revolutionary ideas and he was sacked in 1933. Jan Kamman's photography lessons, however, which Zwart had introduced, were allowed to continue.

Paul Schuitema had really trained to become a painter at the Rotterdam Art Academy, but in the mid-1920s he found graphic design far more appealing. Like Piet Zwart he discovered the possibilities of photography and the combination of photography with typography: 'phototypography'. Due to his use of what were then modern collage techniques, his designs for book covers, posters and advertisements were almost more outspoken than Zwart's, even though around 1930 both artists were producing remarkably similar work.

From 1928 the large Rotterdam firm Van Berkel's Patent offered Paul Schuitema the same opportunities as the cable factory did for Piet Zwart. In his case too the products for which the advertisements and brochures needed to be made – advanced industrial cutting and weighing devices – proved to be an inspiration for creating a progressive new design language. The precision instruments, made of gleaming metal, pre-eminently symbolized a modern society characterized by trade and efficiency. The commissions he received from the Rotterdam printer C. Chevalier in this same period presented Schuitema with the opportunity of becoming more adept at mastering progressive typography techniques. Like Piet Zwart, his socialist persuasions did not stop him making advertisements for commercial enterprises. From 1932 Schuitema designed the cover of the new fortnightly periodical *de 8 en Opbouw*. The large figure of eight, which filled almost the entire cover sheet, was combined with the word *Opbouw* in surprising compositions: different photo collages were added to the basic design for each new issue.

Another confirmed modernist in the field of graphics in the mid-1920s was Gerard Kiljan, who trained at the Quellinus School of Applied Arts in Amsterdam and the Rotterdam Academy of Art. Kiljan's initiatives to reform design education were particularly advanced. In 1930 he set up a new Advertising Department at the Art Academy in The Hague, in which Paul Schuitema was also involved.[39] This was the first Dutch industrial design training course, although initially it was restricted to graphic design. Photography and learning about technical processes and industrial printing techniques were part of the syllabus. Not only was the course pioneering, the democratic teaching methods were quite revolutionary too. The pupils went to work in 'collectives' and the study of historical styles, which Kiljan considered an arbitrary invention, was replaced in The Hague by the new 'objective' subject he had developed called Development of Form. The lessons were modelled on those at the Bauhaus. With Kiljan's aid, a fully fledged Department of Industrial Design was to be introduced much later in 1950.

The Second World War and Post-War Reconstruction

The militancy of many progressive designers had been toned down by the realities of everyday life in the 1930s, but the scope for good plans was to disappear entirely during the German occupation lasting from May 1940 to May 1945. Dutch artists, designers and architects were obliged to join the 'Kultuurkamer' from 1942 onwards, and Jewish artists were no longer allowed to practise their art at all.[40] Those who did not 'sign' were no longer able to work at their profession. Those who did join became members of the Guild of Architecture, Fine Arts and Decorative Craft (Gilde voor Bouwkunst, Beeldende Kunst en Kunstambacht). Art and design from that moment had to comply with the National-Socialist ideology, which in practice made a strong appeal to national Dutch traditions and folk art. Modernism, with its international focus, abstract design language and democratization of technique, was now absolutely forbidden.

Cris Agterberg (Westraven faience and tile factory, Utrecht), two bowls, 1938.

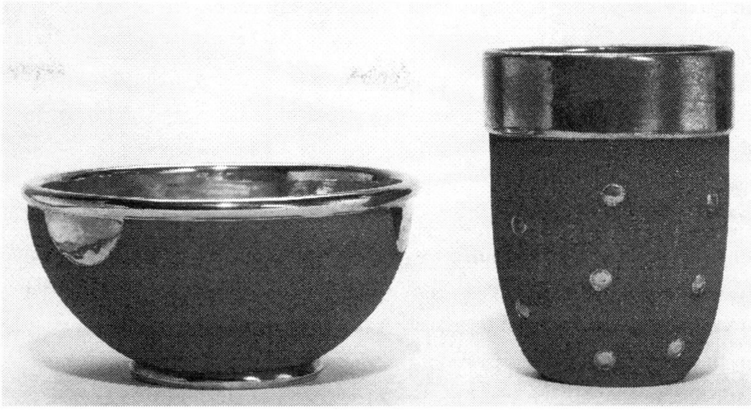

Many artists and designers refused to sign, but some accepted the situation and conceded. Among these was the versatile industrial artist Cris Agterberg. As a member of the advisory body for the Department of Public Education and Arts (Departement voor Volksvoorlichting en Kunsten) set up by the Germans he even wrote reports about the rise and fall of handcrafted art and organized a furniture design competition for the general public: participants had to design a 'healthy and substantial piece of furniture' in which their 'own national character' was well expressed.[41] The department opened the Dutch Art House (Het Nederlandsche Kunsthuis) on the Rokin in Amsterdam, where presentations of decorative craft and applied art were displayed alongside exhibitions of fine art. The emphatically educative character of the institution and its exhibitions was expressed at the exhibition *Tegen Ongezonde Kunst en Wansmaak* (Against Unhealthy Art and Bad Taste) in 1942. Oak furniture, hand-decorated earthenware and hand-made hammered brassware were once again heartily applauded in the Dutch Art House.

During the war the industrial production of consumer goods stagnated. Raw materials became increasingly scarce; factory hands were sent to work in Germany or went into hiding; and factory managers were replaced. The Gispen factory, for example, was more or less forced to work for the occupier: the managers were warned that if they did not comply their workers would be sent to Germany as forced labourers. Once the production of metal furniture had been forbidden – the metal was needed for the weapons industry – the firm switched to making wooden chairs. They also made black-out lamps, constructed in such a way that they could be used in the evenings without taking extra safety precautions. During the winter of 1944–5, when many Dutch people died of starvation, Gispen produced a mini-stove that worked with odd bits of paper and scraps of wood, with the added advantage that food could be cooked on it.[42]

The Netherlands emerged from the war badly damaged and thoroughly shaken.[43] In the summer of 1945 there was a shortage of everything: factories had been bombed or were just ticking over; ports, bridges and roads had been destroyed; Schiphol airport lay in ruins and gas and electricity supplies had almost come to a standstill. Many houses had been destroyed, and to make matters worse the construction industry had stagnated for five years. Owing to the return of 300,000 forced labourers and prisoners from camps in Germany and Poland, as well as tens of thousands of people who had been in hiding, there was a great shortage of housing. Neither did the loss of 'our Indonesia', and also of Germany's status as the most important trading partner, improve the economic situation. Despite all this, morale was high and it turned out that everyone was prepared to work for little in the way of remuneration. Thanks to this post-war reconstruction spirit things got going surprisingly quickly, yet despite this hopeful start, and a show of great solidarity, the road to recovery was long and difficult. In order to boost the economy in a controlled way the distribution system introduced during the war was temporarily prolonged. Many foodstuffs continued to be available only in exchange for rationing coupons.

It was not until aid from the Marshall Plan arrived from the USA between 1948 and 1952, in the shape of billions of dollars of emergency relief supplies and money loans distributed throughout the entire country, that the economy slowly began to recover. Coffee was the last product to come off rationing in 1952. It was not until then that exports and investments began to grow as desired. The rise of the United States as a major world power, both politically and economically, led many industrialists and managers, as well as the Dutch government, to view it as their main model. For most Dutch people America remained a beautiful remote dream, known, if at all, only through movie images.

Not everyone had needed American aid. One of the few firms to thrive during the Second World War had been the Philips light-bulb factory.[44] Through careful tactics and expedient policy the factory had actually come out of the war in better shape than before the conflict. For a long time Philips had not been a purely Dutch concern, since between the wars it had set up new production and sales organizations all over the world. It was partly due to this early internationalization that straight after the war the firm was able to develop earlier initiatives by investing in an expansion of its range of household appliances. Furthermore, since the German firms

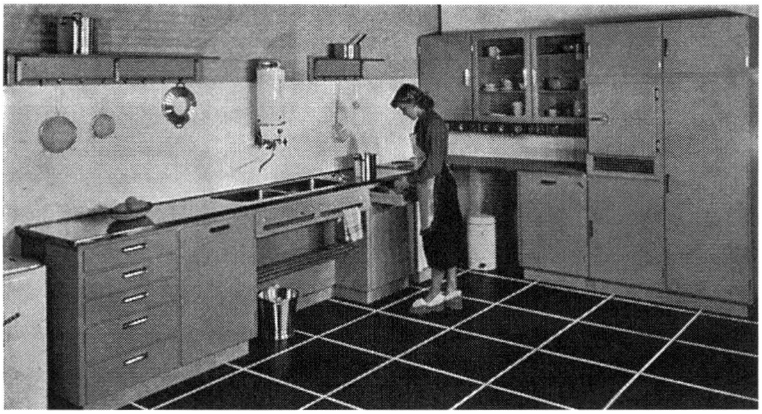

Piet Zwart, Bruynzeel kitchen (1938), reproduced in *Goed Wonen* (1954).

Siemens and AEG had disappeared from the scene for the time being, the products were an instant success.

Most other consumer goods factories took longer to recover, as could be seen at the Salon des Artistes Décorateurs in Paris in 1949, where the Netherlands was able to show only a modest entry. One highlight was the new kitchen by the firm Bruynzeel, which Koen Limperg had started in 1933. Piet Zwart had finished the design in 1938, but the kitchen could only be taken into production after 1945.[45] Here at last, after the initiatives discussed earlier, was the Dutch industry's successful answer to the 'kitchen problem'. The kitchen was the brain-child of director C. Bruynzeel, who, deeply impressed by the working methods in American factories, had transformed his former steam-powered joinery works in Zaandam into one of the most advanced wood-manufacturing factories in Europe. The Bruynzeel kitchen was made up of standardized components, whose measurements had been carefully adjusted to suit the average human body, while the layout was devised after a thorough analysis of all the tasks that had to be performed in a kitchen.

Post-War Idealism: 'Good Living'

The idealism of designers and architects, who before the war would have been counted as belonging to the New Building or New Living circles, proved not to have slackened in the least during the years of war. If anything the opposite was true. During the occupation designers were forced into inactivity, and some even went into hiding, but it gave them time to consider in detail what they should do when peace was once again restored. Even when there was still no end to the war in sight they were speculating about the future position of art and design in articles published in clandestine papers such as *De Vrije Kunstenaar* (The Free Artist).

In practice, the ideals of socially and politically motivated modernism in the 1950s and '60s would be propagated chiefly by the Good Living Association (Stichting Goed Wonen), founded in 1946.[46] The same idealism, however, played a role in the activities of the equally new government-funded Industrial Design Institute (Instituut voor Industriële Vormgeving, IIV), established in 1950. Many designers and architects, indeed, were involved in both Goed Wonen and the IIV. There were various organizational ties between the two and both were active until the late 1960s.

During the war years the VANK was disbanded and plans were forged for a new design organization. Only a few months after the capitulation, therefore, it was possible to found the Applied Artists Federation (Gebonden Kunstenaars Federatie, GKf). This was initiated by Willem Sandberg and Mart Stam. In the everyday idiomatic language of the 1940s the expression 'applied artist' had replaced the names 'industrial artist' or 'artisan'. It referred to exactly the same group, namely to all artists and designers who were designing in either an industrial or craftwork context. The greatest difference between the GKf and the VANK was that it had far stricter entry requirements to ensure that not just anyone could join. In addition to quality requirements, the potential new members' war records played a decisive role in the admission procedure. In addition to Sandberg and Stam, the Board also welcomed Piet Zwart and the graphic designer Wim Brusse. The GKf then affiliated with the Dutch Federation of Artists' Associations (Nederlandse Federatie van Beroepsverenigingen van Kunstenaars), whose structure had also been devised during the war years. Optimistically and idealistically they thought they could unite all artists in this way and that together they would be able to take a firm stand and create a better society.

The energetic GKf created an organization that to some extent was comparable to the pre-war Instituut voor Sier-en Nijverheidskunst (ISN), in that it was intended to advise firms and private individuals on design and more specifically on the designers whose services they could best use. This Aesthetic Advice Office (Bureau voor Aesthetische Adviezen), with Karel Sanders as director, was established in March 1948 in the former building of the Dutch Art House on the Rokin in Amsterdam. For this initiative the authorities were willing to provide financial support.

During the final months of the war, dissatisfaction with previous attempts to educate people on good taste stimulated the furniture designer and salesman A. Bueno de Mesquita to write a report titled *De sociale func-*

tie van de binnenhuisarchitect na den oorlog (The Social Function of the Interior Designer and Decorator after the War). In De Mesquita's view the pre-war living culture of the average Dutch person was sub-standard: either the designers made good furniture that proved to be too expensive and unpopular with the masses or they indulged people's need for status and 'sham luxury' and designed inferior products. The professional view was that this should not be repeated after the war and the only way they would be able to pull it off would be by intensive collaboration between manufacturers, designers, distributors and government. In the end they would have to join together in the interests of good design.

After the war the report stimulated a few initiatives that were supported by the GKf. At first consumers, distributors, designers and manufacturers reorganized their businesses separately. Then on 11 November 1946 the federative Stichting Goed Wonen was established, an organization that was to live up to De Mesquita's ideal. J. Bommer, the social-democratic alderman for the Amsterdam public housing department, became the first chairman of this umbrella organization. Central government took no part in the organization, although it did support them financially later on. However, after several years the four categories of affiliated groups were still unable to agree a coordinated response to recurring problems. For this reason, from 1954 Goed Wonen functioned solely as a consumers' association.

The two important figures from the early years of Goed Wonen were Mart Stam and Johan Niegeman. Stam had been director of the Instituut voor Kunstnijverheidsonderwijs (IVKNO) since 1939.[47] Almost all the older Amsterdam schools in this field merged into this art and design academy in the period between the two wars. In 1967 the school was to be renamed the Rietveld Academy.

Like Stam, Johan Niegeman, who trained as an architect with his uncle H. Th. Wijdeveld, among others, possessed a wide range of talents.[48] He also taught for a while at the Bauhaus, where he first worked with Walter Gropius and where he was later introduced to modern analytical views on design by the radical director Hannes Meijer. After this Niegeman worked in Russia until 1937 on the construction of the new town of Magnitogorsk.

In 1939 Stam asked Niegeman to take over the running of the IVKNO interior design course. The progressive duo together tried to modernize the Amsterdam course and to implement their functionalist design views. In his lessons Niegeman experimented with a method that was inspired by the well-known Bauhaus *Vorkurs*, which aimed at getting students to abandon biased propositions and views on art and beauty. An important aim of this course was that the future designers should be given a clear view of their task and their position in society. Friso Kramer, Dick Simonis, Bé Brand, Cora Nicolaï-Chaillet, Wim de Vries, Kho Liang Ie, Hein Stolle, Jan Vonk and Coen de Vries were just a few of Niegeman's and Stam's earliest students.

Goed Wonen emphasized that it was not primarily concerned with good design or beautiful-looking products but with improving the 'living culture' in a more general sense. Goed Wonen fought against 'tastelessness, material shortages and housing shortages', as first formulated in their aims. It was not things but people who were the central focus. People had to be able to develop their talents as individuals and become 'free and happy'. Differences in class and standards of living did not somehow seem to exist for Goed Wonen, owing to their idealistic post-war reconstruction mindset and the conviction that a harmonious society could now be created in which every person could do justice to his talents. In the utopian society of Goed Wonen the family stood centre stage: in conformity with the ideals then in existence, the woman would be a housewife, managing all the household tasks and mainly in charge of ensuring that the home had a cheerful, cosy, welcoming character.

Goed Wonen endeavoured to reach its goals in various ways, including publishing the periodical *Goed Wonen*, organizing informative sessions in the showroom, giving lectures and courses, and furnishing model houses. The organization even took the step of marketing its own collection of furniture and domestic objects. In the early days the ideals of Goed Wonen were presented most fanatically by designer Wim den Boon.[49] Owing to his role as editorial secretary, the first volumes of the periodical bore the stamp of his personality. However, his tone was so pedantic and patronizing – he became known as the 'minister of the interior' – that it started to irritate the readers and his fellow editors. He was eventually asked to resign.

Goed Wonen appeared from 1948 to 1968 and during this twenty-year period went through a major metamorphosis. At its peak in 1961 there were 7,300 subscribers. In 1968 the name of the periodical was revealingly changed to simply *Wonen* (Living). After a merger with the *Tijdschrift voor Architectuur en Beeldende Kunst* (Journal of Architecture and Fine Art) in 1973 the name was changed again to *Wonen*TA-BK (Living TA-BK). It was now an ordinary but still progressive periodical for architects and designers and had abandoned its old educational mission entirely. From 1986 the journal continued under the name *Archis*.

Since there was little on sale shortly after the war, let alone goods that would satisfy the strict criteria set by Goed Wonen, in the late 1940s and early 1950s a few special Goed Wonen pieces of furniture were put onto the market. 'Minister of the interior' Wim den Boon himself designed a birch-wood stool

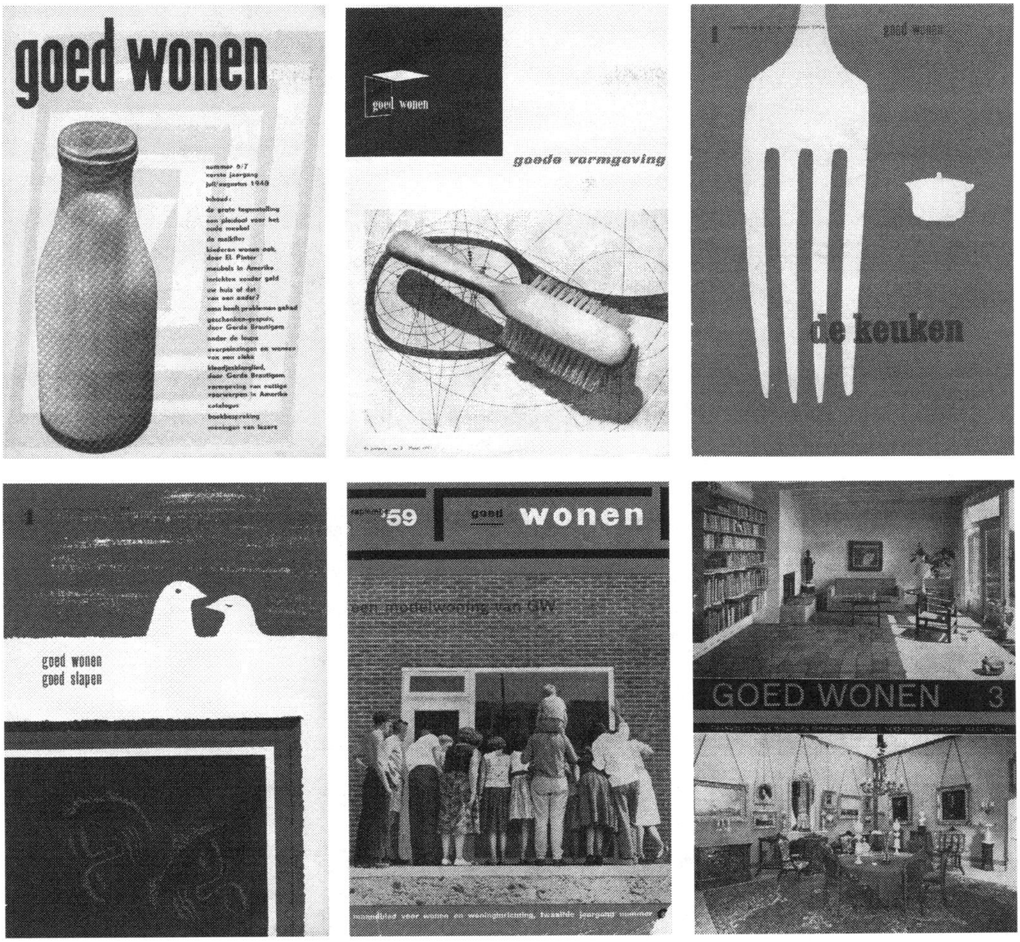

Various *Goed Wonen* covers from 1948, 1951, 1954 (two shown here), 1959 and 1961.

with a curved seat and a backrest made of plywood. Mart Stam developed an oak chair and rush mats, which in terms of design were far removed from his revolutionary free-floating model from 1926. Material shortages, but also people's changed ideas on comfort, played a role. Furthermore, something with a handmade look about it was considered to be a positive quality in these post-war years. Hein Salomonson, who before the war had played an active part on the journal *de 8 en Opbouw*, made a new variation on a familiar theme, the peasant chair, while F. Paulussen was busy designing cane furniture.

In addition to chairs, attention was drawn to cupboards and other 'storage systems'. If interior products were functional, reliable and affordable, and of course could aesthetically pass muster, they were given the Goed Wonen seal of approval. This award was not only granted to products

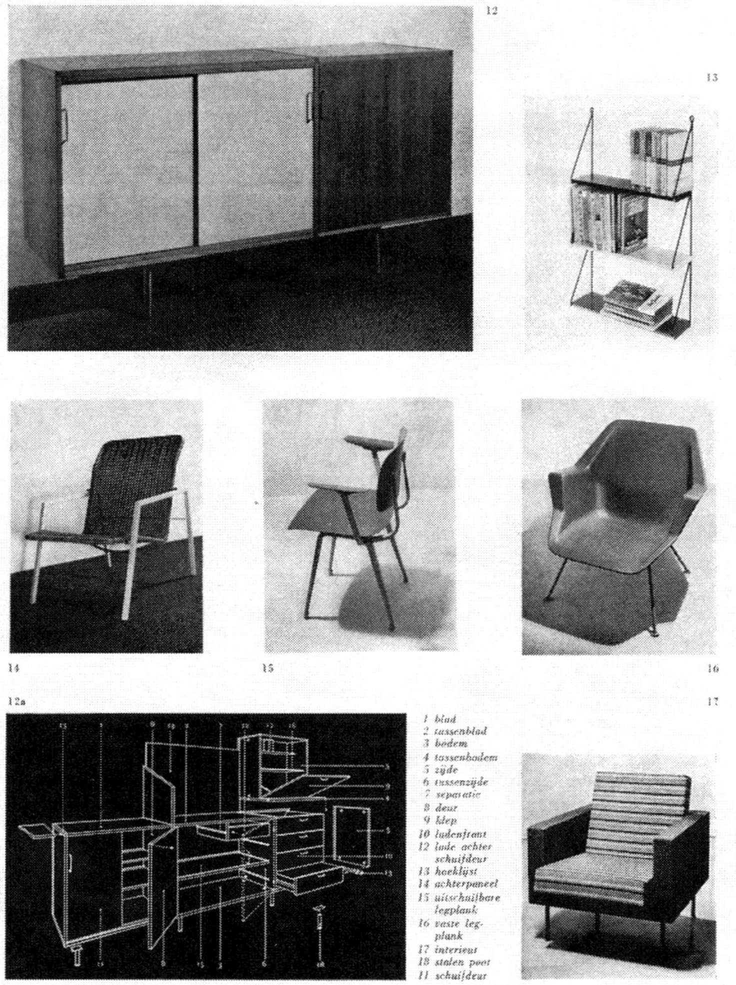

Furniture with the *Goed Wonen* Best Choice label: a cupboard by C. Braakman (Pastoe); a wall rack by T. Reijenga (Pilastro); and chairs by D. van Sliedregt, F. Kramer (Ahrend-Cirkel), W. Rietveld/A. R. Cordemeyer (Gispen) and C. de Vries (Hamer). From *Goed Wonen* (1956).

from Goed Wonen designers but also to other Dutch and foreign furniture and domestic objects. The models designed by Martin Visser for the furniture factory 't Spectrum in Bergeyk, for example, met these strict criteria almost to the letter. As far as ceramics were concerned the original choice was for a simple service by Wim de Vries from the pottery Fris. In 1951 the service Wilma, designed by Edmond Bellefroid for Mosa in Maastricht, was also granted this seal of approval. In the periodical *Goed Wonen* they explained why the service had earned this distinction: the design is easy to hold in your hand; it is agreeable to drink from and furthermore the shape is both 'reserved and festive'. Nevertheless, the editor warned, it was the

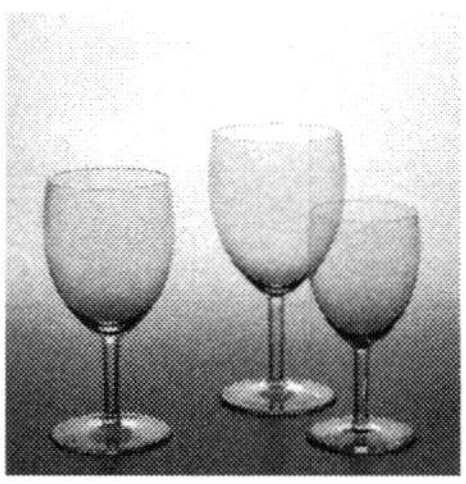

Andries Copier (Leerdam glassworks), 'Gilde' hand-blown wine glasses, 1930.

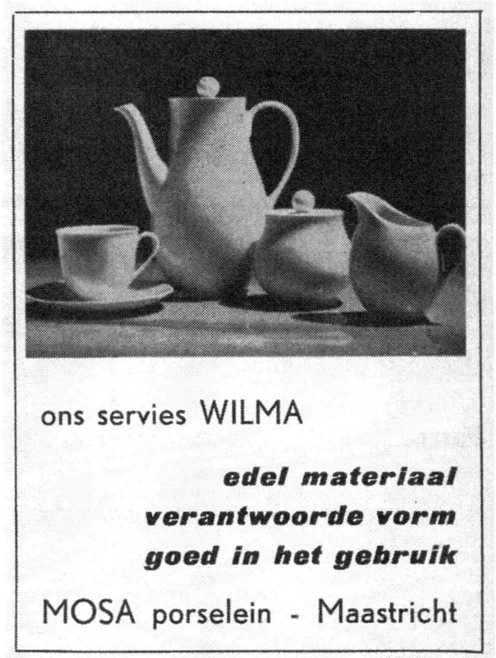

Advertisement in a 1954 issue of *Goed Wonen* for the 'Wilma' service by Edmond Bellefroid (Mosa, Maastricht).

white model that people were supposed to purchase and not the version with 'nondescript decoration', which did after all so weaken the design.[50] In 1957 the aesthetically pleasing Arzberg service 2000 by the German designer H. Gretsch was also awarded this mark of quality, even though it was really too expensive to qualify. For glassware the *Goed Wonen* subscribers were naturally in the right place in Leerdam with its glass factory. The seal of approval was granted to Copier's spherical-shaped vases, his Gilde glasses and a water jug with a set of large drinking glasses, which according to *Goed Wonen* were exemplary for their lack of pretension. Even very simple, trivial domestic products could earn the award, such as the functional dish-rack they selected made by Tomado, and stainless-steel sink tidies by Gero, with their 'exceptionally fine balance between aesthetic concerns and functional design'. Even a doorknob made by the firm Nedap in Amsterdam qualified and was described in the periodical as a 'well-designed object'.[51]

For carpeting the interior *Goed Wonen*, like the pre-war architects from De 8 and Opbouw, recommended using linoleum made by Krommenie: easy to clean, indestructible and available in light, modern colours. The approved curtaining fabrics were made by De Ploeg, but cotton prints by Het Paapje were also among their favourite choices. Electrical appliances, on the other hand, were seldom discussed and apparently had little chance of being eligible for the award: the critical *Goed Wonen* editors thought that radios in the late 1940s were still ugly contraptions, with their 'drawing-room-like appendages'. An exception was made in 1948 for an Erres radio set made by the firm Stokvis, 'a set that does not aspire to being anything more than an electrical appliance'. The editor therefore advised readers to take the gold strips off their old 'far too beautiful' set and 'to have the expensive wooden casing sprayed in the colours white, black or grey'.[52]

The Goed Wonen adherents could see how all these selected 'good' pieces of furniture and domestic articles were supposed to be combined with one another in a special showroom and in the many show houses that were fitted out over the years. The first Goed Wonen show room was set up

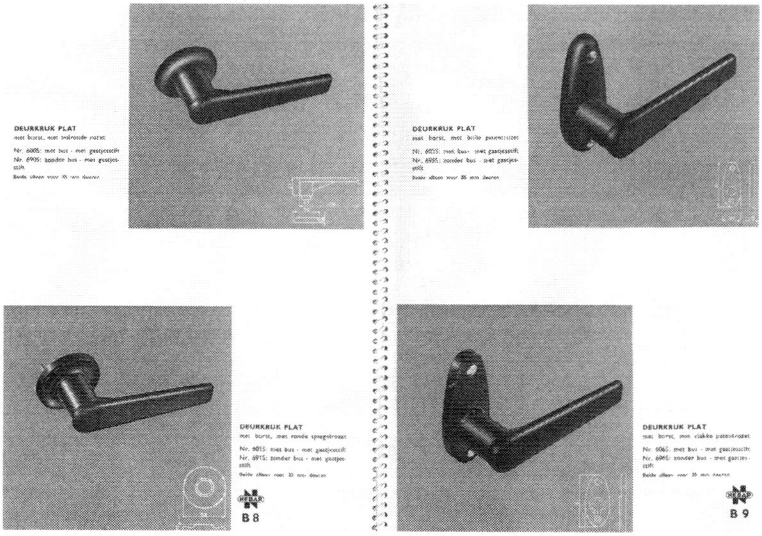

Brochure for the firm Nedap illustrating Bakelite doorknob designs by Wim Gilles, *c.* 1955.

in the property at 56 Rokin, where the Dutch Art House had been established during the war. Each year tens of thousands of people came to look at it and to ask for advice. Bé Brand, who was now married to Johan Niegeman, and Constance Wibaut were in charge until 1957 and taught their visitors the most practical, spacious and charming way to furnish their homes, sometimes with the aid of small-scale models of Goed Wonen furniture. In his 1958 book *Ik kan Wonen* (I Can Set Up House) Johan Niegeman explained again in great detail how people could put together a good interior themselves. In 1967 a second comparable show room was fitted out on the Lijnbaan in Rotterdam.

The Goed Wonen show houses were extraordinarily popular and the Foundation probably reached a great many people in this way. In the twenty years that Goed Wonen was in existence they furnished a total of 75 throughout the country. The smaller public housing units were chosen for them partly because subsidies were available, but also because they were fuelled by idealism. With the aid of these completely equipped through-lounge houses, Goed Wonen suggested ways of creating a sensible, attractive and affordable interior to the residents of these new post-war housing estates. They were not supposed to follow the advice indiscriminately and in fact many did not retain very much at all. This can be seen in a few photos showing how the tenants furnished their new house, which had previously served as a show house. Precious little remained of the light, frugally furnished, modern show interior. The house had been filled instead with large old-fashioned furniture, Persian carpets and curtains in loud-patterned fabrics.

In the course of the 1960s Goed Wonen's stress on moralism disappeared. The increasingly articulate consumer would no longer allow others to lay down the law. Goed Wonen adjusted its viewpoint on other issues too in line with the needs of a changing society. Slowly but surely their exclusive focus on the family interior was widened to make room for the furnishing of houses for the single or elderly. Moreover, it was clear that during the 1960s people had more to spend and that they were constantly hankering after more luxury. The interior was no longer a ready-made space in which people could do little more than shift furniture around. Interior design became interior architecture, and that meant that attention had to be paid to constructional and structural alterations in the available space. At the same time designing furniture came to be seen as more of a task for industrial designers and less as the job of an interior architect.

The IIV and Goede Vorm

The tone of moralistic idealism that was so characteristic of the views of Goed Wonen continued to inspire many designers after the war. For most of them the fierce character of socio-political debate had greatly been toned down, but in their hearts they were still convinced that sober, austere functional design, without too much fussiness, would make people happiest in the long run.

The Netherlands was certainly not unique in this respect. The same body of thought, called *Die Gute Form* (Good Design), had been taught at the Hochschule für Gestaltung in Ulm, Germany, since 1953.[53] This design course tried to resuscitate the Bauhaus ideals. The Swiss artist and former Bauhaus student Max Bill became the school's director. Bill had organized an exhibition in 1949 in Basel titled *Die Gute Form*; in 1957 he wrote a book with the same title. The term had then started to act more or less as a style name. The design method and the ideals of the course given in Ulm were well known in Dutch designers' circles and were also propagated by various organizations as well as by Goed Wonen. These were headed by the government-subsidized Institute of Industrial Design (Instituut voor Industriële Vormgeving, IIV), founded in 1950.

The IIV concentrated initially on the economic side of industrial design (for a more detailed explanation, see chapter Four). The institute was set up by three employers' federations and was subsidized by both the Ministry of Economic Affairs and the Ministry of Education, Arts and Sciences. By arbitrating between manufacturers and designers and by giving aesthetic advice, the IIV would be able to help to improve the quality of the Dutch product and in so doing improve the competitive position of the Netherlands. The central message was that attractive products would sell better and firms that paid attention to design could be confident of greater sales.

A special 'consultative body' of the IIV gave advice to manufacturers – but not for free. The first 'aesthetic advisers' belonging to this agency were Gerrit Rietveld, Christiaan de Moor and Arie W. Verbeek. They judged products in the early 1950s on technical soundness, serviceability, design, and on the price in relation to these aspects. To make sure there were no signs of plagiarism, the materials, colour, decoration and lettering were examined in great detail by these critical gentlemen from the IIV.[54]

In practice, many enterprises wrongly saw the board of the IIV as some sort of 'taste police' and they felt involuntarily judged, and sometimes even treated with contempt. For the Ministry of Economic Affairs this was a reason for slowly cutting off the organization's subsidies. People thought that the IIV was beginning to lose sight of the economic importance of industrial design and was apparently more concerned about elevating taste.

Indeed, when making aesthetic judgements the Foundation could not refrain from passing ethical judgements as well. In 1955 the IIV started to put to the test the products shown at the Dutch Exhibition Centre in Utrecht on the basis of their own critical norms. In addition to the criteria that had been in fashion since the beginning of the century, such as simplicity, functionality and sensible use of material, they now began to keep a keen eye on condemnable, fashionable tendencies. Firms who spent too much time on styling or, in the Board's view, dealt too arbitrarily with 'shapes and colours' were denounced. After all, the dubious practice of 'beautifying' or 'disguising' leaned towards the abhorred, and even more fiercely criticized, concept of fashion. Streamlined American products and their Dutch imitations were dismissed as senseless, materialistic kitsch. For them normal products were quite flamboyant enough. Remarkable in this context, too, is that in the early 1960s they regularly voiced their disapproval of the way that they assumed industrial products were 'unnaturally' or 'deliberately' produced for obsolescence. This too was considered to be a reprehensible and typically American trend.[55]

In 1962 the Industrial Design Centre (Centrum voor Industriële Vormgeving) was established in what was now known as the Beurs van Berlage in Amsterdam. This 'national showroom for good products' was furnished by Gerrit Rietveld and Kho Liang Ie. Visitors to the centre could look at, and

The Industrial Design Centre, Amsterdam, designed by Gerrit Rietveld and Kho Liang Ie, 1962.

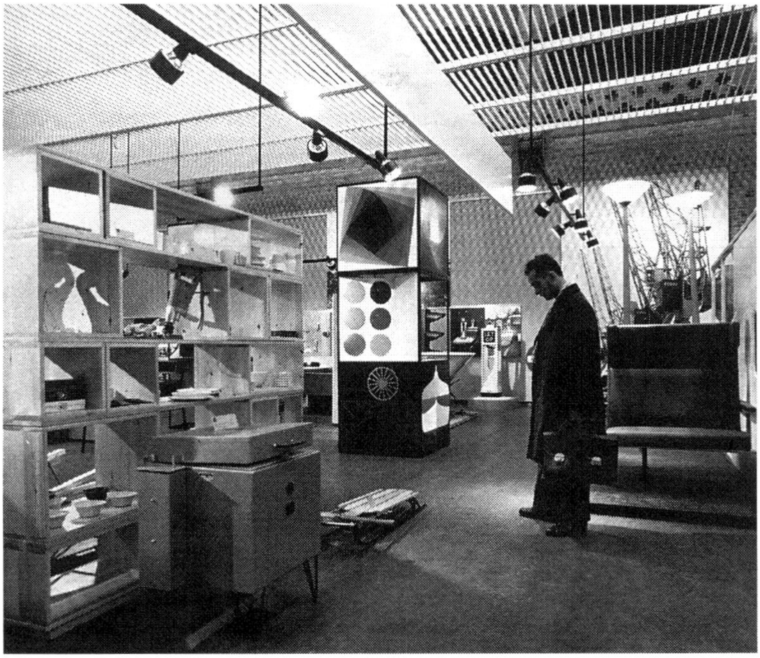

form an opinion on, more than 400 products from 120 Dutch firms. There they could also enter the sort of educational competition that involved choosing the best breakfast set. When assessing the products in the centre the IIV employed 'the recognized qualities of the typical Dutch product': reliability, simplicity and reasonable value for money.

Good Taste from the Museum to the Shop

The permanent exhibition in the Industrial Design Centre, Goed Wonen's showroom, and the various model houses were not the only places where the public could view designed products and be given further training in the moral principles behind them. A few museums and large shops took on an important role in educating the public. In this respect for many years the Stedelijk Museum in Amsterdam played a prominent role, mainly due to Willem Sandberg's involvement.

Modern design had been Sandberg's great interest for many years.[56] He was not only a graphic designer himself, but had also become involved in the VANK long before the Second World War. In 1934 the board of this society had asked him to serve on the committee that prepared the continuously changing exhibitions in the Stedelijk Museum. His influence was shown at once in the exhibitions featuring the Bauhaus artist László Moholy-Nagy and Theo van Doesburg. Sandberg's interest in modern design was aroused mainly through the VANK exhibition *De Stoel gedurende de laatste veertig jaar* (The Chair over the Last Forty Years) in 1935. On that occasion he came to know the architect Mart Stam rather well. As a member of the committee preparing a special Decorative and Industrial Arts section within the Stedelijk Museum, which was later to evolve into a special department of the museum, Sandberg went on to advise on the purchase of contemporary products for the new collection.

During the war Sandberg served with the Resistance. His knowledge of graphic techniques enabled him to forge identity cards in a professional way. Together with artists like the sculptor Gerrit van der Veen and the painter Willem Arrondeus, Sandberg was involved in the legendary assault on Amsterdam's municipal population register on 27 March 1943. He managed to escape and in doing so was the only one who did not pay with his life for this act of Resistance. After this attack Sandberg was forced to spend the second half of the war in safe houses. When the war ended he became director of the Stedelijk Museum, which under his management was to develop into one of the world's most prominent museums of modern art. Despite these duties, Sandberg continued to be active as a graphic designer. His unremitting interest in 'good' design was shown in his enthusiastic support of the many initiatives in this field.

During the 1950s the Stedelijk Museum organized a few exhibitions with an extremely pronounced educational and moralistic tenor.[57] *Mens en Huis* (Man and Home), for example, was put together in 1952 by J. W. Janzen and Wim den Boon, *Goed Wonen*'s former 'minister of the interior', who had just relinquished his editorial post. At the exhibition there were mainly 'good' and 'naturally' designed products to be seen, including sports equipment and tools. The furniture exhibited, including the crate chair by Gerrit Rietveld and the butterfly chair by Jorge Ferrary-Hardoy, was made of natural materials such as wood, cane, canvas and leather. A few interiors were also furnished using 'good' products that had been deliberately put in the 'wrong' place; all were, of course, purposefully provided with a text giving a full explanation. To check whether an interior was 'good', the organizers gave the visitor what they thought was a simple rule of thumb: if an ordinary milk bottle – or a wooden crate of apples – contrasted too much with the surrounding interior, then the visitor had definitely been looking for his furnishings in the wrong place.

The exhibition *Wonen en Wonen* (Living and Living) was organized in 1954 by the Stedelijk Museum in close collaboration with Goed Wonen.

Here a historic retrospective was presented of the development of the interior over the preceding fifty years. The idea behind it was to convince the visitor that opting for modern products was inescapable. In addition there was the tried and tested, pre-war formula of placing a traditionally furnished interior opposite a modern one so visitors could then compare the two. The young curator Hans Jaffé thought it was the museum's responsibility to bring good design closer to the public in this way. On display at the exhibition were simple, down-to-earth, industrially produced but attractive domestic articles that almost everyone could afford. To lower the psychological barrier that prevented people from entering a museum, Sandberg ordered the installation of a balustrade all round the outside of the museum so that passers-by could easily look inside.

Gerrit Rietveld, Frits Eschauzier and Wim den Boon had been involved in organizing the first post-war educational exhibition a few years earlier in 1951 in the Gemeentemuseum in The Hague. Its controversial title, *Kunst en Kitsch* (Art and Kitsch), attracted a remarkable number of visitors and reactions. Nevertheless, it overshot the mark. In the 'wrong' Kitsch department too many people recognized too many products from their own interiors and, to make matters worse, in the Art section they were appalled at how austere, bare and uninviting 'right' things could be.[58]

Between 1920 and 1970 a few large shops and department stores played an exceptional role in educating the general public on good and bad taste and in promoting interest in modern design. When it came to good design, the owners of these enterprises combined their business acumen with just the right amount of idealism. Before the war the firms in question were mainly the furniture companies Metz & Co. in Amsterdam and Pander and Bas van Pelt in The Hague. These were the shops that had something to offer consumers with a taste for the progressive. If money were no object, customers could go to large renowned specialist shops like Focke & Meltzer in Amsterdam or Jungerhans in Rotterdam that also sold high-quality Dutch and foreign goods, but these were less idealistic in their approach to good modern design. In the large cities there were also a number of smaller craft shops and art dealers that sold sound Dutch design, such as 't Binnenhuis in Amsterdam, which continued under the management of Jac van den Bosch until 1929. Other well-known design shops were De Distel in Rotterdam and De Zonnebloem in The Hague. Some industrial artists even ran small sales outlets from their homes, including Cornelis van der Sluys with his business, De Opbouw, in The Hague, and Cris Agterberg with his shop in Utrecht.

After the war Metz & Co., though not Pander, managed to regain its leading role in the field of good design after a few difficult years.[59] Henk de Leeuw,

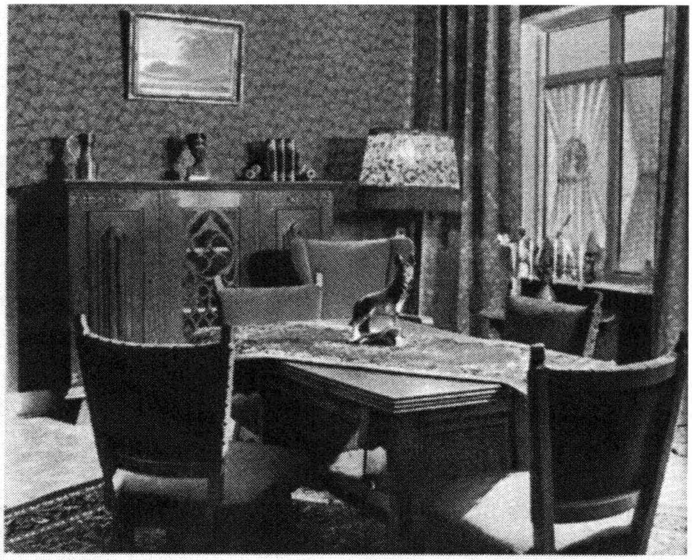

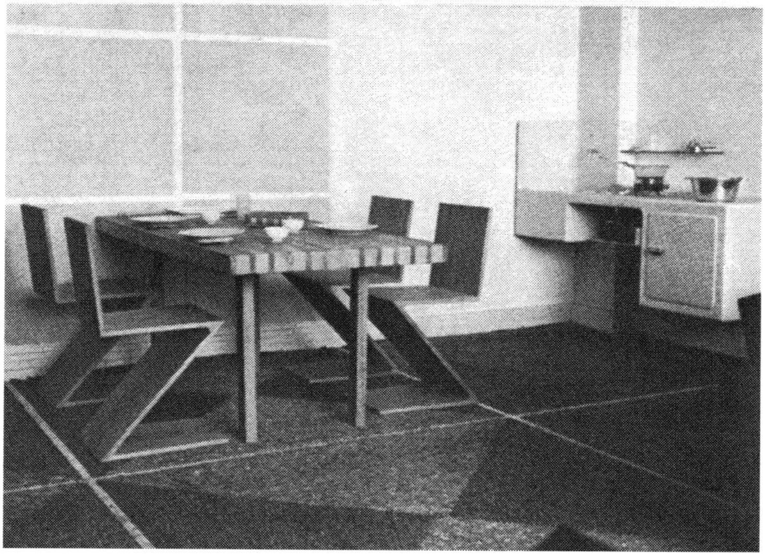

A 'wrong' and a 'right' interior installed at the exhibition *Kunst en Kitsch* ('Art and Kitsch'), in the Gemeentemuseum, The Hague, 1951.

son of Joseph de Leeuw who originally established the firm, took over the management and followed his father in combining idealistic missionary zeal with great commercial talent. Just as before the war, Metz commissioned designs from Dutch and foreign designers, while at the same time producing foreign designs under its own name with exclusive sales rights in the Netherlands. Bart van der Leck and Gerrit Rietveld continued working for Metz, and Sonia Delaunay's fabric designs were once again put into production.

The tradition of regularly organizing instructive exhibitions in shops was resumed and continued until the early 1970s. The first of these post-war presentations in 1949 was devoted to the theme 'Cane in the Home', in keeping with the material shortages at that time.

But gradually Metz's role changed from educator to taste creator and went on to become a trendsetter. Henk de Leeuw was one of the first in the Netherlands to become interested in new Italian furniture, so chairs by Gio Ponti, Carlo Pagani and Franco Albini were taken into production by Metz. He also obtained the exclusive sales rights in the Netherlands for glassware by Venini. De Leeuw managed to do the same for the designs by Poul Kjaerholm, Alvar Aalto and Harry Bertoia. At the Metz store, the Dutch public was often introduced for the first time to new designs by Charles and Ray Eames, Florence Knoll, Eero Saarinen and Arne Jacobsen. Even the Barcelona chair by Mies van der Rohe (1929) was first seen at an exhibition arranged at the Metz store in 1960.

A firm that could be compared to Metz in terms of its aims and organization was My Home, run by Bas van Pelt.[60] After it was established in 1931 in The Hague, this firm quickly evolved from a cross between a craft shop and an interior design office into a shop for modern home furnishings by Bas van Pelt himself and other designers. During the 1930s the organization branched out from The Hague and opened showrooms in Maastricht and Enschede, and launched a new shop in Amsterdam. After the war Bas van Pelt and Metz were often the major sales outlets in the Netherlands for the work of front-ranking foreign furniture designers. Bas van Pelt also had new designs made under his own production label and he continued to have close ties with Goed Wonen. The shop maintained excellent contacts with the departments of applied art at the Stedelijk Museum in Amsterdam and the Gemeentemuseum in The Hague. It was clear that the board of directors at Bas van Pelt, and similarly at Metz & Co., felt that their enterprises had a cultural and social mission to fulfil beyond an emphasis on commerce alone.

After the war the important furniture department of the De Bijenkorf department store, founded in 1894, started to play a role in spreading the message of Goede Vorm (Good Design).[61] In 1947 they decided that from then on the Amsterdam branch would concentrate entirely on modern, contemporary furniture. The complete range of traditional furniture was ruthlessly discarded. The furniture department's new chief buyer, Martin Visser, appointed that same year, was responsible for this revolutionary action, basing it on educational motives. Visser was in no doubt that modern, well-designed furniture would have a civilizing effect on the Dutch buying public. The manager of De Bijenkorf shared his belief that there was a need for a

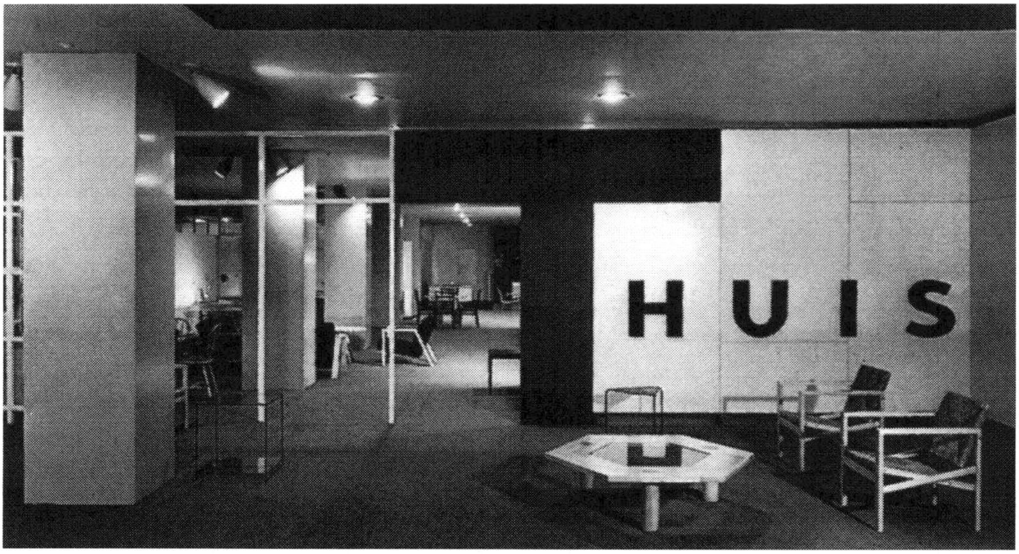

Aldo van Eck, Martin Visser and Benno Premsela, one of the *Ons Huis, Ons Thuis* ('Our House Our Home') exhibition in De Bijenkorf, Amsterdam, 1953.

modern progressive furniture store with a slightly less expensive range than the collections at Metz & Co. and Bas van Pelt. De Bijenkorf stated categorically that they wanted the furniture to be progressive but not extravagant.

In 1949 Benno Premsela was employed by De Bijenkorf to present modern furniture and other articles in an attractive way. A couple of years later he became responsible for the complete presentation of the department store, including the window displays. Premsela and Martin Visser then organized the famous *Our House Our Home* (Ons Huis Ons Thuis, OHOT) presentations at De Bijenkorf, which – like the artistic window displays – became a household word in Amsterdam. De Bijenkorf's cultural mission even extended to organizing art exhibitions, building up an art collection and holding lectures in the shop.

The OHOT exhibitions showed just how much Visser and Premsela were taken with Goed Wonen ideals. At the first presentation, which tackled the furnishing of cramped accommodation, they presented samples of practical, folding and stackable furniture. The designs were simple and the wood, wickerwork and cane used in their fabrication gave them a sober and natural air. In addition to the Dutch makes of 't Spectrum, UMS Pastoe, Wagemans & van Tuinen (Artifort) and Gispen, they also displayed Scandinavian furniture. Curtain material from De Ploeg and the hand-printers Het Paapje and linoleum from Krommenie completed the interiors displayed in modern Goed Wonen style. Paintings by Karel Appel and Corneille that hung on the walls further underlined Visser's enthusiasm for culture. It was the first time that work by members of the Cobra group could be seen in the Netherlands, despite its controversial reputation. In 1953 De Bijenkorf commissioned paintings by Appel and Constant as the basis for designs for curtain material, which they had printed by Het Paapje and Herman Hart's textile-printing workshop in Amsterdam.

After Martin Visser left in 1954 to become a designer for the furniture factory 't Spectrum, De Bijenkorf continued its progressive, modern policy, although by then the moralist attitude had been greatly toned down. The close ties with *Goed Wonen* did not fade away, as is shown by the coverage they still received in this periodical. All in all, between the 1950s and the 1970s there was not much to choose between De Bijenkorf, Bas van Pelt and Metz. Through their collections, and in particular through their special exhibitions, activities and close ties with the Stedelijk Museum and the Gemeentemuseum, all three stores played a major role in popularizing the modern interior and good modern design as a whole.

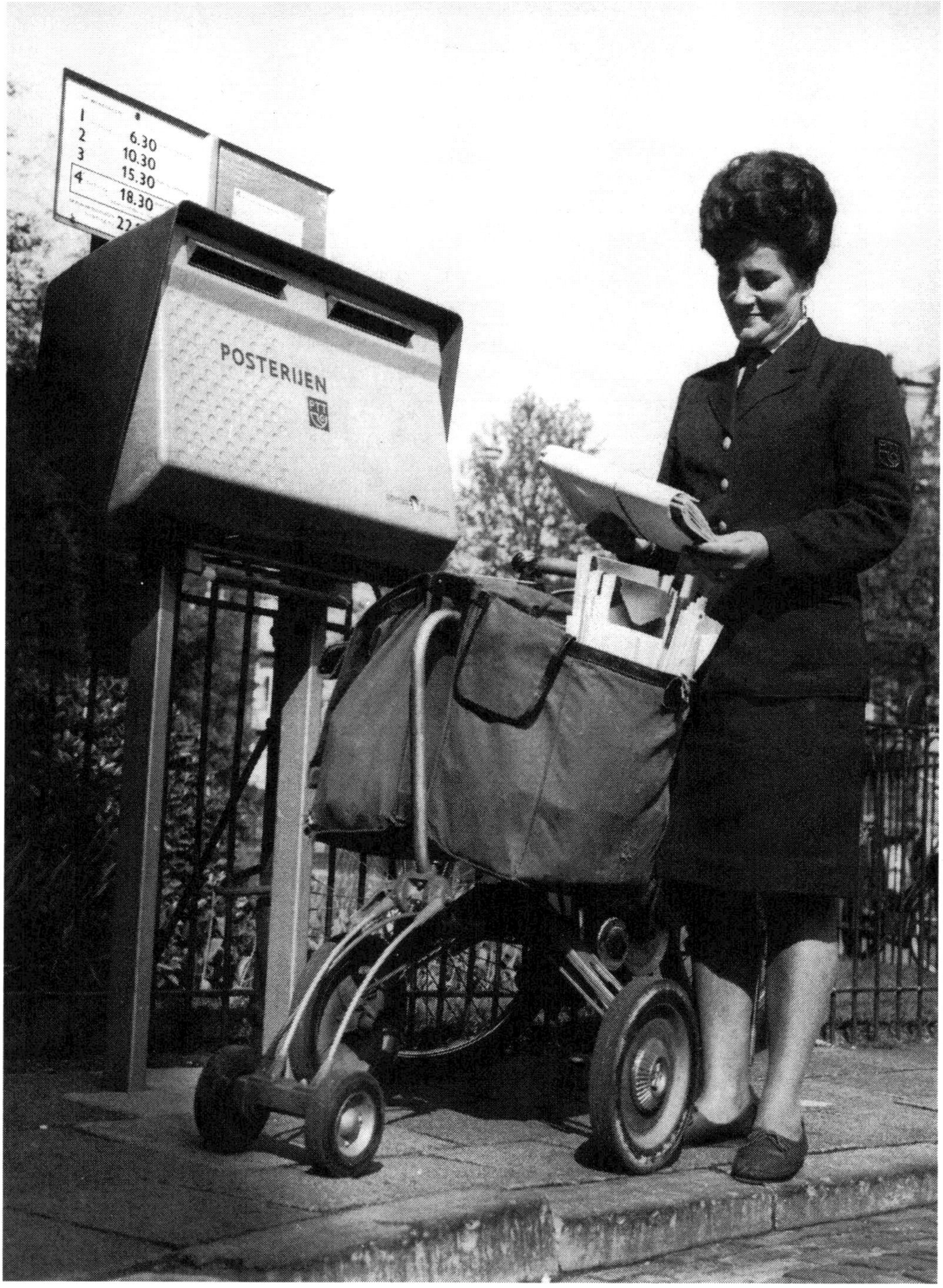
1
2 6.30
3 10.30
4 15.30
18.30
POSTERIJEN

4

Design as Profession, 1945–80

In the post-war decades industrial design in the Netherlands shifted from its relatively marginal, idealistic and artistic status into an important economic and social instrument. On the factory floor, draughtsmen, model-makers and engineers made way for professionally trained industrial designers. Independent industrial designers and design companies also managed to acquire a strong position in the cultural and economic realm. Complementing them, a small but active group of traditional Arts and Crafts practitioners, potters, weavers, textile printers and jewellery-makers carried on working as usual in their own studios and workshops. Industrial activity doubled between 1948 and 1962, productivity reaching a peak in the 1960s that has never been equalled since.[1] The government stimulated this development as best it could, though its priority was to create jobs.

An increasing number of manufacturers in this period began to see design as a vital link in their product development process and made it an important part of their policy. Sometimes this stemmed from idealistic convictions, but to an increasing degree it was driven by economic, or purely commercial, motives. Design education profited from this surge of interest and expanded quite substantially, resulting in a growing number of qualified professional industrial designers. These new designers were no longer solely interested in domestic objects and interior decoration; their field was extending from the simplest of domestic articles to agricultural equipment, medical apparatus, street lamps and railway carriages. Advertising and corporate identity took off in a big way too and provided plenty of work for the many new graphic design companies. By the 1980s design was firmly established. Spread over five museums, the exhibition *Holland in Vorm* (Dutch

Emil Truijen and Rob Parry, double pillar box, 1956–60.

Design) in 1987 featured design from the post-war years. The catalogue is an exceptionally rich source of the history of this flourishing discipline in the Netherlands in the third quarter of the twentieth century.[2] This chapter focuses on the design policy of the period, looking at how the business community, designers, government and educational institutions created the necessary conditions to redesign the Netherlands. In addition to the inescapable discussions on style, and the social or artistic calibre of the design, one of the most pressing questions in this period was how the designers' services could be most efficiently deployed.

The Government's Role and Designers' Initiatives

Straight after the war initiatives to promote modern design policy originated in designers' circles rather than via the state, and in a few cases came from private firms. In March 1945, two months before the liberation, a summary of a detailed report by the designers Paul Schuitema and Piet Zwart and the economist Jan Bouman appeared in the underground paper *De Vrije Kunstenaar* (The Free Artist).[3] This set out concrete plans for an industrial procedure that would be suitable for producing domestic objects once the war was over. Their report contained the first serious plan, written by experts, for the introduction of industrial design in the Netherlands. These three considered design to be an important and fully fledged discipline of national economic import, capable of a wide-reaching social impact. After the war they sent their report to the government, proposing that in the new structure it would be responsible for the coordination of the social, economic, technical and aesthetic aspects of design. This also held for future industrial design courses, which in the report's terminology was referred to as 'design engineering'. Here too they had interesting recommendations for the authorities, including the advice that design should be taught at technical schools as well as in art academies.

In response to this report, in 1945 the government immediately installed a Committee for Industrial Design, but this aroused little enthusiasm among the business community. Manufacturers seemed to be terrified of the idea of compulsory measures being imposed and were equally worried about artists becoming too influential. But Zwart, Schuitema and Bouman were not satisfied either: only one designer, Willem Gispen, was asked to join the Committee.

After Karel Sanders established the Aesthetic Advice Office in 1948, very slowly the Dutch government began to show some real interest in design on pragmatic grounds. Although industrial production was getting back into its stride, the deficit in the balance of payments meant that far more would have to be exported in future. On reflection, they decided that design really did seem to be able to make a contribution. That is why, after all, a central Industrial Design Foundation (Stichting Industriële Vormgeving) was set up at the end of 1949 on the initiative of Sanders and the three most important employers' organizations, reluctantly supported by the Ministry of Education, Arts and Sciences and the Ministry of Economic Affairs. The existing association of entrepreneurs, the Bond voor Kunst in Industrie (BKI), amalgamated with the new organization in July 1950 and a national Instituut voor Industriële Vormgeving (IIV) became a reality.[4]

The BKI had celebrated its twenty-fifth anniversary in 1949 with a large exhibition in the Stedelijk Museum in Amsterdam titled *Goed maar mooi* (Not just Good but Good-looking). Here it became clear that the BKI entrepreneurs were working with renewed dedication. The exhibition was no longer about art and industry going hand in hand as such, or about artistic decorative objects, but about industrial design in the modern sense of the word. In addition to furniture and decorative products for the living room, there were also electrical appliances on show, from radios and gramophones to sewing machines and vacuum cleaners.

Once again, on Sanders's initiative the industrial designers joined forces in 1952 to form the Circle of Industrial Designers (Kring Industriële Ontwerpers, KIO). Most product designers did not feel at home in the GKf, that is if they were allowed to join it at all, since the federation had a tough entry policy: new members were strictly vetted by a selection committee. The GKf, whose official name was the Society of Practitioners of Applied Arts (Vereniging van Beoefenaars der Gebonden Kunsten), had all kinds of members from widely varied backgrounds, including many graphic designers and craftsmen still working in a traditional way. On the other hand, the left-wing character of the organization, the prominent part some of its members had played in the Resistance and its focus on Amsterdam made it exclusive at the same time. In 1948 a number of graphic designers, mostly from Rotterdam and The Hague, who were keen to work more commercially established the Society of Advertisement Designers and Illustrators (Vereniging van Reclameontwerpers en Illustratoren, VRI). Machiel Wilmink, who had already founded the professional journal *De Reclame* before the war, was their first chairman.[5] From inside the GKf another professional organization was established in 1959, the Netherlands Industrial Designers Federation (Nidf), but this 'group of seven' was not very influential, although those involved included renowned designers like Willem Gispen and Piet Zwart.

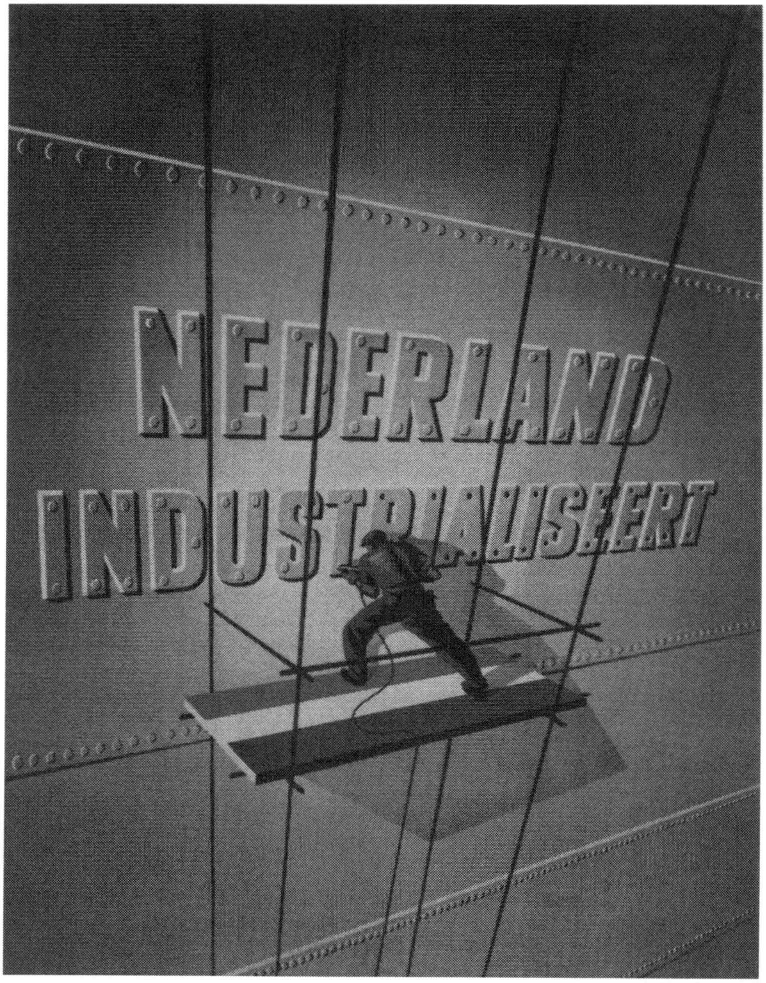

Wladimir Flem, poster proclaiming 'The Netherlands are Industrializing', c. 1948.

The IIV

The words 'Increased purchasing power through industrial design' appear on the front of a brochure designed by Karel Suyling and printed for the IIV[6] in 1952, and it goes on to say: 'if a product looks better it sells better'. The board thought it knew exactly what fell into the category 'better': 'Good design demands: the highest level of functionality, a dependable structure [and] an attractive appearance . . . By improving these three characteristics the Dutch product is bound to command a strong position when compared to its foreign competitor.' They included an alphabetical list of more than fifty products, from earthenware, glassware and radio sets to refrigerators,

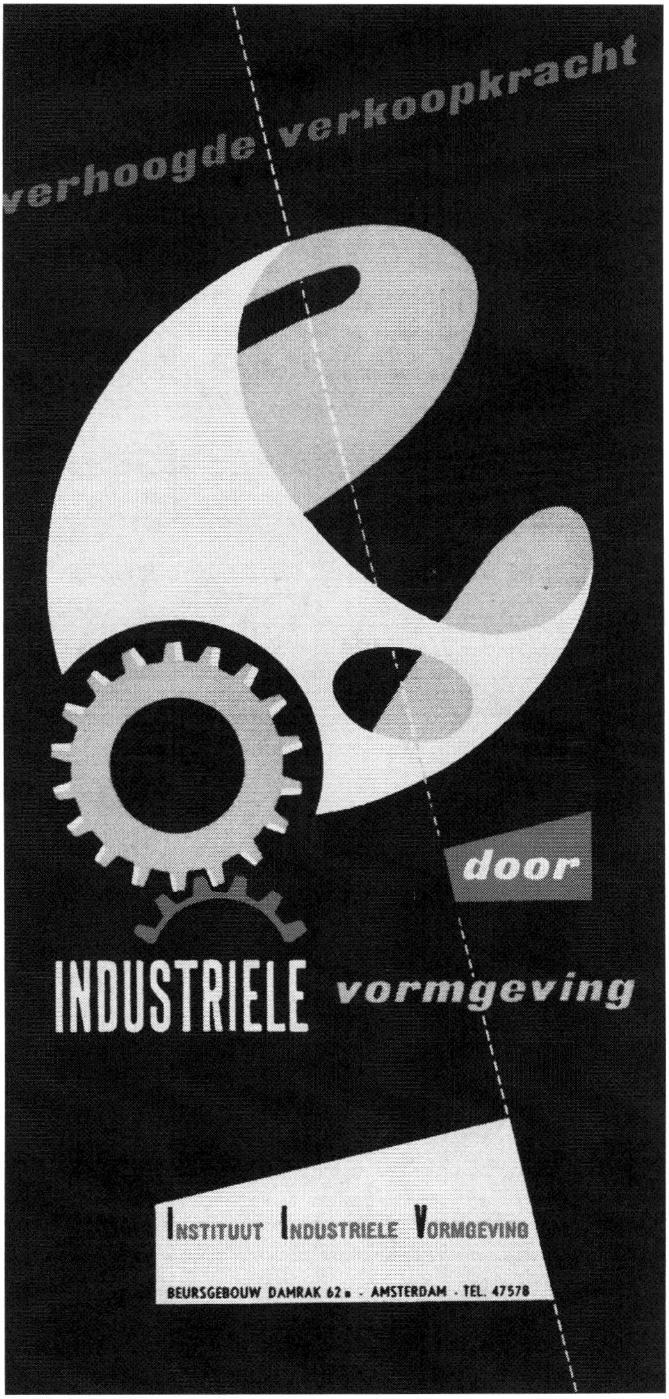

Karel Suyling, brochure for the IIV, 1952.

sports articles and washing machines, demonstrating the range of designers' skills to enterprises that were still in the dark about what they had to offer and how broad this type of professional expertise was. 'Once you realize that the assistance of an industrial designer is as important for your company as that of your economist, your technical engineer, your sales manager or your lawyer, you can approach the IIV, an organization working in this new field in the Netherlands.'

The IIV considered mediation to be its most important task. Its aim was to bring firms into contact with suitable designers and to this end it built up a comprehensive documentation system providing information on each designer's past projects and specialities. The Institute played an important role in the 1950s and '60s, seeing itself as stimulating 'brisk and free traffic' between industrial firms and designers. Its enthusiasm proved infectious, bringing together designers and firms, and organizing informative meetings and excursions, while readily passing on its knowledge and experience to government, industry and industrial design courses through policy memoranda, brochures and informative exhibitions. The variety of congresses and symposia the IIV organized made a major contribution too.[7] Foreign celebrities were brought to the Netherlands to give lectures and the IIV members exchanged their expertise and know-how with foreign sister organizations. Under this flag, the IIV invited speakers like Henry Dreyfuss and Walter Dorwin Teague from the USA. From 1952 onwards they published the *Maandbericht* (Monthly News; later the *IV-Nieuws*, 'Industrial Design News'), a newsletter in which manufacturers and designers were kept informed of all new developments.

Theo Ruth (Wagemans & Van Tuinen (Artifort), Maastricht), 'Congo' Easy Chair 1001, 1952.

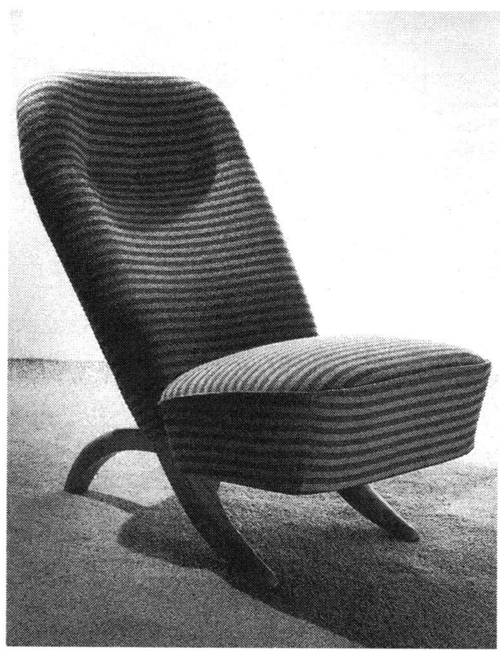

The exhibitions, which followed one another in rapid succession, reached an ever growing public. They were organized in the IIV's own premises on the Rokin (from 1954 on the Herengracht), but they also took the form of special presentations at trade fairs, where they featured a specific segment of the market, such as furnishing fabrics, electrical appliances and kitchens. The IIV also held exhibitions in which they collaborated with foreign organizations, as well as mounting displays for specific firms. Auping, Artifort, Philips, Sikkens, Mosa and Stokvis, for instance, were given the opportunity to show their new designs, and even

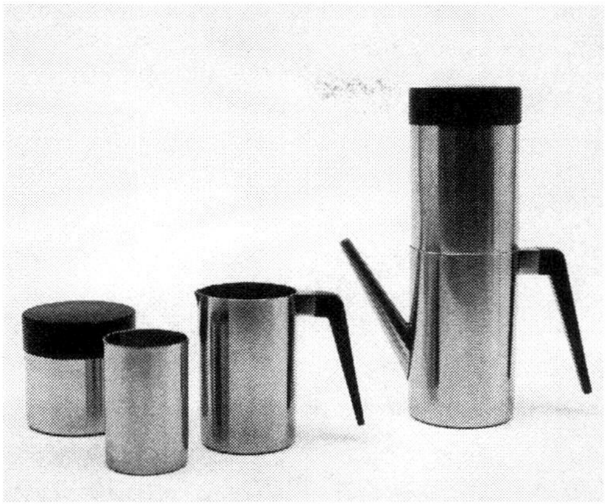

Dick Simonis (Gero), stainless-steel coffee-service, 1959.

the Dutch branches of Olivetti and Braun were invited to take part in these presentations.

One of the first events to show that the effort put in by the IIV was actually paying off was the Triennale in Milan in 1954. The Institute coordinated the Dutch entry and saw to it – with extra support from the Ministry of Education, Arts and Sciences – that some thirty firms took part. A few Dutch designs that were later to become famous were first presented at this event. It was here that the public would become acquainted with the ingenious Revoltstoel, made by De Cirkel and designed by the young Friso Kramer, son of the Amsterdam School architect Piet Kramer.[8] For the first time, instead of round tubes Kramer used U-shaped steel tubes, which were not just cheaper but could be used in a more varied way, allowing for more creativity in the design. It was in Milan that Wagemans & Van Tuinen (Artifort) from Maastricht showed the extraordinary Congostoel designed by Theo Ruth. Among the Triennale winners were the Leerdam glassworks, for their Gildeglas designed by Copier, and Gero from Zeist for their cutlery and stainless-steel pans by Dick Simonis. The potteries, which had been duty-bound immediately after the war to devote a considerable quantity of their raw materials to the production of standard consumer durables, had extraordinary success with their newest designs. The important and long-established Sphinx and Mosa factories presented attractive services by Pierre Daems and Edmond Bellefroid. The smaller firms Fris and Sint Maarten Porcelein won high praise for their pottery designed for everyday use by Wim de Vries and Han Knaap.[9]

To everyone's amazement, the economic returns from this event proved to be high. It would appear that the Dutch had become so accustomed to regarding this sort of exhibition as primarily a cultural affair that they were almost surprised to find that orders had been placed by foreign buyers.

In 1956 the IIV participated in an exhibition in the Stedelijk Museum in Amsterdam. Most of the exhibition, which had the unambiguous title *Industrial Design*, consisted of a presentation of German design, coordinated by Wilhelm Wagenfeld, and Italian design, organized by Marco Zanuso. In addition the IIV filled three small galleries with a display in which information was given on industrial design as part of a company's production process. The Revoltstoel by Friso Kramer and earthenware by Edmond Bellefroid served as models.

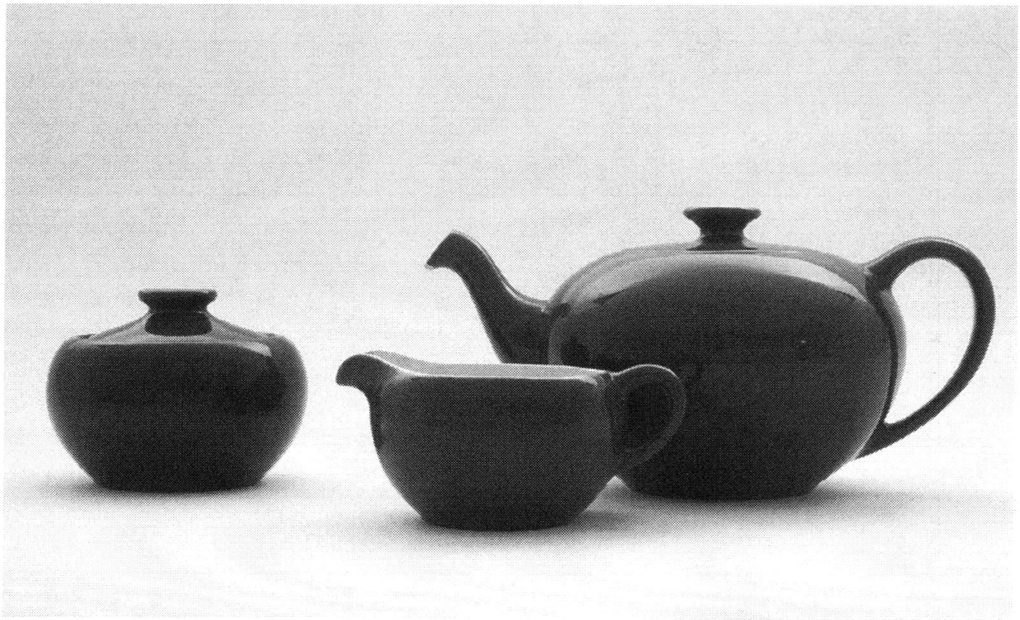

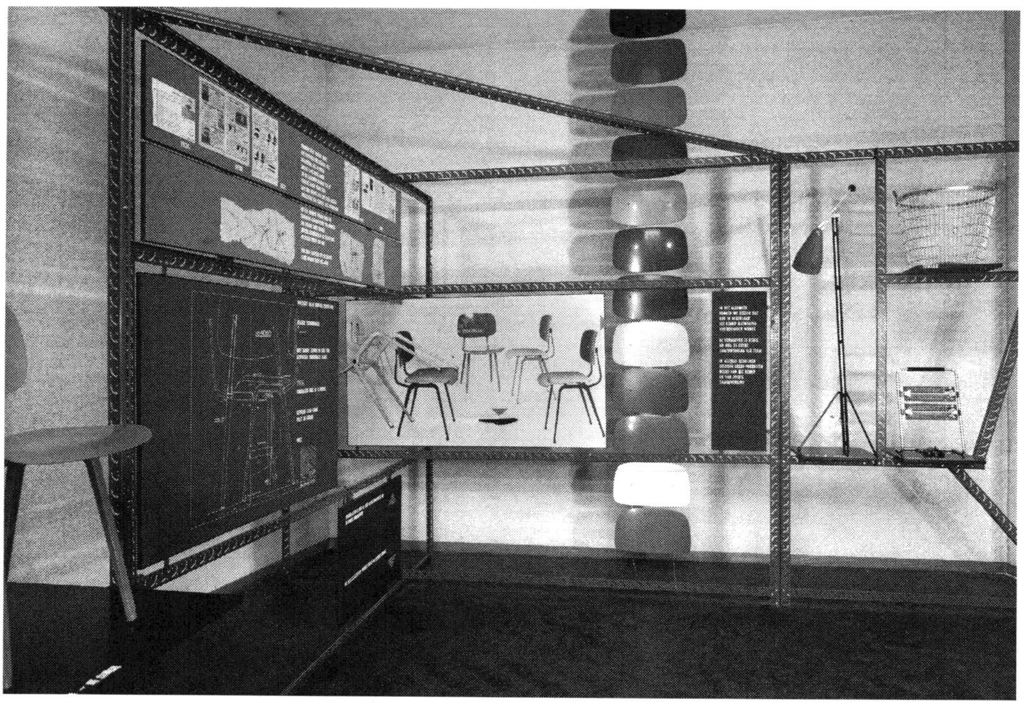

Wim de Vries (Fris, Edam), 'Edam' tea service, 1949–52.

Another successful publicity stunt was the tour of a number of factories and design companies that the IIV organized in 1956 for a large group of Dutch journalists. This resulted in scores of newspaper articles with headlines such as 'Industrial design, a weapon to be deployed on the free market'.[10] These articles also boasted about the great technical advances being made in Dutch industry and the 'substantial scientific investigations' conducted by Dutch designers.

In 1957 the IIV joined the International Council of Societies of Industrial Design (ICSID). Two years later, at the first congress of this international organization in Stockholm, one of the Netherlands' representatives was L. C. Kalff, who had been designer-in-chief for Philips for many years. In his speech to the congress Kalff typified Dutch design as 'reliable, simple and inexpensive'. Despite this characterization, which was neither spectacular nor original, the speaker was highly commended. The 'serious Dutchmen' were held in great respect, as could be read in the report on the congress in the IIV's monthly review *Maandbericht*.[11]

At the second ICSID conference in 1961 in Venice the current state of the art of design was displayed in two hundred photographs of new products from sixteen different countries. As well as once again showing cutlery by Gero designed by Dick Simonis, and glasses from Leerdam by Andries Copier, the Netherlands displayed advanced technical products such as a fertilizer distributor by Wim Rietveld, Gerrit Rietveld's son, an amperemeter by J. Wouda, a tramcar by Friso Kramer and Jaap Penraat, a sun lamp from Philips and even an aeroplane, the famous Fokker F27 Friendship designed by H. C. van Meerten.

The cheerful public façade of the IIV concealed many conflicts behind the scenes in which money played the crucial role. One problem was that the government had been living under the illusion that the Institute would become financially independent in the short to medium term, and that its contribution to the IIV's funding could then be considerably lowered. This proved not to be the case. In practice the commercial firms associated with the Institute were not always happy with the idealistic advice they were given. The ongoing criticism levelled by the IIV often left firms feeling patronized and discredited, with the impression that the IIV's criticism was more to blame for curbing their economic prosperity than for stimulating it. Of course, government subsidy had never been granted to the Institute with this scenario in mind. It had been motivated by the need to advance industrial activity, working on the assumption that design was a stimulating instrument, not just to be supported as an end in itself. The government had no intention of frustrating industry by

A room at the *Industrial Design* exhibition at the Stedelijk Museum, Amsterdam, showing the Revolt chair by Friso Kramer, 1956.

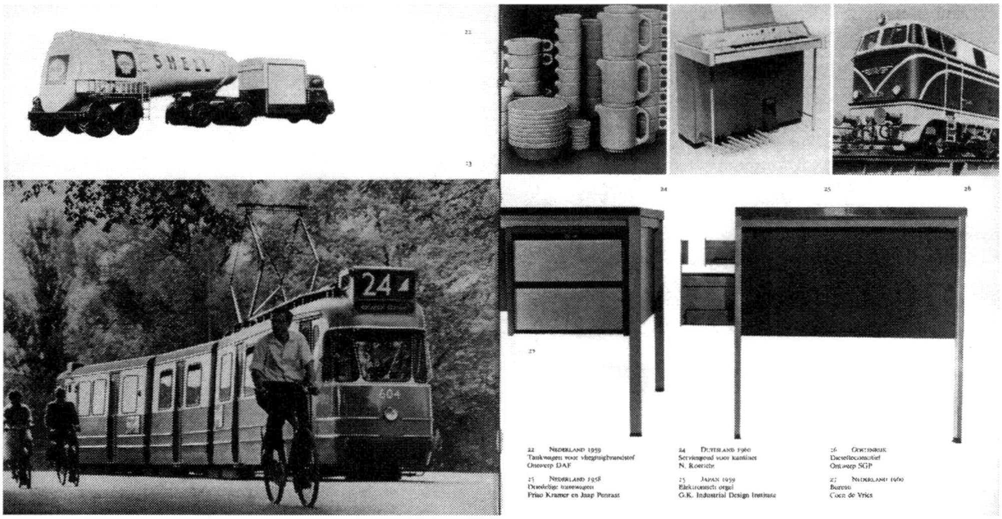

Vorm ('Design') brochure, published by the IIV and the Sikkens firm, printed on the occasion of the ICSID conference in Venice, 1961, including designs by W. Rietveld en J. Penraat (tramcar, 1958), Daf (tanker, 1959) and C. de Vries (steel desk, 1960).

supporting the IIV's strict design norms and for this reason it reduced its contribution to the budget.

The conflicts, the financial problems and the various reorganizations, accompanied by an equal number of resulting policy changes, led in 1961 to a structural change of course. A new, independent national Industrial Design Council (Raad voor Industriële Vormgeving) sprang to life, with representatives from industry, commerce, consumers, industrial designers and education. This 30-member strong Council had to operate and promote design across a far broader front than the IIV had ever done. The Council had two executive bodies at its disposal, the existing IIV, which continued to help the more than two hundred affiliated firms to find suitable designers, and a new institution yet to be established, the Industrial Design Centre (Centrum voor Industriële Vormgeving, CIV), which would liaise with consumers and the retail trade. In this Centre they planned to organize frequently changing presentations of well-designed Dutch industrial products. A Selection Committee, consisting of figures who enjoyed the trust of all parties concerned, would select these products on the basis of the Council's established norms and guidelines. In this way they thought they could take a more independent and objective stance. The British Council of Design was taken as a model, although at the official installation of the Council it was observed, somewhat wryly, that the British organization received approximately two hundred times more financial support than its Dutch counterpart.[12]

Off to America

In the 1950s America was the great ideal for Dutch entrepreneurs and consumers, particularly as far as modern design was concerned. Even the Ministry of Economic Affairs had recognized soon after the war that industrial design, as practised in the United States, could play an important part in stimulating the economy. This led to a ministerial Committee for Increased Productivity, in close consultation with the director of the IIV, Karel Sanders, putting together a select group of designers in 1953. The chosen few were allowed to familiarize themselves with the American situation in some depth, at the Dutch government's expense. In addition to Sanders, the happy few included Wim Gilles, René Smeets, Jaap Penraat, Karel Suyling and the journalist Rein Blijstra.[13]

Until then Wim Gilles had been a designer at the metalwork factory of Diepenbrock & Reigers in Ulft (DRU) in the east of the country.[14] This firm was a model for many factories that had only just taken their first serious steps in the field of design. Although it had been started in the eighteenth century as an iron foundry based on traditional craftsmanship, its first steam engine was installed in Ulft in the mid-nineteenth century and the firm had grown to more than 600 workers at the start of the twentieth century. From far back in the company's history new models for garden benches, letterboxes and enamel pans had been 'moulded' by a small group of model-makers. The firm's success was based on this long-standing tradition, together with a close eye on the products being produced by its competitors.

In 1948 the director of the DRU, J.A. Ingen Housz, took the initiative to alter radically the design process in his factory. To implement this he took on the young mechanical engineer Wim Gilles, who subsequently used his own judgement to introduce a modern design methodology based on market research and a self-developed system of product analysis.[15] After this Gilles thought up a new design methodology involving a logical, well-reasoned protocol. This 'mathematical organization of form' would mean that the outer appearance of the product would no longer be determined by the subjective, artistic preference of an individual, but would be the result of an objective, verifiable, more or less scientific process.

One of the results of Wim Gilles's innovative ideas first saw the light of day in 1954: a whistling tea kettle made from enamelled steel plate. The revolutionary materialization of the kettle, which was to become a famous design, was described in 1955 in the new *Technical Winkler Prins Encyclopedia* as a typical, practical example of a modern, process-based approach to industrial design, demonstrating the attention paid to technical detail and its functionality. The fact that they selected an ordinary kettle, rather than a more traditional object like a chair, a vase or a carpet, reveals a great deal about contemporary thought regarding design developments in the Netherlands.

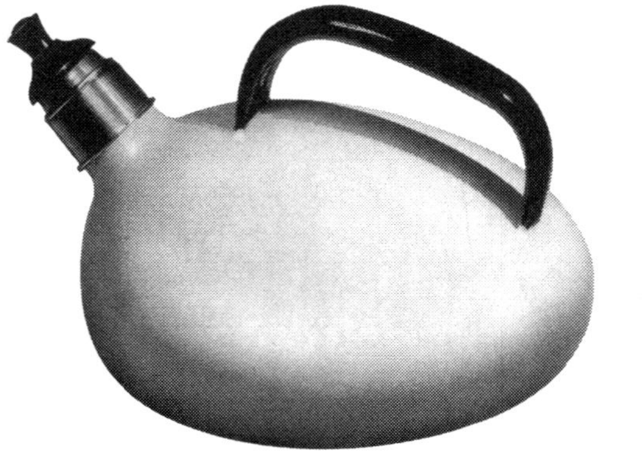

Wim Gilles (DRU), enamelled sheet-iron kettle, 1954

René Smeets, who was also invited to join the study tour of America, had become director of the new industrial arts school in Eindhoven in 1950.[16] The idea was that he would concentrate mainly on design education. Before he became director in Eindhoven, Smeets had worked for a few years as a self-taught designer at the ceramic factory Russel-Tiglia in Limburg and had also served as an officer in the army. The evening class in industrial design at the school in Eindhoven was meant to spur on industry in the Brabant region. Unlike industrial-design teachers between the wars, such as Piet Zwart, Gerard Kiljan and Mart Stam, who in their courses propagated collaboration with industry for idealistic reasons, Smeets was mainly inspired by pragmatic considerations. To his way of thinking, the economy and the needs of the Brabant enterprises were the main concerns, and it is no coincidence that Louis Kalff, the designer-in-chief at Philips, was involved in the creation of the school. Firms mostly required 'attractive'-looking articles, which a large number of people would love to own, so increasing the turnover. 'Attractive' to Smeets did not mean in the first place 'sober and honest', but rather 'beautified' or even 'decorated'.

Freelance designers were represented in the America group by Jaap Penraat and Karel Suyling.[17] Penraat had been trained during the war years by Mart Stam and Johan Niegeman at the IVKNO in Amsterdam. The fact that Penraat had also spent those years forging papers and identity cards to successfully smuggle more than four hundred Jews out of the Netherlands only came to light years later. Towards the end of his life this Dutch Schindler was internationally decorated for his act of heroism. At the time of the American tour Penraat was already one of the most progressive Dutch designers, interested in technique and user-friendliness and in the development of entirely new technical products. One of these was a new tramcar that he designed with Friso Kramer, but he was also one of the first in the Netherlands to introduce the open kitchen. He emigrated to America in 1958, so obviously the country must have made a good impression on him during the tour.

Karel Suyling was self-taught. He worked mainly as an advertisement and packaging designer and dedicated himself to the emancipation of his discipline. From 1955 to 1970 he designed many advertisements, including those promoting Citroën in the Netherlands.[18]

The designers travelled around North America for six weeks, visiting design courses in Boston, Cleveland, Chicago, Illinois, Cincinnati and New York. They also looked in on fifteen design companies, including the large offices of Henry Dreyfuss, Walter Dorwin Teague and Raymond Loewy, and the large design departments at Kodak and General Motors. The world the Dutch designers entered was totally different from the one they were used to back home. Not only were such large offices unknown phenomena in the Netherlands, but the group was also impressed by the professional, business-like character of the American offices and envious of their commercial success. In the United States design had been a completely accepted and respected link in the production process for many years, whereas in the Netherlands it was unknown on such a large scale. Such phenomena as consumer research, product analysis and product presentation were also new for the Dutch visitors, at least when taking into consideration the professional manner in which it was organized in America. Moreover, the industrial designer proved to be involved in the whole process, from the formulation of the design commission to the presentation of the new article to the consumer. The Dutch group viewed the highly regarded, versatile, well-trained and commercially driven American designers with a certain amount of jealousy.

However, the Dutchmen did not fail to notice that all the American success stories had their drawbacks. They were particularly critical of the fact

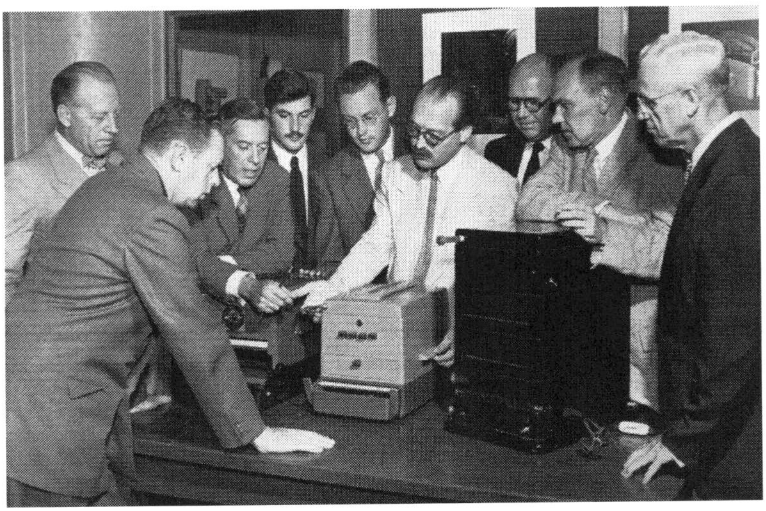

A Dutch delegation of designers visiting the design office of the Eastman Kodak Company, Rochester, New York, during its study tour in 1953. From left to right: K. Sanders, R. Smeets, T. G. Clement (of Kodak), K. Suyling, W. Gilles, J. Penraat, Robertson, R. Blijstra, unnamed Kodak employee.

that in the United States too much attention was paid to things that people back home considered unimportant, or would even have condemned, such as superficial styling and the slavish following of trends – something still detested in the Netherlands. They were also undecided about the new phenomenon called marketing. On the one hand they really saw the commercial benefits it brought, but on the other they felt that yielding to consumer demands conflicted with upholding an 'objective' view of good form, which still completely governed Dutch thinking about design.

When they returned to the Netherlands each participant wrote a report on his experiences. Jaap Penraat and Karel Suyling concentrated on the position of the freelancer in America; Wim Gilles and Karel Sanders analysed the relationship between industry and designer; and René Smeets wrote his report on his experiences in American design education.

Two years later, under the auspices of the IIV, three Dutchmen were invited to join an international party of enthusiasts on a tour of America: G.C.J. Schoemaker, director of Inventum, a factory making electrical appliances, and the designers Wim Rietveld and J. Wouda.[19] They were introduced to a relatively new branch of the design profession during their trip, the 'medical advice officer', an area of work that some American designers already appeared to be engaged in. This was the Dutch design world's first introduction to the new discipline later to become known as 'ergonomics'.

New Training Courses

Well-trained, contemporary designers, a prerequisite for implementing a goal-orientated, modern design policy, were still scarce shortly after the war. In the design schools – then still usually called Arts and Crafts schools – the traditional crafts were still the main focus. There were only a couple of exceptions, including the Academy in The Hague where, under Cor Alons's management, a department of Interior Design and a department of Advertising, under the supervision of Gerard Kiljan, had both been in operation since 1934, and where something resembling a modern approach to industrial design was in evidence.[20]

The other exception was the New Art School (Nieuwe Kunstschool) in Amsterdam, also founded in 1934 by the former Bauhaus teacher Paul Citroen; but this course did not survive the war.[21] Those who had taught at this progressive, private, non-subsidized school included the architect Alexander Bodon, graphic designer Hajo Rose, weaver Käthe Schmidt and photographer Paul Guermonprez, the last three of whom were former

Bauhaus pupils. Research on colours, materials and structures were the main educational themes, modelled on the *Vorkurs* established by Johannes Itten at the Bauhaus. One of their first pupils was Benno Premsela, who was later to become an influential interior designer and one of the foremost authorities on Dutch design.

From the early 1940s Mart Stam and Johan Niegeman tried to modernize the approach to education at the Instituut voor Kunstnijverheidsonderwijs (IVKNO) in Amsterdam.[22] They taught their students to focus their attention on people's needs and to tackle design commissions in an analytical and systematic way. They condemned the sort of educational approach based on the artist's ego and which focused on artistic-minded expression. In their eyes artisanal design was outdated, although they did still concentrate mainly on traditional interior design.

Changing course proved to be easier said than done. A disillusioned Mart Stam had already left by 1947 and Niegeman could not manage to put together a syllabus that fell in line with modern industrial society. Disenchanted, he left in 1955. Nevertheless, a number of very promising designers had graduated under his inspiring supervision: his pupils Friso Kramer, Jaap Penraat, Kho Liang Ie and Coen de Vries unquestionably numbered among the most progressive industrial artists in the Netherlands in the 1950s. It was not until 1960 that, in addition to the department of Interior Design, a fully fledged department of Industrial Design was set up in Amsterdam under W. J. Jaarsveld.

The study tour of America in 1953 stimulated a few important reforms in the educational programme at the Eindhoven college. René Smeets incorporated his findings in the syllabus for an entirely new daytime Industrial Design course, which started up in 1955 and ran parallel to the evening course. This was indeed the first specialized School of Design in Europe. In the curriculum the main focus was on intensive collaboration with industry. After a general foundation year, the five-year course had three specialization profiles: product design, publicity and product presentation, and textiles. The fourth year was set aside for internships so students could gain hands-on experience with professionals in the workplace. The promising Wim Gilles was taken on as a teacher by Smeets; he rose to the position of director from 1970 to 1973.

The Hague Academy of Fine Art had started a 'weekend course' on industrial design in 1950, partly at the request of the Ministry of Economic Affairs. This course was meant for people already working in the field, such as young adults who had completed Technical School and were attached to a firm as a structural engineer. It was for this reason that the lessons were given on Friday afternoons and Saturdays. In addition to the traditional, creative design subjects, they also taught new subjects related to the new study of ergonomics, product analysis and visual communication. Gerard Kiljan became the course coordinator. Apart from his teaching, this designer executed several commissions of his own: he was, for example, responsible for the design of the Bakelite telephone (1955), produced in hundreds of thousands by the firm Heemaf in Hengelo. In these post-war years he also designed Joy lemonade bottles, including the labels, in an attempt by the Hilversum soft-drink manufacturer Koster to compete with Coca-Cola.[23]

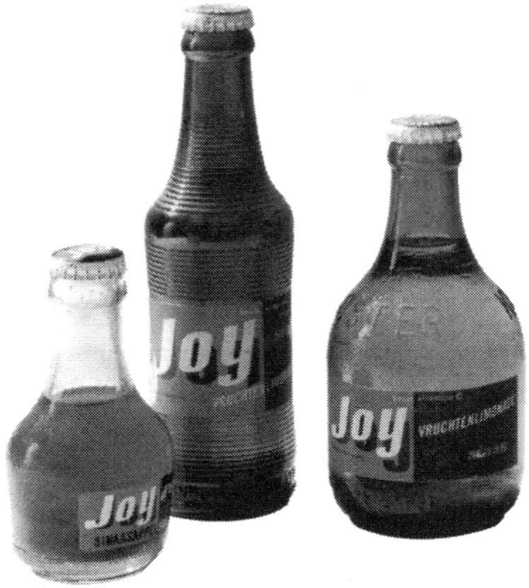

G. Kiljan (Wed. Thijssens & Zn/Joy) three bottles with labels, 1948 and 1960.

A group of prominent designers, including Cor Alons, Willem Gispen and Gerrit Rietveld, were brought in to teach on the new course in The Hague. They also asked the young Kho Liang Ie and Friso Kramer, both of whom had only just graduated from the IVKNO in Amsterdam, to teach there. The first graduates from The Hague included Joop Istha, J.C. Berkheij, Joop van Osnabrugge and Wim Rietveld. Joop Istha developed into a versatile designer of home and technical appliances and boasted a large international network. From 1975 to 1990 he was professor at Delft Technical University. Johannes Berkheij specialized in medical equipment and for a few years ran a design company with Joop Istha. One of his successes was the cylindrical gas heater (1968) designed for the firm Etna in Breda. Joop van Osnabrugge also became a versatile designer of consumer products, including electrical appliances, kitchens and stoves. Finally, as well as designing chairs and lamps, Wim Rietveld turned his attention to technical devices, agricultural equipment, lorries and trains.[24]

Meanwhile, the Dutch government was pressing for the creation of an industrial design course at a higher, more scientific level. Delft Technical University seemed the obvious place for it, an idea that was supported by leading designers who had already presented a case for such an institution before the war and who had done their best to professionalize their discipline. The most well-known champions were Mart Stam, Wil Sandberg and the glass designer Andries Copier. But back in Delft they were still not interested in the idea. The projected new department would have to be set up by staff from the departments of Architecture and Mechanical Engineering, but at the time they could see no point in establishing a course for 'semi-artists'. In the Architecture department Professor M. J. Granpré Molière still held sway, a traditionalist who was highly critical of modern industrial mass production. It was only when the architects J. H. van den Broek and F. A. Eschauzier became professors in Delft that the tide turned, and even then it was to take until 1964 before the first two students in Delft could begin studying industrial design. G. J. van der Grinten, who had done all the preparatory work in Delft over the years, including stressing the economic importance of such a course, was to become the first Extraordinary Professor of Industrial Design in the Netherlands.[25] In comparison to the design courses given at industrial design schools, the new course in Delft was focused more on technique, on subjects relating to man and society, and on design methodology.

Design Policy in the Factories

The young industrial designers who had been trained in Amsterdam, Eindhoven, The Hague and later Delft were able to find work easily due to the flourishing economy and the fact that industrial design as a field was becoming increasingly regarded in the 1950s and '60s. They obtained commissions quite easily or found part-time jobs as designers for a few days a week. Those commissioning their work came from a wide range of sectors as a growing number of manufacturers came to realize that it was in their own interest to pay more attention to design. The demand for designers was often greater than the supply.

The evolving electrical household appliances sector in particular provided a great deal of employment.[26] It was in this sector that the products of a fully developed design policy came to the fore. The vacuum cleaners and irons that could be found in most households were in dire need of replacement and consumers were also starting to show an interest in coffee grinders, hair driers, mixers, electric cookers, sewing machines, refrigerators and fully automatic washing machines. The enormous increase in wages and the abolition of luxury tax in 1955 gave sales of these attractive items a boost: between 1957 and 1964, for example, the number of families owning an electric washing machine increased from 31 to 83 per cent, while those with black-and-white televisions increased from 8 to 68 per cent.[27]

The foremost symbol of progress was the modern kitchen, equipped with the newest home appliances. The popular Polygon Newsreel commentary, shown at the beginning of every cinema programme, reported on the American kitchen in 1954, 'the kitchen of the future', in which, at the touch of a button, a housewife could prepare a delicious meal with the greatest of ease. Those in the audience who dreamed of such a kitchen could view something similar for real in 1957 at the exhibition titled *Het Atoom* (The Atom) in Amsterdam. This exhibition, and a handful of national demonstrations of comparable ambition, combined economic information and industrial propaganda with entertainment. The largest and most impressive display was staged under the name *E55* in Rotterdam. The main theme in this case was the resurrection of this seaport town, which had been razed to the ground during the war.

The most familiar names among more than two hundred factories in the Netherlands that were making electrical home appliances in the 1950s and '60s are Philips, Van der Heem (under the brand name Erres), Inventum, Indola, Holland Electro, Daalderop and Ruton. This is where most of the design activity was taking place. Their design policy gradually changed during these years from one of pragmatism, mixed with a considerable amount of idealism, if not downright paternalism, to a purely commercial policy based on market research.

The Royal Electrical Appliances Factory Inventum in Bilthoven, established in 1908 as Inventa, had already produced great numbers of electrical irons, hot plates and electric heaters even before the First World War. The Amsterdam Municipal Electricity Company, which was then engaged in competitive warfare with the Municipal Gas Company, had a policy of promoting the sale of electrical appliances to stimulate the use of electricity. As a result of this campaign, by 1916 there were already more than 15,000 electric irons in everyday use there and by around 1920 Amsterdam was the 'most electrified city' in the world. Far more thought, however, was given to the technical innovations applied to these products than to their design, for on the whole it was products made by large foreign companies, such as the German firm AEG, that set the standard. The design idiom used for the modern models in the range was international commercial Art Deco.[28]

Only one designer from Inventum's pre-war years has been recognized, Arie W. Verbeek from Rotterdam, who designed a minimalist electric heater in 1929 that went into production in 1932. This handsome appliance has found a place in the design collections of various museums. Verbeek was one of the first designers with a firm belief in the need for this new discipline to be applied to industrial mass-produced articles in the Netherlands.

After the war the management at Inventum began to start thinking in a more contemporary and structural way about design. Their electric heaters were subsequently modernized by Wim Rietveld. The first result of his study tour of America (accompanied by the director of the firm Schoemakers) was

Arie Verbeek (Inventum, Bilthoven), electric heater, 1929.

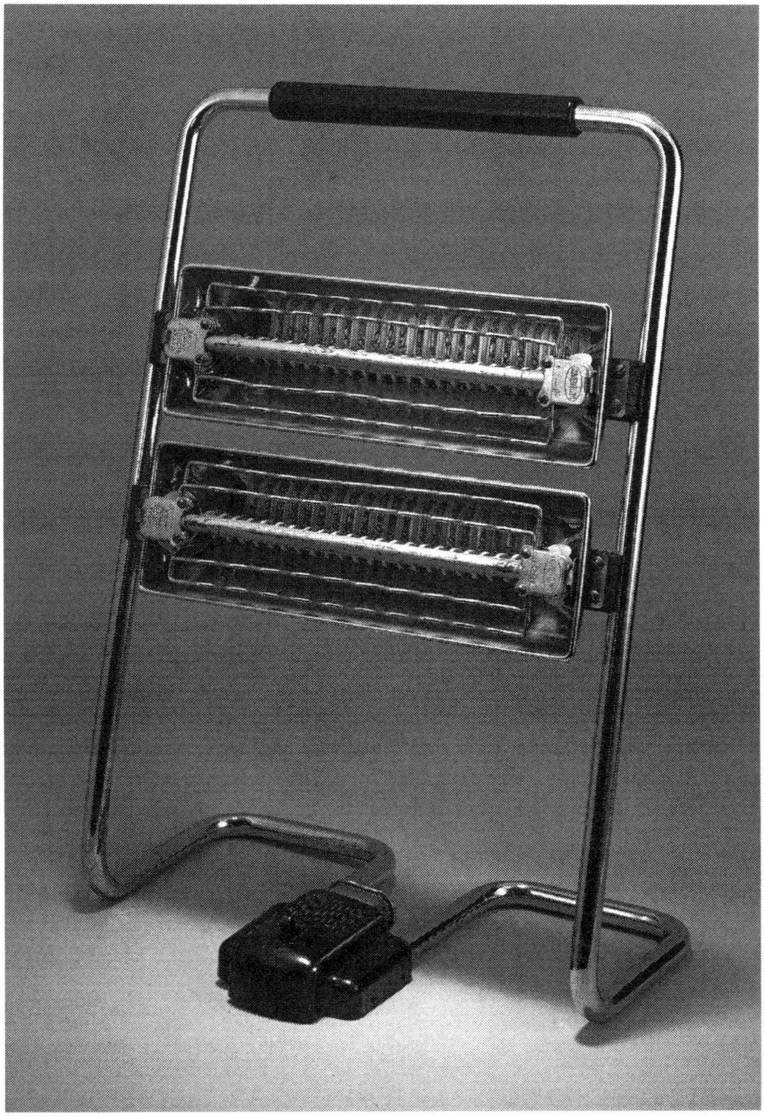

that the factory stopped playing safe by simply extending the range of products. Instead, it reduced the core collection and made it available in more than one colour. In this way the production process could be organized far more efficiently and cheaply, reducing the price of the appliance and increasing turnover. Marketing was introduced only after Rietveld left Inventum. This involved recording consumers' current needs and assessing their future expectations. On the basis of the results of this study it seemed advisable to expand the range of products on sale. What in essence had been a paternalistic outlook towards design, and a great faith in a rational approach to good form, was now slowly but surely being abandoned. The management plumped for a more commercial design policy that would allow them to react more quickly to shifts in fashion and lifestyle trends. Market segmentation became a household word. The new strategy was accompanied by a much more dynamic advertising policy. Designer Joop van Osnabrugge proved to be willing to go along with this method of work, which was much more lucrative from a business viewpoint.

Philips in Eindhoven had already felt the need to consider product design at an earlier stage to bring it into line with company policy. In the mid-1920s, in addition to light bulbs, the firm began to produce radio sets.[29] Unlike electrical home appliances, radios were products that were more likely to be given pride of place in the living room, which meant that more care had to be taken about the way they looked. In 1925 Louis Kalff took over as head of Philips's own advertising studio. In those years this department was also responsible for the 'aesthetic supervision' of new products, as well as designing posters, packing and stands at home exhibitions. In 1926 Kalff designed the first shell-shaped Bakelite loudspeaker, available in several colours, which could enhance a living room in the same way as a work of art. The first radio case followed in 1927. Kalff was put in charge of the new 'Artistic Design Team' in 1930, consisting of a small international group of designers exclusively engaged in product design. The number of radio models was soon enlarged and production increased at a fast rate: by 1932 a million Philips radios had already been sold.

Philips started to think about other products as well. In those days radios were typically seasonal articles, so the search was on for alternative Bakelite products that could be sold all year round and not just in the autumn and winter. Soapboxes, sewing cases and even toilet seats proved to fit the bill, but the electric shaver, developed by the engineer A. Horowitz, was the greatest hit, despite grave initial misgivings on the part of the Philips Management Board. The Philishave, the first electrical dry shaver with rotating blades (still in its original form with one shaving head), was presented to the public in 1939 at the Utrecht Spring Trade Fair.[30] After the war the Philishave appeared in a streamlined version, made from a white, Bakelite-related synthetic material. The design acquired the nickname *Eitje* (Small Egg). During the war years a second rotating head was added and since 1966 the Philishave has been manufactured with three shaving heads.

After the war the design team at Philips was replaced by the new department called 'Appliance Design'. In choosing this name they wanted

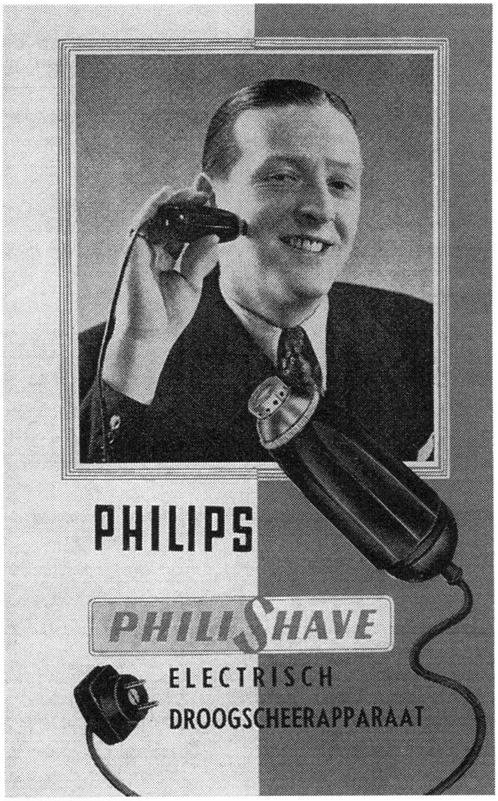

Advertisement for the Philishave dry-shaving method, 1939.

to stress Philips's intention to produce an ever increasing number of electrical appliances. After the shavers, radios and record players, the range of products was enhanced with portable radios, vacuum cleaners, televisions, coffee grinders, irons, sun lamps, spin dryers and all sorts of professional appliances.

Rein Veersema was one of this new department's most successful designers. This status meant that he was greatly appreciated by the management, but also led to differences of opinion with Kalff. In 1956 Veersema was put in charge of the design office and design policy was structurally modernized. It was his aim to give the totally divergent forms of all the various Philips products one 'face': the 'Philips family' look. What by then had become a large international enterprise had to be given a corporate image. Moreover, Veersema introduced ergonomics as a fully fledged part of the design process. A more methodical and interdisciplinary design policy was gradually developed at Philips, in emulation of the shining example of the rival German company Braun, where Dieter Rams was at the helm from 1960, and of the design vision of the Hochschule in Ulm, where the goal was to create timeless design. Unfortunately, Veersema's policy was only partially successful and the firm certainly did not achieve the desired commercial results. Veersema was succeeded in 1965 by the Norwegian Knut Yran, who had more commercial insight and was far less convinced of the universal validity of the design rules propagated in Ulm. It was not his ambition to make timeless designs: he was more interested in making products for the future, while also being a great admirer of commercial American design. Yran introduced a new design system at Philips, consisting of a clearly defined design-track, running from project briefing right up to final delivery of the packed product to the retailer. In this way the uniformity of design of Philips products was considerably strengthened by an ever expanding and ever more important Philips design office, known as the Concern Industrial Design Centre (CIDC). In 1981 Bob Blaich took Yran's place, while he in turn stepped down in 1991 to make way for Stefano Marzano. Thus for decades Philips design policy was delineated by foreign designers, not by Dutch.

In the course of the 1960s this increasingly large and powerful firm, originally based in Eindhoven, swallowed up all the other Dutch firms operating in the electrical appliances market. One of the last to be taken over by Philips was Van der Heem. After the war important work was still being carried out there by the designer Piet van der Scheer and later by Joop Istha. From the start Van der Scheer was willing to take into account the wishes of the sales department as well as those of the technical engineering department. In the 1950s and early 1960s Van der Heem was inspired by American advertising methods and went even further by experimenting with the most advanced marketing methods. But, at the same time, it continued to believe in the benefits of good design. In its slow but sure efforts to ensure that good design would become generally accepted, it analysed consumer behaviour, and modern sociological theories on top-down dynamics for innovation were translated into commercial strategies. In spite of all that, Van der Heem lost its independence in 1969 and the Erres brand name was to become Philips's second trademark.

Modernization in the Furniture Sector

Like the producers of electrical home appliances, the furniture industry profited from the great demand for replacements after the war and the explosive growth in prosperity.[31] At first everything this branch of trade produced was sold quite effortlessly, even if no attention was paid to design. This period of booming business, however, also meant that the factories devoted far too little time and effort to modernization. The hundreds of generally small family firms were satisfied with their profit margins and neglected to invest enough of their earnings in the renewal of extremely antiquated machinery – just as little time and money was spent on marketing. A 1968 study examining the state of affairs in the furniture industry, commissioned by the Dutch Economic Institute (Nederlands Economisch Instituut), concluded that the situation had scarcely altered in thirty years. Not surprisingly, when trade fell off it proved to be fatal. The sector could not stand up to foreign competition, which was able to supply the goods faster and make products more suited to the demand. For that matter the demand was still mainly for traditional, classic models in a style that those in progressive circles denigrated as *balpoten* ('legs-on-balls') style or *old finnish* (a Dutch corruption of the English word 'old-fashioned').

In the 1950s and '60s there were really only three factories that pursued a modern, interesting design policy for consumer furniture: 't Spectrum,

Martin Visser
('t Spectrum, Bergeyk),
'Riethoven' metal and
rattan chair, 1959.

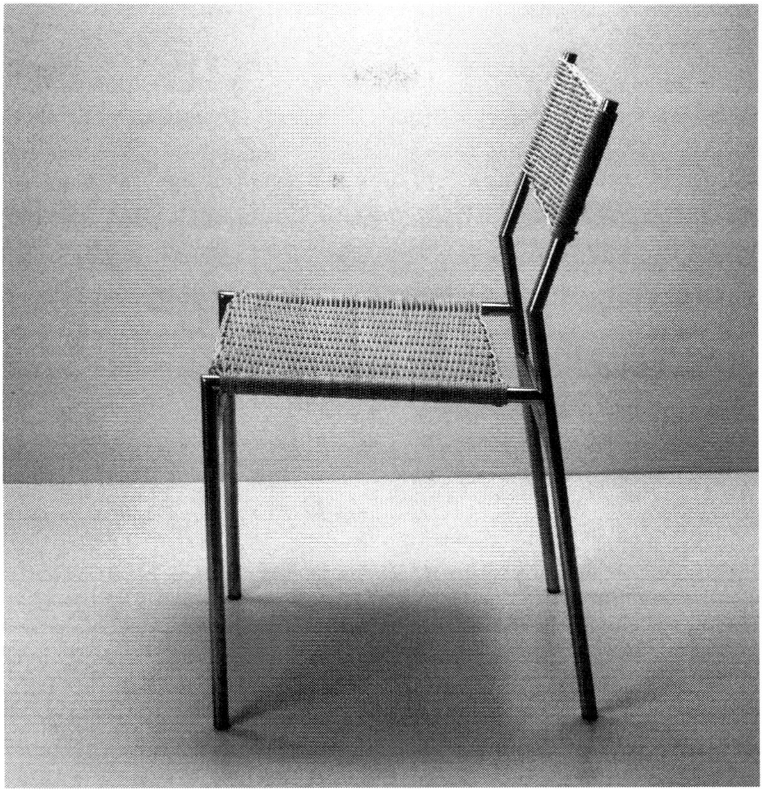

UMS-Pastoe and Wagemans & Van Tuinen, later called Artifort. 't Spectrum in Bergeyk was started during the war as a daughter company of the weaving mill De Ploeg.[32] It was very idealistic, just like its parent company, and the employees had a large say in company affairs and a share in the profits. The purpose was to make timeless designs that everyone could afford. For a few years post-war material shortages restricted them to making small pieces of furniture and domestic objects, but by the mid-1950s 't Spectrum had progressed to larger items. The firm's preference for good but sober design had changed little over the years. In the new statutes that the factory drew up in 1957 they included a clause stating that the furniture had to comply with the 'demands of good taste'; in other words, it should be timeless, functional, affordable and reliable. The use of natural, traditional materials, and a design idiom associated with Scandinavian furniture, contributed to this image. It is not surprising that furniture from 't Spectrum was extremely popular in Goed Wonen circles. From the outset the firm collaborated intensively with professional designers. Martin Visser, who transferred from De

Bijenkorf to 't Spectrum in 1954, was the most familiar and the most productive, working in the modernist style demanded by the management board.

As time went on 't Spectrum design became more refined and elegant. From about 1970 the firm accepted more variation in form and material. The consumer's increasing demand for more luxury and comfort, and for domestic objects to fit their social status, was not ignored. Even 't Spectrum, however, could not escape the consequences of an economic crisis. The decline in purchasing power, growing competition from cheaper foreign furniture and the change in popular taste resulted in the firm being wound up in 1974. The enterprise was, however, to be continued in a different form. In 1988 Spectrum Furniture (Spectrum-Meubelen) was established, a firm that to this day markets furniture that is exceptionally contemporary in its design.

In the 1930s the Utrechtse Machinale Stoel- en Meubelfabriek (UMS), founded in 1913, was one of the firms that had introduced modern furniture into its range, mainly for commercial reasons.[33] Cees Braakman, son of the works manager, was already their most important designer. From 1945 UMS had to be rebuilt virtually from scratch after being badly hit during the war, providing them with an opportunity to focus exclusively on contemporary design. Inspired by American working methods and the design of Charles and Ray Eames, Braakman brought a totally new line onto the market under the brand name Pastoe. Much of the firm's products were made of plywood, sculpted into the desired shape using an advanced technique called high-frequency compression. Although the basis for this modern Pastoe design policy was not in line with the principles propagated by the Goed Wonen Foundation in all respects, their spokesman nevertheless praised the Pastoe storage cabinets in the periodical *Goed Wonen*.

Wagemans & Van Tuinen in Maastricht had been making furniture since the 1920s, but it was not until the 1950s that the firm became well

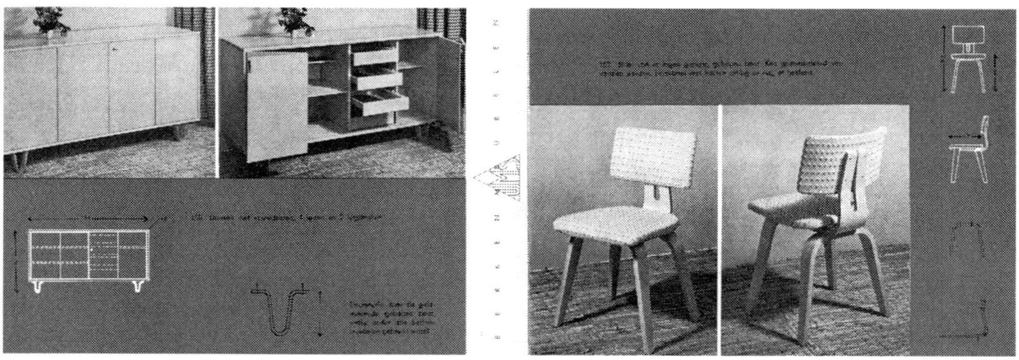

Brochure for *Berkenmeubelen* (birch-wood furniture) presenting designs by Cees Braakman (UMS/Pastoe, Utrecht), 1951.

known for its modern design products under the brand name Artifort.[34] Unlike 't Spectrum and UMS, Wagemans & Van Tuinen reached an international market with the Artifort brand. The firm's first success outside the Netherlands came when a chair by the industrial designer Theo Ruth was shown at the Triennale in Milan in 1954. A completely new design policy was introduced when Kho Liang Ie was taken on as a consultant. Kho, born in the Dutch East Indies and of Chinese ancestry, had been trained by Johan Niegeman at the IVKNO as a furniture designer and interior designer. After completing his studies he began his career as a public relations official and organizer of exhibitions on interior design for the Goed Wonen Foundation. For a while he also edited the periodical *Goed Wonen*. In the mid-1950s he entered into an alliance for some years with the graphic designer Wim Crouwel, after which his international career as a furniture designer expanded enormously.

Thanks to the design policy outlined by Kho, within a few years Wagemans & Van Tuinen became one of the leading modern furniture factories in Europe. The most important reason for this success was that Kho started the tradition of collaboration with foreign designers. On his initiative the French designer Pierre Paulin was appointed to the Artifort design team in 1959. The British designer Geoffrey Harcourt followed in 1962. Artifort's colourful, organically shaped 'sit-sculptures' were totally different in character from the strictly functional furniture made by 't Spectrum and Pastoe, making Dutch critics cautious in their judgement. But it was Artifont's elegant, 'trendy' and fashionable furniture that slotted so well into the international market.

The Dutch furniture industry went through a very rough period in the 1970s. The sector was starting to pay the price for its past preference for short-term profits. Only the factories that had concentrated in time on modern furniture, and had also renewed their production and product-development methods, managed to survive the onslaught. The rest were wiped out by foreign competition. It was not until the early 1980s that the Dutch furniture industry managed to recover. In addition to Pastoe and Artifort, firms like Castelijn, Montis, Gelderland and Rohé brought updated furniture onto the market from young designers like Pierre Mazairac, Karel Boonzaaijer, Gerard van den Berg, Gijs Bakker, Axel Enthoven and Aldo van den Nieuwelaar. The furniture on display in the exhibition *Dutch Furniture 1980–1983* in the Rotterdam Bouwcentrum heralded a new heyday for the industry. The firms just mentioned combined forces when it came to presentation and marketing activities, using the Dutch Design Centre in Utrecht as their forum.[35]

A few Dutch factories continued to enjoy success in the field of office furniture.[36] Gispen, Ahrend, De Cirkel, Oda and Staalmeubel BV are the most important names, but a great many more enterprises were active in this sector. Gispen amalgamated with Staalmeubel BV in 1966; De Cirkel and Oda became a part of Ahrend in 1967. The post-war economic revival, expansion and large-scale reorganization of clerical work eventually gave this sector an enormous boost in the 1950 and '60s. Clerical staff doubled in number and the surface area of offices grew up to the mid-1970s by a staggering factor of three. Furthermore, in those years 'open-plan' office design was introduced, resulting in a demand for new furniture of a quite different nature, such as desks and cupboards that could be connected up in different ways so that the large open workrooms could be shared in a flexible way.

In the 1950s Gispen increasingly concentrated on office furniture.[37] This reputable factory, moved to Culemborg in 1935, was forced to do without Willem Gispen, its director and designer-in-chief, after 1949. He felt that his factory obligations left him too little time for his own creativity. However, this certainly did not mean that interest in design disappeared at Gispen's. The firm's most important designers over the next two decades were Wim Rietveld and then Anton Cordemeyer. Both were given plenty of opportunity at Gispen's to experiment with new techniques and materials. Rietveld introduced compressed laminated wood in combination with metal components secured by rubber discs, a fastening technique developed by Eames. Together with his father Gerrit Rietveld, in 1957 he developed for Gispen the Mondial desk chair, which had a folded-metal support and a seat made of synthetic material.[38] In the following year,

Wim Rietveld (De Cirkel), 'Piramide' adjustable hall chairs, 1960, as illustrated in Nico Verhoeven, *Doelmatigheid van industriële vormgeving*, 1962 (IIV brochure).

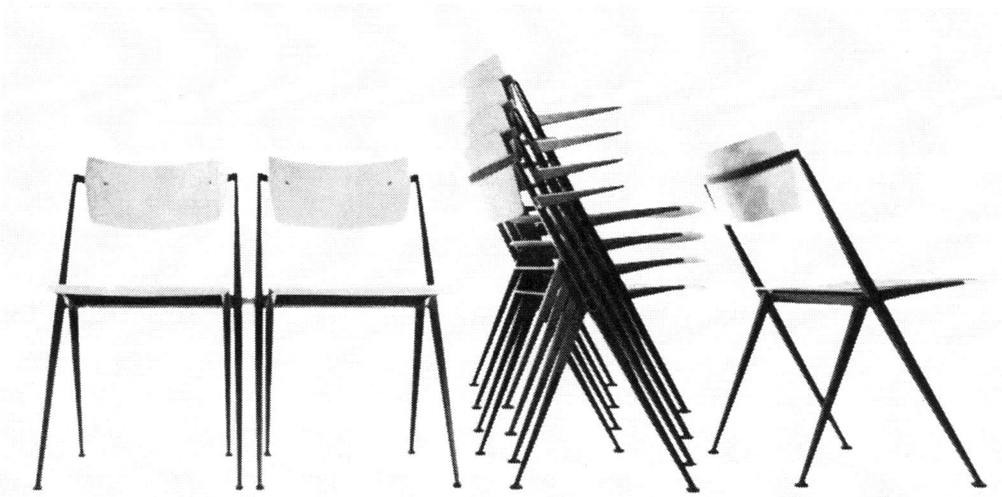

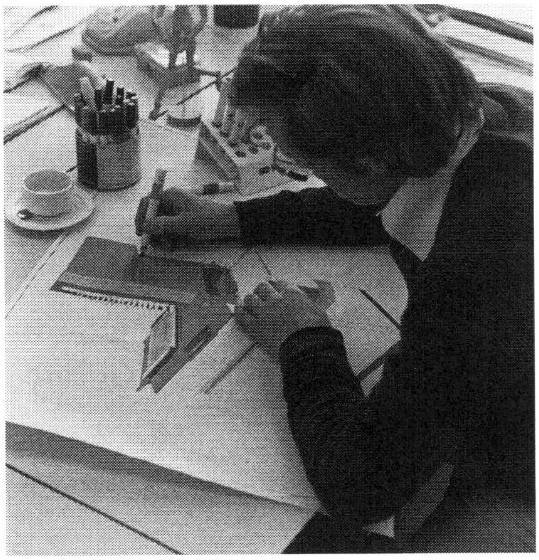

Designer in action at the Océ van der Grinten design studio, Venlo, 1982.

with his successor Anton Cordemeyer, he developed the first synthetic bucket-chair in the Netherlands, a design that well suited an office seating area without looking out of place in a home interior.

Willem Gispen founded the furniture firm Kembo in 1953, selling his new designs for office and school furniture as well as for living-room furniture, with typically organic shapes and displaying clear Scandinavian, and later also Italian, influence. They were produced in several different factories.

In the 1950s and '60s Ahrend, which had already produced office furniture before the Second World War, had the most ambitious design policy of them all.[39] The firm's most important designers were Friso Kramer and, once again, Wim Rietveld. The products Ahrend brought onto the market during those years were actually manufactured by De Cirkel. A particularly pioneering and successful example was Friso Kramer's Revoltstoel, which is still in production today. For many people the methodical and innovative way in which this chair was planned and developed acted as a model for how industrial design should be incorporated in a factory's total policy. The Revoltstoel was followed by the Resultstoel and by various designs for school furniture, desks and drawing tables. The director of De Cirkel, Jan Schröfer, commissioned Wim Rietveld to design the Piramidestoel, a variation on the Mondial chair he had earlier made for Gispen. In 1972 Friso Kramer developed his successful Mehes system for Ahrend, a brand name created by the acronym for 'mobility, efficiency, humanization, environment and standardization', the keywords reiterating the essence of well-designed office furniture.

In addition to furniture, there was explosive growth in demand for new business machines and office requisites. Océ van der Grinten in Venlo developed into a flourishing firm producing photocopiers, whose success was partly due to the clever design of these machines. Louis Lucker, who graduated in 1963 from the Eindhoven Industrial Design Academy, became the first designer to be given a permanent job on their staff. Scores of others were to follow. The design department at Océ soon became one of the largest design offices in the Netherlands. These technical machines, designed with great care and attention to ergonomic principles, have won international awards on many occasions.[40]

Directors with Ambition

Not all the highlights in Dutch industrial design in the 1950s and '60s were the result of a conscious policy. A few success stories, like Daf and Tomado, were more the outcome of a special nose for business, or possibly the result of a handful of factory managers' romantic ambitions.

In the mid-1950s Hub van Doorne and his brother Wim, who had been director of Van Doorne's Automobielfabriek (Daf) since the 1930s, realized a long-cherished boyhood dream.[41] They succeeded in making the first truly Dutch passenger car since the loss of the famous Dutch Spijker in the 1920s. Daf's lorry production lines had been successful since before the Second World War. The triumph of this new small car, however, was not so much based on its unusual design as on important technical innovations made possible by Hub van Doorne's 'variomatic', a revolutionary automatic gear-change system. Jan van der Brugghen, structural engineer at Daf, took care of the technical innovations and Wim van den Brink, originally an aeronautical engineer, designed the bodywork. The result was a typically Dutch car: sober looking, functional, easy to operate and without any fussy, unnecessary styling features – no chrome strips or luxurious accessories. The most unusual element was the raised headlights on the low bonnet.

The presentation of the first Daf 600 at the annual automobile show in the RAI in Amsterdam in 1958 was a sensational event. Thousands of visitors attended, including the international press, and 4,000 cars were sold straight off. Yet despite its undeniable success with the customers, Daf was all too soon to gain the hackneyed image of being a 'silly little car' for women and old-age pensioners. As years went by the simple design was adapted to try and improve its unexciting image, while also adjusting quite a lot of the technical specifications. In 1961 the more luxurious Daffodil made its entry on the market and in 1966 the Italian designer Giovanni Michelotti was hired to modernize the car, ultimately to no avail: in the mid-1970s Daf was taken over by the Swedish firm Volvo. Nevertheless, in Born, Limburg, cars were still being designed by the Dutch: Volvo's present chief designer in Gothenburg, Fedde Talsma, was educated at Delft Technical University and has divided his time between Sweden and the Netherlands for more than twenty years.

Just like the 'silly little Dafs', articles produced by the firm Tomado have become icons of the period.[42] Many 'baby-boomers' can remember the bookshelves that cheered up their bedrooms in their teenage years. The simple, black steel-wire shelf supports could be attached to the wall with a screwdriver with relative ease, and once in place the metal shelves (in the

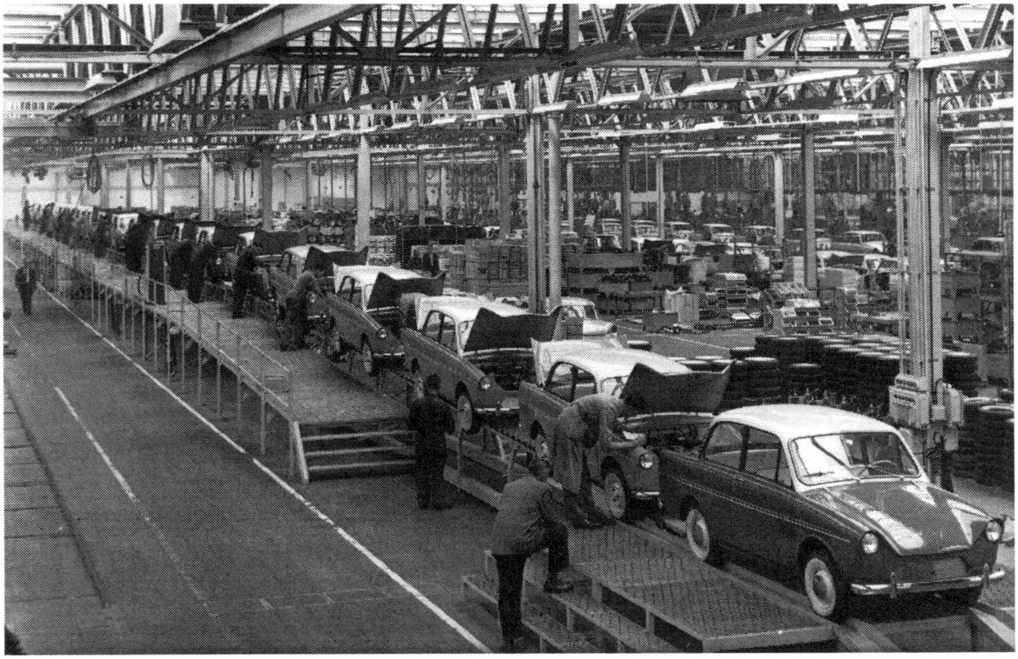

A row of Dafs on the assembly line at Eindhoven, *c.* 1960.

colour scheme of your choice) could simply be clicked on. Tomado (an acronym for Van der Tocht's Mass articles Dordrecht), was established in 1923 by Jan and Wim van der Togt and started life as a factory for simple household articles made of steel wire. The early concern the brothers held regarding the functionality of their products was demonstrated in 1933 when they developed a new practical dish-drainer in close collaboration with the Nederlandse Vereniging voor Huisvrouwen (Netherlands Union of Housewives). The firm's heyday came after the war, when Tomado met an enormous need by quickly bringing onto the market dish drying-racks, colanders, bookends and other household products. These included bottle-lickers (to get the last remnants out of glass yoghurt bottles), soap-whisks (round perforated-metal soap-holders with handles for whisking around left-over bits of soap in water) and jam pot-holders (used to hold jam pots when filling them from a pan of hot jam), articles that would puzzle consumers today, who are no longer so thrifty, or perhaps so domesticated. The almost frivolous design of the light and modern-looking wirework undoubtedly stimulated sales. Nevertheless, it was recommended by the reliable organization Goed Wonen. Tomado became a household name for all practical-minded housewives. In the late 1950s the firm started to provide some of its products with a thick synthetic coating;

the Tomado dish-drainer then became a Tomestic dish-drainer. Unfortunately, this factory did not survive either: in 1971 it was taken over by the Belgian firm Bekaert and is presently owned by Metaltex, based in the Czech Republic.

Plastic's triumphal onward march had started early in the Netherlands.[43] Bakelite had been introduced for household goods back in the 1930s in the form of radio cases and gramophones, but soon lemon squeezers, ashtrays, soapboxes, insulating grip-handles, door handles and toilet seats made from this modern material also came onto the market. In the 1950s new synthetic materials were introduced, encouraging an unusually large number of Dutch firms to start actively experimenting. Initially the still inadequate knowledge available on the new material and its processing techniques, combined with the pressure to produce as much and as quickly as possible, led to poor quality and did not contribute to well thought-out design. Nonetheless, plastic's advance was well under way. After trying thermosetting plastic, which proved in practice to be a difficult material to work with, as the 1950s progressed they moved on to using soft thermoplastics. Moreover, injection moulding was introduced as a manufacturing technique in addition to compression. Here too America paved the way. The expensive moulds necessary to manufacture synthetic objects were often obtained second-hand from America or Germany – a procedure that inevitably did not stimulate well-considered or progressive design policy.

Following the increasing market for household goods, electrical home appliances and toys, the demand for plastic camping articles also grew throughout the 1960s. Eventually synthetic materials would be accepted everywhere: in the living room, at the office and in the world of leisure and

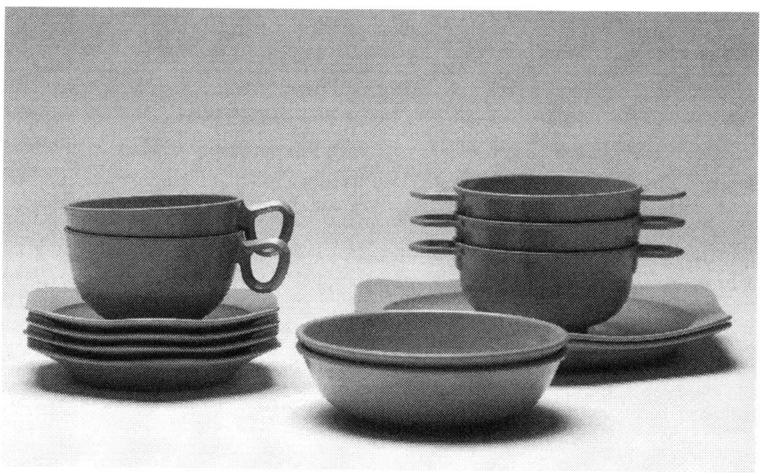

Andries Copier (Van Nifterik, Putten), parts of a plastic (melamine) dinner service for KLM, 1946.

entertainment. Ashtrays, wastepaper baskets, lamps, clocks, typewriters, radios, long-playing records, advertising and packing materials, glasses, disposable cups and much more were all made of plastic and sold well.

Of the scores of plastics processing firms founded in the 1950s only a few proved to be viable after adopting a successful and progressive design policy. One such is Mepal, which has been based in Lochem since the beginning of the 1960s and merged with the Danish firm Rosti in 1993. In 1963 the management initiated a test project and contracted a group of well-known designers, among them Piet Zwart, Coen and Wim de Vries, Charles Jongejans and Dick Simonis, to develop a new range of storage boxes.[44] Tiger and Curver were two other Dutch companies making high-quality synthetic household products that had been meticulously designed.

New Design Companies

It was gradually becoming clear that designers could make an important contribution to a company's economic prosperity. As the prestige of Dutch designers rose they were able to establish professional design companies modelled on famous American companies such as Loewy, Teague and Dreyfuss, which they had come to know through journals, international congresses and, in a few cases, from personal visits or internships. Emile Truÿen, for example, trained in the early 1950s at the Interior Design department of The Hague Art Academy. After completing his studies he went to the United States, where he took up a post teaching at the Pratt Institute in New York. Once back in the Netherlands, Truijen began an association with Rob Parry, whom he had known since his time in The Hague and who had recently been working for Gerrit Rietveld.[45]

The economic recovery in this period provided the two with many commissions, including one from the PTT (Dutch Post Office) for the twin letterbox (1957) that was later to become so familiar. They produced an ergonomically acceptable design made partially from plastic, notable for its attractive bevel-edged contours and an ingenious system for keeping local and national mail separate. The result was so good that the box remained in continuous service right up to the end of the century. Truijen and Parry, however, had already gone their separate ways in 1958.

Truÿen's next move was to set up the design company Tel Design in The Hague with Jan Lucassen in 1962 (the name is derived from the initial letters of Truijen en Lucassen).[46] In 1961 Lucassen had been one of the first to graduate from the new Industrial Design Academy in Eindhoven, where Truijen was teaching at the time. Tel Design aimed to cover all design

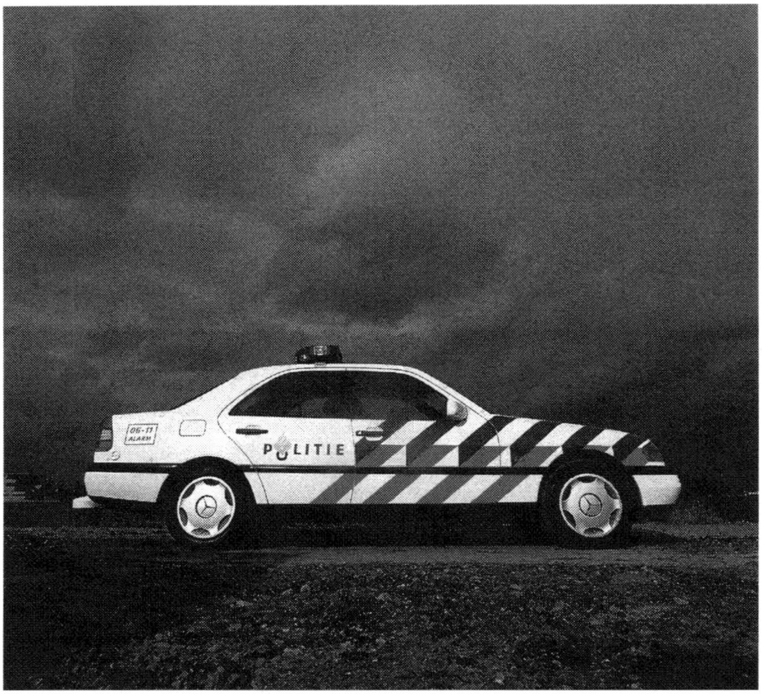

Studio Dumbar for the Ministries of Justice and Home Affairs, police car livery, 1993.

disciplines and all lines of work, and the two designers' ambition was to work in the commercial way Truijen had become familiar with during his stay in America. When Tel Design was commissioned to design a new house style for the NS (Dutch National Railways) in 1967, they took on the graphic designer Gert Dumbar as a third partner. Dumbar had been trained at the Royal College of Art in London and, with his flair for visual communication, was able to give a refreshing new lease of life to Tel Design. The firm in its original form closed in 1976, but the partners carried on working independently as they went their separate ways: Emile Truijen became a professor at Delft Technical University in 1977 and Gert Dumbar established Studio Dumbar, which remains active to this day. As well as its designs for the Dutch National Railways, Tel Design carried out pioneering work for the new discipline of public relations and for house styles. Studio Dumbar has become well known both in the Netherlands and abroad for many inventive logos and publicity campaigns, such as those created for the V&D department store, the ECI book club, the sometimes ridiculed livery of Dutch police cars, and the recently announced commission to supply a uniform house style for all departments of the Dutch national government.[47]

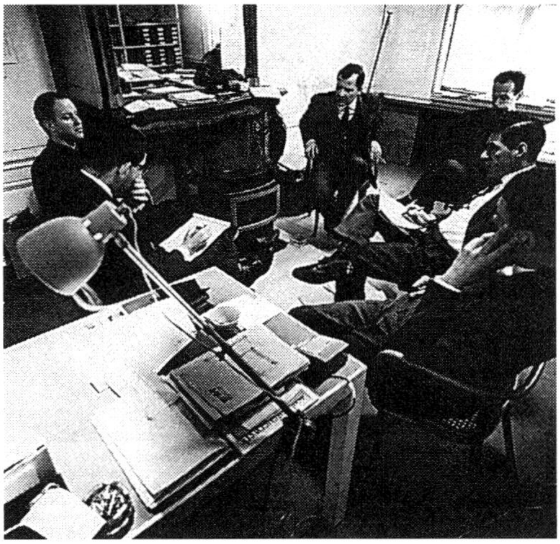

The founders of Total Design at their new design studio in Amsterdam.

Another new company, the Associatie voor Total Design BV, was established in 1963 by Wim Crouwel, Benno Wissing, Friso Kramer and the brothers Paul and Dick Schwarz.[48] Their aim was to offer services on every facet of design, from stamps to exhibitions. With a prestigious office address on the Herengracht in Amsterdam, they recruited 'junior designers', such as Ben Bos, appointed business managers (namely the Schwarz brothers) and engaged clerical staff. Total Design presented itself as an international, professional and modern organization that was not modelled on the commercial American design companies, but on such studios in Great Britain as Fletcher Forbes Gill and on some of individual designers in Germany and Switzerland with whom they were in touch. Furthermore, the humanist design philosophy taught at the Hochschule für Gestaltung in Ulm was very important for their method of work. At Total Design they worked in a business-like, professional way, with Paul Schwarz as accounts manager maintaining contact with the clients. At the company's offices the commissions were dealt with by separate design teams, and this rational division of work enabled them to be handled as efficiently as possible. Preferably they designed along rational and established lines on the basis of a rigorous grid.

Wim Crouwel had trained as a painter at the Groningen Academy of Art, but started his career in 1952 designing exhibition stands. His introduction to Swiss typography in these years was decisive for the further course of his career. The clarity and the logic of functionalism from the 1920s and Swiss typography, which continued to build on this tradition, was an continuing source of inspiration. In 1956 Crouwel worked for a while with the interior designer Kho Liang Ie. One of their joint design projects involved planning stands for the exhibition *Het Atoom* (1957) in Amsterdam, one of the optimistic post-war reconstruction events (see above). They were also responsible for a series of arresting exhibition stands, notable for their austere, subdued, minimalist design, commissioned from such firms as Auping, De Bijenkorf and Linoleum Krommenie.

When Total Design was set up in 1963 Benno Wissing already had an adventurous career behind him.[49] He trained to become an artist at the Rotterdam Academy of Art. After the war, lured by the attraction of communism, he stayed for a while in Yugoslavia and Czechoslovakia. He showed his social commitment in 1946 by becoming involved in Groep R (Group R), an artists' organization that put collaboration first and aimed at abolishing art with a capital A. He supported himself by designing stands and décors and doing graphic work, making close to a hundred posters and catalogues for Boijmans Van Beuningen Museum in Rotterdam. In Wissing's unrelenting need to encourage a more anonymous form of 'visual communication', his most important sources of inspiration were El Lissitzky and Moholy-Nagy, as well as the Dutch designers Zwart and Schuitema. Corporate photography books, a popular medium in those years, were a good example of this visual form of communication. In this genre Wissing designed books such as *100 jaar Grasso* (100 Years of Grasso, 1958), in which he visualized the hundred years of history through which this 's-Hertogenbosch engineering factory had been active using a wide range of typographical resources, complemented by contemporary photographs by Violette Cornelius.[50]

The third partner in Total Design was Friso Kramer. Unlike Crouwel and Wissing, Kramer was involved exclusively in product design and never in graphic design. Just like his associates, however, he supported the principles of functionalism and rational and analytic design methods. The three designers of Total Design each brought their commissions and clients to the new joint company. Wissing continued to design the print work for the Boijmans Van Beuningen Museum, Crouwel did the same for the Van Abbe Museum in Eindhoven and for the Stedelijk Museum in Amsterdam, whilst Kramer brought his order portfolio for Ahrend into the alliance. Within a few years the wide-ranging mode of operation that Total Design had initially envisaged had already proved unrealistic. Designing industrial products, Friso Kramer's speciality, did not take off as well as he had predicted within the confines of the company and, somewhat disappointed, he left it in 1968. Benno Wissing soldiered on until 1972, but became increasingly irritated by the hierarchic way in which the company was organized and the resulting compartmentalization of responsibility.

The main stable factor at Total Design continued to be Wim Crouwel, who developed into the 'face' of the company. His commitment, pragmatic, professional attitude and social skills – not least the ease with which he was able to communicate with his clients – ensured that Total Design continued to exist, albeit in an ever changing structure. The company changed its name to Total Identity in 2000. At its height the studio had forty members of staff. Permanent staff, in addition to Crouwel, included Ben Bos, Daphne Duijvelshoff and Jolijn van de Wouw.[51] Other well-known designers, such as Paul Mijksenaar, Jurriaan Schrofer and Anthon Beeke, were attached to Total Design for shorter periods of time.[52]

Their first major client was the Coal Trade Association (Steenkolen Handels Vereniging, SHV), which asked Total Design to develop a house style for them, as well as a distinctive logo for PAM domestic fuel oil, petrol and liquid gas, which the SHV was launching on the market. The result of these commissions, which were mainly worked on by Benno Wissing, was a coherent programme of carefully designed logos, letter headings, packaging materials, annual report layouts and calendars, including the lettering on the goods trains that transported the oil and the total look of PAM petrol stations.

Randstad, a temporary employment agency, approached Total Design in 1967 when the agency was still fairly new. Not only did Randstad want a completely new house style, but they also asked the designers to think about how the organization's image could be improved. At the time, temporary employment agencies were considered to be employers without a great deal of social conscience, and they met with widespread disapproval. This commission was carried out entirely by Ben Bos, the most important part of it being the logotype, which was closely related to the revolutionary New Alphabet produced at the time by Wim Crouwel.[53] The combination of the logotype and the style of the letters used for the name Randstad was so well chosen that it is still in use 40 years later, and has not dated at all. This project was followed by logos for the Rabobank, the Rotterdam Ahoy hall complex, De Gruyter supermarket, Makro Cash & Carry, Het Spectrum and Kluwer publishers, the National Investment Bank, the Dutch Municipal Building Fund, the Holland Festival and many more.

Both Total Design and Tel Design are quite rightly associated with the modern house-style concept. The differences between the two were sometimes barely visible, but on the whole Total Design's work was more austere and minimalist, Tel's more expressive. The house styles of what were at the time two of the Netherlands' best-known supermarket chains, namely De Gruyter's, designed by Total Design (Ben Bos), and Simon de Wit's, designed by Tel (Frans van Mourik), nicely illustrate the difference between the two.[54]

In the past some organizations had tried to bring a certain uniformity to their company's printed materials by means of a specific logotype or a distinguishing graphic style. Two examples, both discussed earlier, were Jacob Jongert's printing work for Van Nelle's coffee, tea and tobacco factory and Paul Schuitema's work for Van Berkel's patent office. As far back as the seventeenth century the VOC (Dutch East Indies Trading Company) had used its own logotype. Nonetheless, Tel and Total Design, and the many new Dutch graphic design companies that were to follow in their footsteps

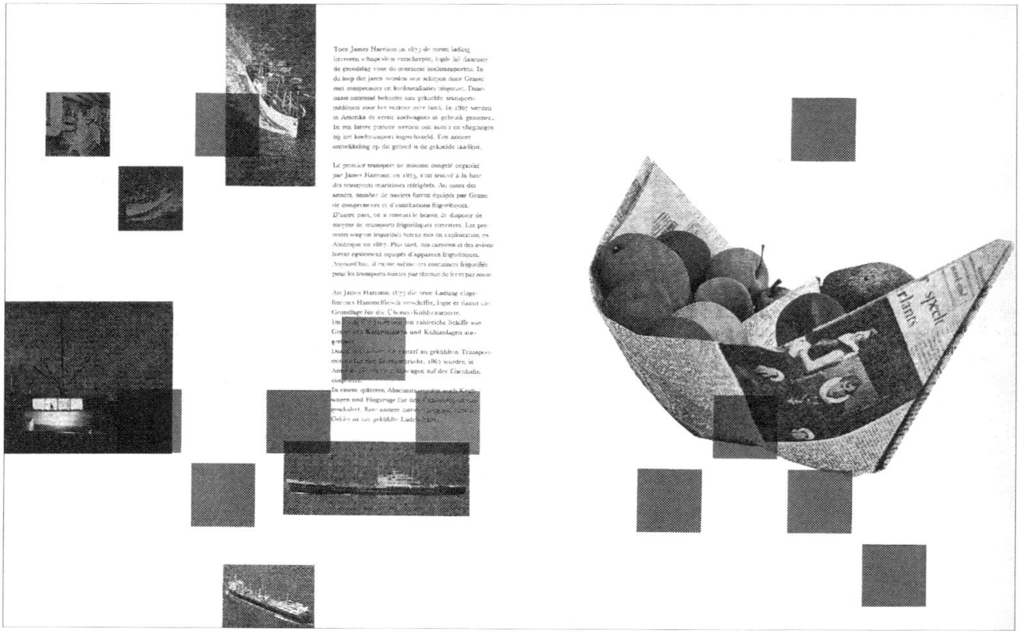

A double page in Benno Wissing, *100 jaar Grasso*, 's-Hertogenbosch (1958), with photographs of Violette Cornelius.

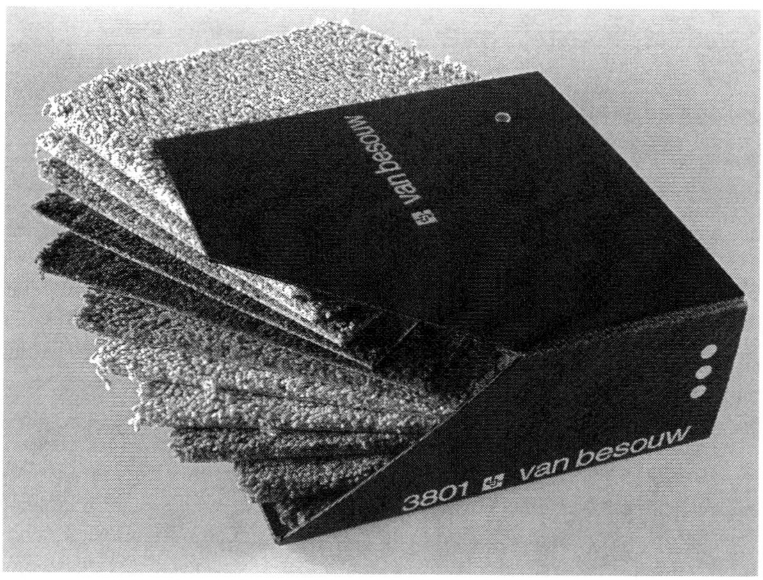

Marijke de Ley (Studio Premsela Vonk for Van Besouw, Goirle), samples of the cotton strap-carpet of 1970.

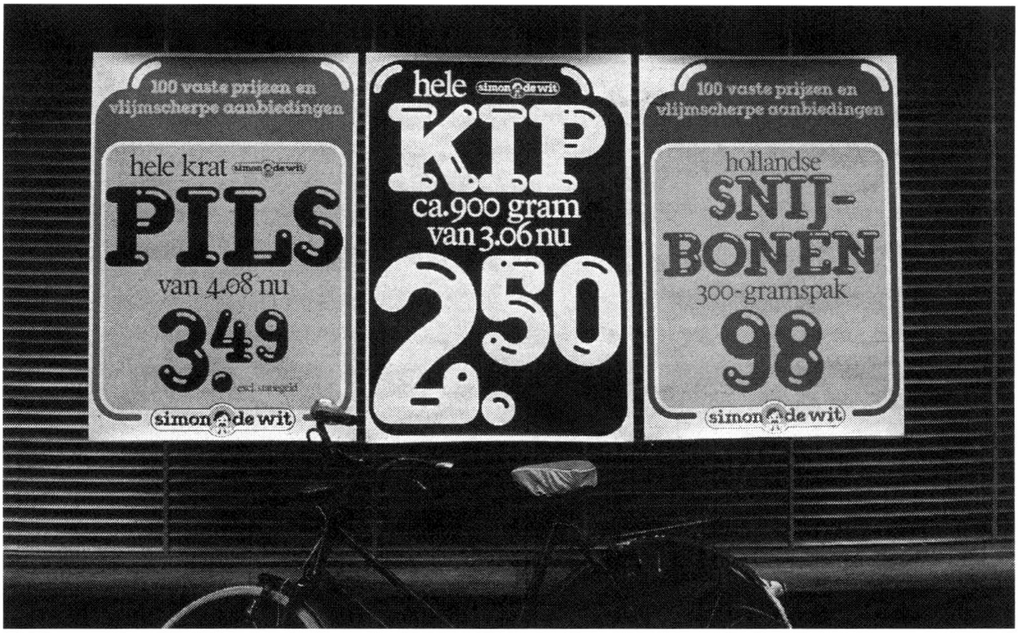

Tel Design, three posters for Simon de Wit, 1970.

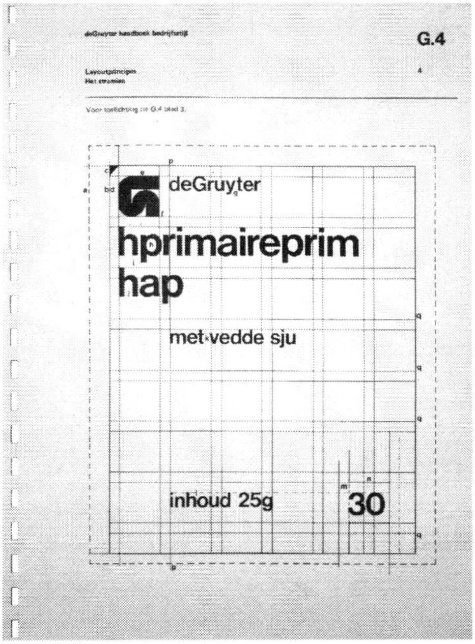

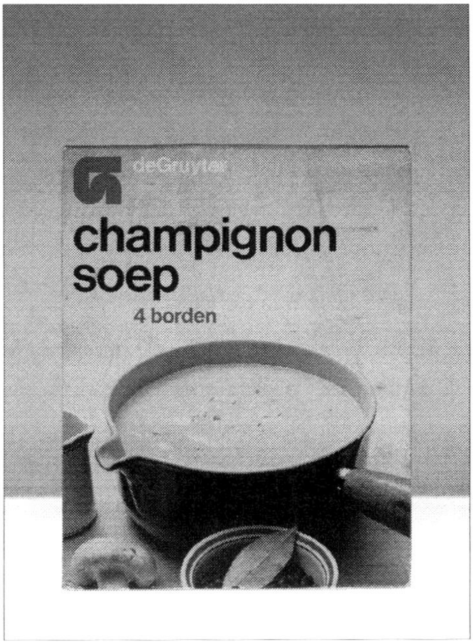

Total Design, design and final result for De Gruyter soup packaging, 1971.

in the 1960s and '70s, broadened the concept of house style to embrace the phenomenon of corporate identity, or corporate image.[55]

A large, but different sort of design company was Premsela Vonk's design studio in Amsterdam. Benno Premsela and Jan Vonk had regularly worked together since 1956 and, straight after Premsela left De Bijenkorf department store in 1963, the two of them set up a joint company.[56] In contrast to Total Design and Tel, this studio did not concentrate primarily on graphic design, but rather on interior design with a special interest in textiles.

At first the commissions were for exhibition show houses and their interiors. Between 1963 and 1969 they attended to a large share of the presentations in the Industrial Design Centre (Centrum Industriële Vormgeving) in the Beurs in Amsterdam. Together they also supplied a new 'look' for the furniture factory Pastoe, by designing modern, contemporary catalogues (with photographic work by Jan Versnel), showrooms and exhibition stands.

Benno Premsela's work for the carpet factory Van Besouw in Goirle began in 1967 after the firm's director Jan Mes had been introduced to Premsela through the IIV. The regular visits Premsela made to Goirle were the preface to an upheaval in the existing traditional world of carpets. Premsela questioned everything: not just the designs, colours, materials and techniques, but also the way advertisements were made, the way they presented their products at home fairs and in the shops, the role of the consumer and, not least, the call for major investment. Thanks to Benno Premsela, from then onwards Van Besouw's carpet factory considered modernization to be a social obligation.

For the Premsela Vonk studio staff designing was synonymous with product development, which entailed giving advice on an enterprise's commercial policy. The designs for Van Besouw were characterized by their simplicity, a certain air of timelessness and high quality. The cotton bouclé carpet developed by Marijke de Ley in 1970 was a pioneering product. A second major change in this field was thought up by Diek Zweegman, who devised a system by which flax could be worked in with cotton. Another designer from the Premsela Vonk studio, José de Pauw, was awarded the Kho Liang Ie prize in 1980 for her furnishing fabrics for Vescom, Auping and the German firm Gerns & Gahler. In 1988 Premsela Vonk merged with the graphic design studio BSR in The Hague and a few years later BSR Premsela Vonk changed its name to Eden.[57]

Benno Premsela's influence was far-reaching, not only as a designer but also as a critical member of a wide-ranging selection of advisory committees; these governed almost all fields of cultural life in the Netherlands from the 1960s to the '90s. It was no accident that Benno Premsela's nickname was 'the Pope of Art', fulfilling the same function as F. J. van Royen and H. P. Bremmer had done before the war. His name lives on today in the Premsela Foundation, the national organization for design and fashion.[58]

Quangos

For many commercial establishments, restyling printed matter and using a new modern logo was an effective way of attracting custom. The Dutch authorities and many quangos followed suit, the most important being the State Printing Office (Staatsdrukkerij) and the State Publishing Office (Sdu).[59] Straight after the war, under the supervision of P. Knuttel, an optimistic decision was taken to make everything printed by the government readable and pleasing to the eye. J. F. van Royen's heartfelt condemnation of all state printing matter as 'ugly', reiterated three times in 1912, was about to be remedied at last. The Staatsdrukkerij appointed a design team of their own for this purpose, which at its height had a staff of twenty. Between 1955 and 1988 prominent roles in this team were played by Ton van Riel, Karel Treebus, Gertjan Leuvelink, Jelle van den Toorn Vrijthoff and Irma Boom. In 1976 Hein van Haaren became director and also head of the Design department at the Sdu. The firm had developed into an organization that young designers were keen to work for and where experimentation was encouraged. Sometimes, however, the Sdu's role was restricted to implementing the designs made by independent designers or studios. Through the years commissions were granted to Piet van Trigt, Jurriaan Schrofer, Gerard Wernars, Pieter Brattinga, Rob Schröder, Lies Ros and Esther Noyons, to mention just a few. Type designers like Gerard Unger and Bram de Does also worked regularly for the Sdu, but on occasions they also brought in large companies like Total Design, Tel and BRS.[60] In 1988, however, the Sdu was privatized, and from then on the various public bodies were allowed to place their orders with the company of their choice.

Quangos often commissioned work from large well-known design companies as well. Tel and Total Design made designs for Schiphol airport, the Dutch National Railways, the Bank of the Netherlands, urban transport companies, the National Broadcasting Foundation, the Ministry of Waterways and Public Works, and the PTT (the National Post and Telegraph and Telephone Company). All these large projects had an enormous influence on visual culture in the Netherlands. As a result, the hundreds of powerful, clear, and usually simple, graphic designs that came into circulation from the 1970s onwards were familiar, even in the remotest parts of the country. They are responsible for the widespread view shared by many foreigners that the whole of the Netherlands has been supplied with well-thought-out design.

With the advent of jet airliners in the early 1960s, and the enormous increase in air travel brought about by burgeoning tourism, the construction of a new and modern national airport became an inevitability. The interior and furniture designer Kho Liang Ie was asked to take care of the interior of the complex of buildings to be designed by the architect M. Duintjer, while Benno Wissing of Total Design was employed for the signposting. This would be a gigantic and ambitious commission. Kho enlarged his own company by employing the interior designers Nel Verschuuren and Tinus van de Kerkhof.

Kho and Wissing did not rush headlong into the assignment. Instead, by way of preparation, they made a study tour of other large airports. Based on these experiences they decided that creating a restful environment should be given top priority. Travellers were often tense, uncertain about where to go and in a hurry, so a simply designed, orderly space could do a great deal to improve their frame of mind. Hence a tranquil, light colour scheme was chosen for Schiphol. Only the signs designed to lead travellers in the right direction were permitted to have a bright, contrasting colour. The walls were covered with white wall-tiles produced by the firm Mosa in Maastricht and sheets of white Formica supplied by various Dutch firms. The white lamellated ceiling was a variation on the ceiling that Kho had developed earlier with Gerrit Rietveld for the Industrial Design Centre in the Beurs in Amsterdam. In the spacious lounges they placed simple, sharply outlined but comfortable lounge chairs and couches made by Artifort, rounded off with small tables, wastepaper baskets, large plant containers and telephone booths. The check-out counters were also redesigned. Arie Jansma designed a simple concept for the shops in the waiting area in the form of cubes that could be rearranged at will.

Signposting as a separate discipline was then still in its infancy. Benno Wissing developed a system whereby large yellow signs, hung high above the heads of the travellers, indicated the main directions. All were produced in large simple letters, in both English and Dutch. Secondary information was given on smaller green signs. No pictograms were used, other than arrows.

At the opening in 1967 the result proved very satisfactory. In an international framework, Schiphol was considered to be one of the most beautifully designed airports in the world: 'Schiphol puts passengers first', wrote the British trade journal *Design* in just one of the positive reactions.[61] Since then, of course, the airport has been partially modernized and extended several

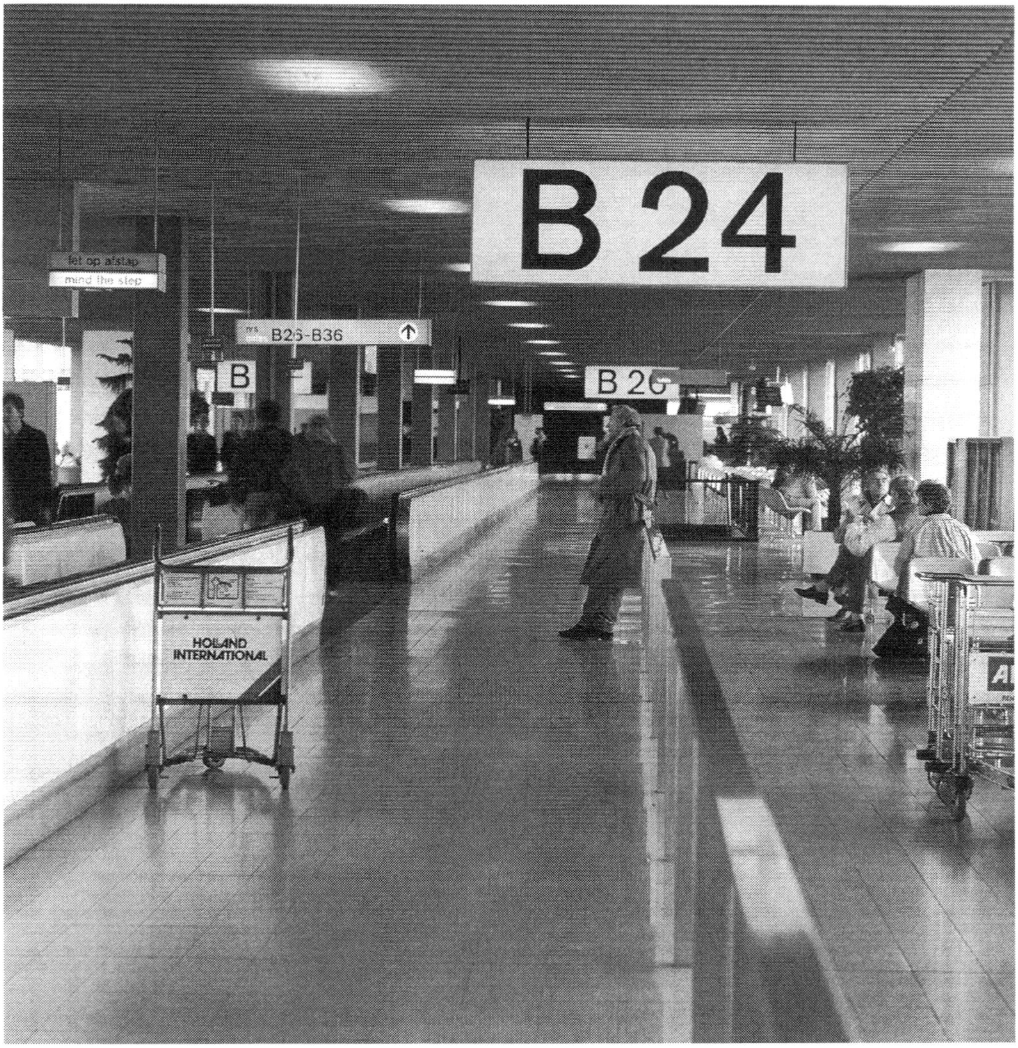

Benno Wissing (Total Design), signage at Schiphol Airport, 1967.

times, for example, by Jan Benthem from Benthem Crouwel Architects. Even after Kho's death in 1975 the company Kho Liang Ie Associates was involved in these adaptations. In the mid-1990s Bureau Mijksenaar adapted Benno Wissing's original signposting and finally added to it a series of pictograms. The basic idea and the characteristic ambience of Kho and Wissing's design, however, seem to have been preserved.

Not only airline passengers arriving at Schiphol become directly acquainted with the Dutch government's internationally famous policy of stimulating design in public spaces. Those travelling through the

Netherlands by train or car – or even by bike – are welcomed by several eye-catching design projects also initiated by the government.

Between 1937 and 1995 the Dutch Railways (NS), which originated after a merger between several private railway companies, was entirely financed and directed by the Dutch state. In the course of the 1960s it became clear that the NS would have to change its old-fashioned, official image if it were to have any chance of competing with the steady increase in car traffic. A modern house style was needed to give it a new, more contemporary character to rejuvenate the organization and make it more dynamic.

Tel Design was commissioned to do the job in 1967. They thought up a new colour for the trains, a logo, a new lay-out for the railway timetable and even new signposts for the stations, including a series of pictograms. It had the desired effect. The modern, powerful logo designed by Gert Dumbar, showing a combined double arrow pointing in two directions, is still in use today and does not look at all outdated. The new colour for the passenger trains, for which, to everyone's surprise and to some people's indignation, he chose a warm chrome yellow, is still also considered to have been an excellent choice. With this fresh, original colour, the Netherlands wanted to make a clear statement and impress the international rail transport community. For many it is still a treat to see these yellow trains, preferably under a blue sky, travelling through the flat green landscape. The new railway timetable format was worked on by Gertjan Leuvelink, who like Dumbar had come to strengthen the Tel Design team in 1967. The NS's own design department run by Siep Wijsenbeek concentrated in this period mainly on the modernization of the rolling stock and the interiors.

The changes made to the railway had a considerable impact, as a national railway company touches on everyone's lives, young or old, rich or poor. It was undoubtedly very significant for the position of the discipline in the Netherlands that the corporate policy of a large, nationwide organization like the NS should provide so much scope for modern design.

Those who travel by car, rather than train, are confronted with signs installed by the ANWB (Dutch Touring Club), the national equivalent of the Automobile Association.[62] They provide the sorely needed clarity to deal with the complicated, overcrowded Dutch road network, even though the role Dutch designers played on the roads was less pronounced than on the railways.

The first signs produced by the ANWB, which was founded in 1883, were installed as early as 1894. The ANWB, then a private organization, devised and funded everything itself since the Dutch government did not see any point in providing such a service at the time. It was not until 1966 that the

ANWB introduced the same type of lettering on all the large blue direction indicators on the motorways – the first attempt at uniformity in almost 70 years. The font chosen was an adaptation of that also used along American highways. Once it was adopted, graphic designers in the Netherlands were up in arms.[63] Formally speaking, they objected because they thought the signs were not easy to read, but in fact the true reason was that they would rather have seen the commission go to a Dutch designer. A 'lettering committee' from the GKf complained officially about them but to no avail, and the signs remained a thorn in the flesh of many Dutch graphic designers. In 1975 a Signposting Conference was organized at Delft Technical University, where designers, traffic experts and signpost-makers discussed the ANWB signs. But once again the graphic designers' complaints were not generally sustained. It was to take until 1994 before the design company NPK Industrial Design was commissioned to alter the ANWB signs. For the letters they called upon the help of the most prominent type designer in the Netherlands, Gerard Unger. Working on the basis of new insights on legibility, and not deviating too much from the old letter type, he finally adjusted the much criticized signs in such a way that they could also be computer-generated.

Those travelling by bike through the Netherlands are served by the special smaller ANWB signs, on which the destination is written in red. The traditional ANWB 'mushroom' road markers are also still in use. These direction indicators were designed back in 1919 by an architect named J.H.W. Leliman and are still popular in the Netherlands. Since then these low, angular concrete signposts, painted white with metal caps, have risen in number to a total of 5,000 spread over the whole of the Netherlands. Recently they have started to be replaced with a similar, but lighter, design made from synthetic material. The traditional design has proven to be so popular that it was recently nominated in a competition held to select the Best Dutch Design Product. Its popularity undoubtedly owes much to feelings of nostalgia.

Dutch Money and the PTT

Until the introduction of the Euro in 2002 Dutch money had a high profile all over the world.[64] Since the war great care has been taken in the design of banknotes and coins. Paper money is printed by Joh. Enschedé & Zonen in Haarlem. The final responsibility for its distribution rests with the president of the Netherlands Bank. Coins are struck by the Netherlands Mint in Utrecht, accountable to the Minister of Finance. New designs for both coins and notes were created by the winners of contests.

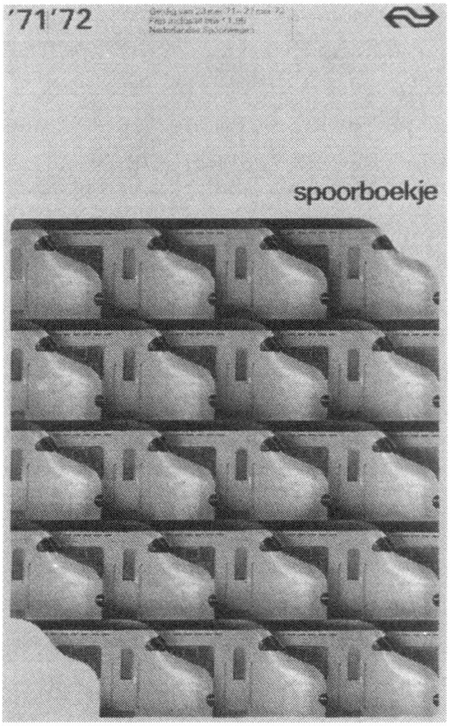

G. J. Leuvelink (Tel Design), Dutch Railways timetable, 1972–3.

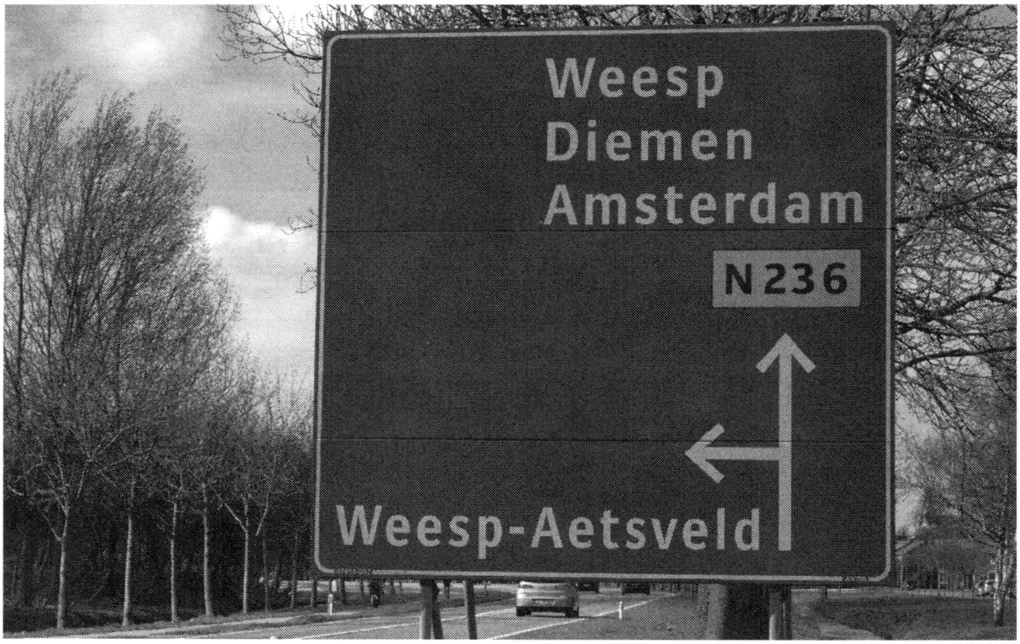

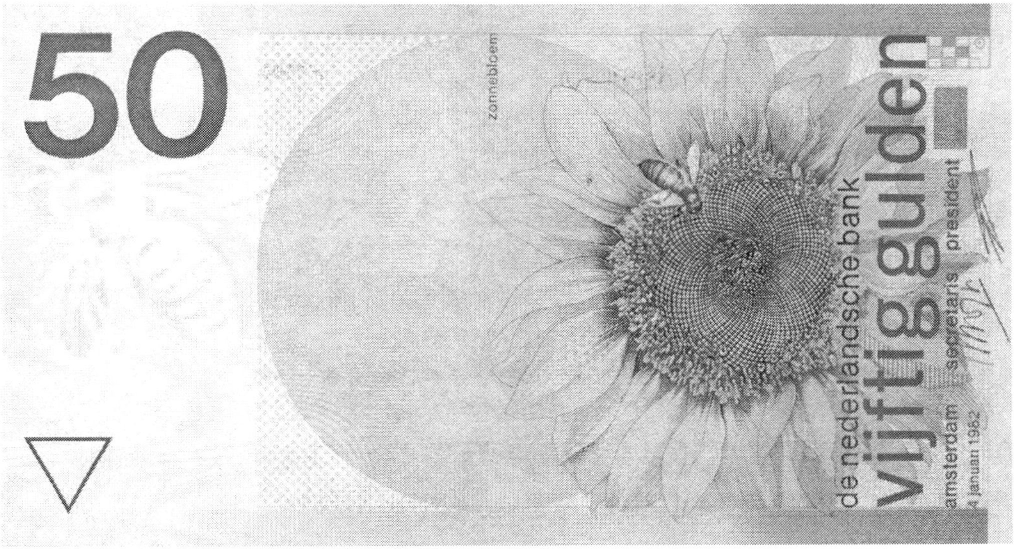

R.D.E. Oxenaar and J. J. Kruit (De Nederlandse Bank), 50 guilder bank note, 1982.

W. J. Rozendaal drew the first new Dutch banknote shortly after 1945. The design was not very spectacular for the time and the final result was even less satisfactory due to various technical problems experienced in the printing process. J. F. (Eppo) Doeve was then selected in 1950 from a group of five designers to make a completely new series featuring well-known Dutch historical figures. In order to avoid printing problems, this time his sketches were completely worked out by the Joh. Enschedé staff. The result was that these notes remained fairly traditional. It was not until the graphic designer R.D.E. (Ootje) Oxenaar was commissioned to design a new series in the 1960s that this policy was reversed. He was recommended to the Netherlands Bank by Karel Schuurman, who at the time was the PTT's Aesthetic Adviser and already knew Oxenaar from his postage stamp designs. His series of notes with highly stylized historical portraits, executed in bright colours, of the country's 'Hall of Fame' was extremely refreshing. However, this series was to be followed in the 1980s by an even more talked-about sequel, the revolutionary, colourful notes to the value of 50, 100 and 250 guilders with illustrations of a sunflower, a snipe and a lighthouse. Hans Kruit also contributed to the design of this series. The traditional portrait was abandoned for the first time. Over the years Oxenaar had acquired a great deal of knowledge about the extremely specialized printing process used to manufacture banknotes. This 'secret weapon' enabled him to induce the staff at Joh. Enschedé to execute practically all his stylistic and technical innovations. This was a considerable feat considering

NPK industrial design/ Gerard Unger, ANWB signage, 1994–7.

the innumerable safety regulations that the Netherlands Bank had to comply with, which by that time had grown into a thick wad of specifications. The series was completed between 1988 and 1997 by three equally distinct designs by the designer Jaap Drupsteen. Thus designing banknotes had grown to be a great deal more than just supplying a new picture. It had become part of a democratic and professional process, whereby Oxenaar and Drupsteen had been successful in reserving sufficient space for creativity and humour despite all the technical obstacles and safety regulations. Although these achievements were admired in other European countries, there are unfortunately few signs of this erudition on the new Euro notes.

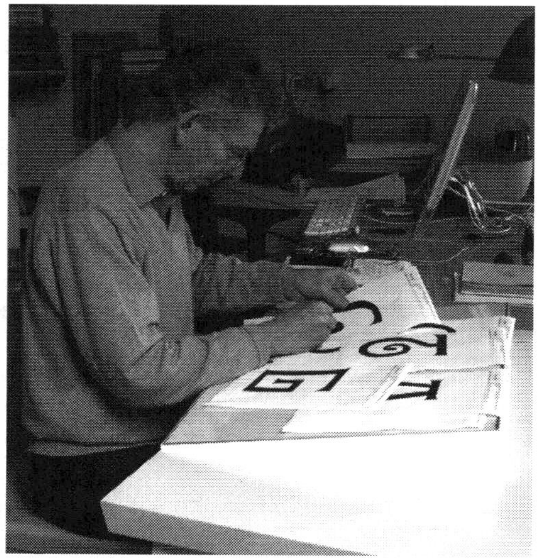

Type designer Gerard Unger at work, 2007.

No fewer than nine designers were allowed to compete in the 1980 contest for the new Dutch coin design. The chosen design was not by a graphic designer, but by an industrial and jewellery designer, Bruno Ninaber van Eyben.[65] His coins had a distinctly modern look about them, combining a highly stylized portrait of Queen Beatrix with an abstract motif that denoted the value of the coin. The coins themselves were simple, original and well thought out, even though the system denoting the value of the coin was not easy to fathom. The letters and the numbers on the coins were then made more legible with the help of the type designer Gerard Unger. These coins too were replaced in 2002 by the far less spectacular Euro coins.

The influential role played by the PTT as commissioner of the most wide-ranging designs, from postage stamps up to post offices, has already been sketched meticulously in numerous publications.[66] Much of the Netherlands' high reputation in the field of design is based on this work. The book *Design is geen vrijblijvende zaak: Organisatie, imago en context van de ptt-vormgeving tussen 1906 en 2002* (Design is not a Non-committal Business: Organization, Image and Context of PTT Design between 1906 and 2002), published in 2006, not only runs through all the facets of this 'success story' once again, but also scrutinizes them critically. The end result is that a number of persistent myths clinging to this historical account have now been called into question. One myth that has been unmasked is that, in the author's view, Jean François van Royen, the man who is usually mentioned in the same breath as pre-war PTT design policy, did not in fact delineate a distinct PTT design policy at all. His main aim is alleged to have been to implement a socially minded policy for the benefit of the artists. What mattered most to Van Royen was ensuring that commissions were distributed honestly and generously and designers were provided with a source of income in what for many of them was a difficult period. His own preferences were not the crucial factor. Moreover, Van Royen allowed others to advise him at great length, mainly listening to the artist Willem van Konijnenburg and the art critic, and later curator of the Kröller Müller Museum, A.M.W.J. Hammacher.

Van Royen died in 1942 in Camp Amersfoort, where he had been imprisoned by the Germans on the grounds of his alleged involvement in a campaign against the Kultuurkamer (see chapter Three). After the war his work was taken over by the Department of Aesthetic Design (Dienst Esthetische Vormgeving, DEV), run in succession by Willem Frederik Gouwe, Christiaan de Moor, Karel Schuurman, Hein van Haaren, Ootje Oxenaar and Marie Helène Cornips. This department was not just an important commissioner of work, but also showed itself to be a powerful player in design culture in a broader sense. The DEV acted as a mediator when commissions for the PTT were being handed out, but it also advised other institutions including, as mentioned above, the Netherlands Bank. Furthermore, the department played a role in design education and adjudicated at design competitions. It also determined which artists should be brought in to make decorative artwork in, or close to, new post offices and other PTT buildings. In 1951 the Netherlands government put into operation the 1% Regulation, specifying that one per cent of the building costs for Government buildings had to be spent on art. In those years the DEV also built up its own art collection.

Among the various commissions distributed by the DEV, those for new postage stamps were always the favourite, and the most prestigious. Postage stamps were the PTT's and the Netherlands' visiting card. Chris de Moor, aesthetic adviser from 1951 to 1963, was so fascinated by postage stamps that he wrote a book about them in 1960. In it he discussed the twelve commandments governing postage stamp design – twelve aesthetic, technical and practical tips and rules to be observed when designing stamps.

In addition to the standard stamps (the definitive series) showing the cost of postage in numerals, which were in continuous use for years on end, special new series were produced regularly. The children's stamps (which cost a little more than the postage due, so that the extra money could go to a children's charity), the summer stamps (with a summer theme for holiday postcards) and the various commemorative series were annually recurring projects. From the 1930s onwards these started to function as a sort of mini-poster, the stamps together forming a composition measuring at the most 8 square centimetres on which, within the specified technical and functional limits, the PTT gave graphic designers a free rein to exhibit their creativity and originality.

The first post-war definitive stamp was made by Jan van Krimpen, the typographer who worked for the printer Johan Enschedé & Zonen. It was a quiet, classical, ornamental execution of this commission. In 1976 Wim Crouwel designed the second post-war definitive series. In his design he tried to make a stamp that was modern but also as neutral as possible. This design was based on his favourite working method as well, omitting all references to tradition, penmanship or even emotion. The issue of this stamp, about the same time as the publication of the telephone book that Crouwel and his staff at Total Design had created, met with strong opposition. Critics did not share the view that the designs were modern and functional but condemned them for being uninteresting and paltry. The last thing they would have called them was neutral. In the next chapter we shall look in greater detail at the consternation this design caused. For that matter, just as much fuss was made in 1981 about the stamp with the queen's head on it drawn by artist Peter Struycken, and with lettering by Gerard Unger. With the aid of computer technology, which was then still in its infancy, Strucken abstracted Queen Beatrix's portrait using only separate round dots; the result failed to win everyone's favour.

By 1970 the PTT had developed into such a complicated organization that the management decided to rejuvenate its image totally. A large-scale PTT house-style operation headed by Ootje Oxenaar was initiated. The two major rivals at the time, Total Design and Tel Design, were asked to submit plans. In the end, thanks to Hein van Haaren's mediation, they opted for a unique joint project involving both renowned design teams. Both had their proposals ready in 1978, but it took until 1981 before everything had been adopted throughout the organization. Brochures, postage stamps, books of stamps, diaries, telephone books, work wear, company vehicles – everything and everyone was supplied with the new logo, in the new colours with the new lettering. In 1988 Studio Dumbar was commissioned to revise the house style of the newly privatized KPN (Royal Netherlands Post), successor to the PTT.

In the following year the DEV was transformed into the Art and Design Department. There was a storm of protest from the art world in 2002 when the KPN closed down this department. It was a sign that in the meantime the KPN had become completely business-like and commercial. This brought to an end a long tradition of design idealism: the cultural and social role of this former state enterprise's design department seemed to be played out.

Most other Dutch companies had already been forced to revise their design policies in the 1970s for economic reasons. For ordinary commercial firms, their unswerving belief in the great cultural, social and economic significance of industrial design was at an end. The 1970s saw the advent of renewed discussion on the benefits of design and the social position of the designer. Room was created for an entirely new interpretation of the discipline and the role of its practitioners.

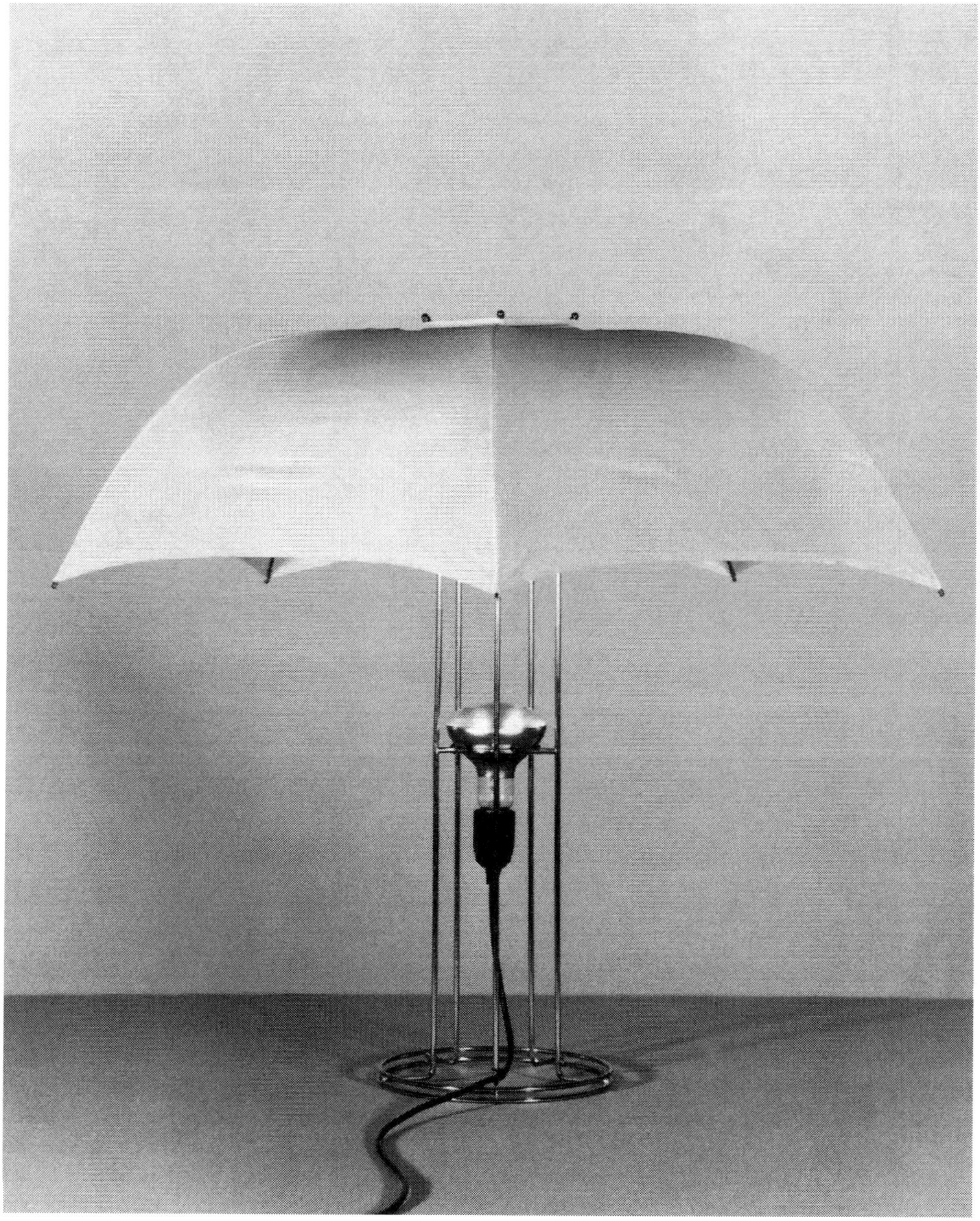

5

Design for Debate, 1960s to the Present

In the closing decades of the twentieth century design, including fashion, became an inescapable theme. The debate took place in the columns of trade journals, during meetings held by professional associations and organized interest groups, in educational circles, at trade fairs and exhibitions; but there was seldom, if ever, any consensus of opinion. Moreover, the discussions were not restricted to professionals, because at the time public interest in modern design was on the increase. In recent years even the popular media have been writing a great deal on the subject. A recent highlight in this nowadays public debate was the selection of the 'Best Dutch Design' in 2006. People had increasingly to take a point of view, or pass judgement, on questions involving design. Thus the contemporary style of the familiar telephone book, the cheerful-looking banknotes or the postman's new uniform became the stuff of everyday conversation. The selection of consumer goods available to the public multiplied to such an extent that even buying a new washing-up brush could prove to be tricky: no longer was there just the familiar wooden one sitting on the shelf – all of a sudden there were three more cheerfully coloured plastic ones for us to choose from. And every office clerk became aware that an office chair, in addition to being comfortable and functional, was supposed to be designed in adherence to 'ergonomic' principles.

Magazines on houses and gardens, fashion and lifestyle appeared in ever increasing numbers and their circulation grew. In the same way, for many people in the Netherlands a day at the large furniture malls has become a favourite outing. In recent years Dutch people seem to have become preoccupied with the creation of their own style and ambience, aided by a constantly growing range of fashionable articles and consumer goods. Moreover, they are prepared to spend a great deal of money to

Gijs Bakker, umbrella lamp, 1973.

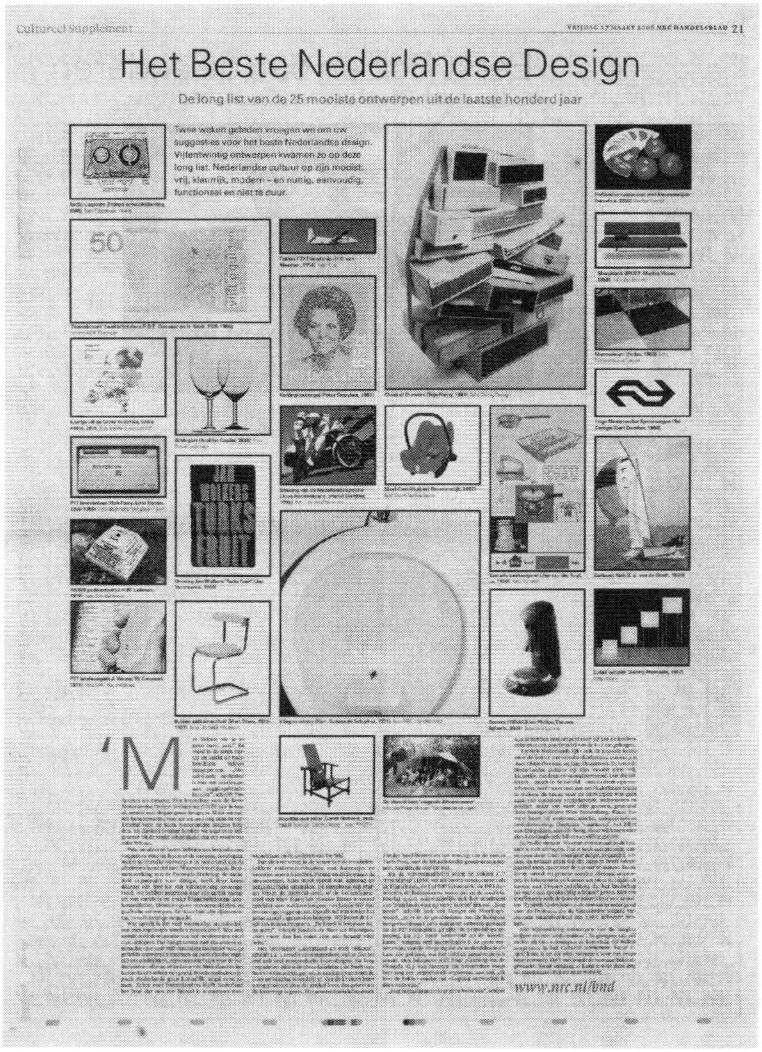

Cultureel Supplement 21

Het Beste Nederlandse Design

De long list van de 25 mooiste ontwerpen uit de laatste honderd jaar

Twee weken geleden vroegen wij om uw suggesties voor het beste Nederlandse design. Vijfentwintig ontwerpen kwamen zo op deze long list. Nederlandse cultuur op zijn mooist: vrij, kleurrijk, modern – en nuttig, eenvoudig, functioneel en niet te duur.

www.nrc.nl/bnd

'Het Beste Nederlandse Design', NRC, 17 March 2006.

achieve a specific look. Since the beginning of the 1980s, largely thanks to the Swedish home furnishing store IKEA, those with less money to spend have seen many attractive modern products come within easy reach. In these years the HEMA, the most typically Dutch of all chain stores, successfully changed its main objective from 'good and inexpensive' to 'good, inexpensive *and* well-designed'. To highlight its broader objective it has started to hold a popular annual design competition.

A number of recurring themes have dominated the design debate among professionals in the last thirty years, the most important being the modernist canon. Critics have increasingly condemned the rational and functional style for being 'boring, anonymous and lacking in emotion', and in its place have advocated design showing more 'expression, significance and passion'. In the process, the old discussion about the value of handicrafts was rekindled, as was that regarding the relationship between design and art. Others preferred to distance themselves from these questions and emphasized the commercial, interdisciplinary and problem-solving character of design. For this reason they heatedly advocated the use of the term 'designing for industry' rather than just 'design'. Another persistent topic was whether design should be used as an instrument in economic policy: is it permissible for increased production and turnover to be design's most important goal? This theme became all the more urgent in the 1970s due to the worldwide economic recession and interrelated environmental problems. Thus design has become part of a wider political discourse, once again with a moral component. Two additional themes have emerged more recently, the consequences of automation and globalization.

In the 1980s the debate was provided with a new historical dimension by art historians who began to reflect upon the subject of design and its history. Exhibitions like *Industry & Design* (1985) and *Dutch Form* (1987) presented an overall picture of the most important twentieth-century Dutch products and producers. Dutch museums, with their burgeoning collections, aspired to a role that went much further than just propagating good form. The number of books on Dutch interior decoration and design has grown and the journal *Jong Holland* (Young Holland) has been providing space for scholarly studies on these themes since 1985. Historical awareness was also fed by contributions from abroad: American pop culture, Italian postmodernism and the publications of the architectural

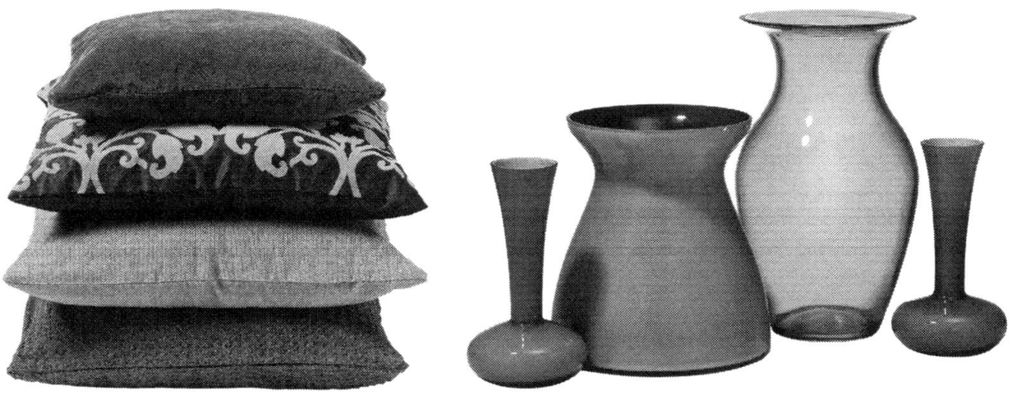

Pillows and vases from the HEMA, 2006.

theorists Robert Venturi and Charles Jencks all reached the world of Dutch design in the 1980s. The result was a critical and sometimes ironic appraisal of their own history and principles.

At the end of the twentieth century the debate sometimes seemed to have become an end in itself: the more discussion a new product provoked the better it was perceived to be. In the mid-1980s this was taken to such extremes that it led to a fierce polemic about the alleged absence of professional design criticism and high-quality discussion in the Netherlands. It had evolved into a debate for debate's sake or, rather, a debate about the shortcomings of the ongoing debate.[1] The Rotterdam Design Prize, a national design competition begun in 1993, focused more on generating discussion than on selecting the best design. In that same year the government saw the new National Design Institute's most important task to be stimulating debate.

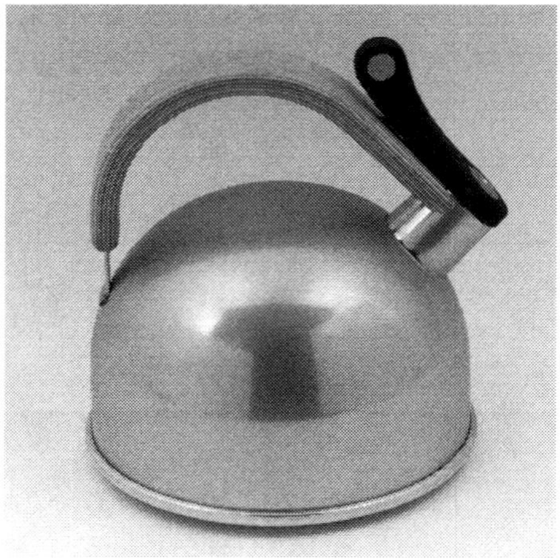

Nicolaï Carels, *Le Lapin* kettle, winner of the HEMA design contest, 1989–90.

Critique of the Modernist Canon

The exhibition *My Room*, organized in 1960 in the Stedelijk Museum in Amsterdam, was one of the first occasions at which modernism, which until then remained unassailable, was called into question. The exhibition was arranged for the annual presentation of the firm Rath & Doodeheefver's new wallpaper collection. To mount the exhibition the director of the wallpaper factory, J. F. Rodenberg, had enlisted the help of the leading designer Kho Liang Ie, the critic and sculptor J. J. Beljon, who was also the director of The Hague Art Academy, and the graphic designer Gerard Wernars. In organizing this presentation, as becomes clear when reading the catalogue written by Beljon, the three of them had very deliberately decided to put the previously sacrosanct canon up for discussion. Beljon thought it was time that people learnt how to break the rules and that each person should be given the chance to put together 'My room' to suit his or her own taste. It was high time for the reintroduction of the scope to 'express the human element', and for the nurturing of 'decorative passion', because they were in danger of 'losing a wealth of emotions and human warmth'.[2]

Nobody was surprised at Joop Beljon's views on the subject. Three years earlier his appointment to the Academy of Art in The Hague had also sounded the death knell of the Advertising department, once so progressive. Under

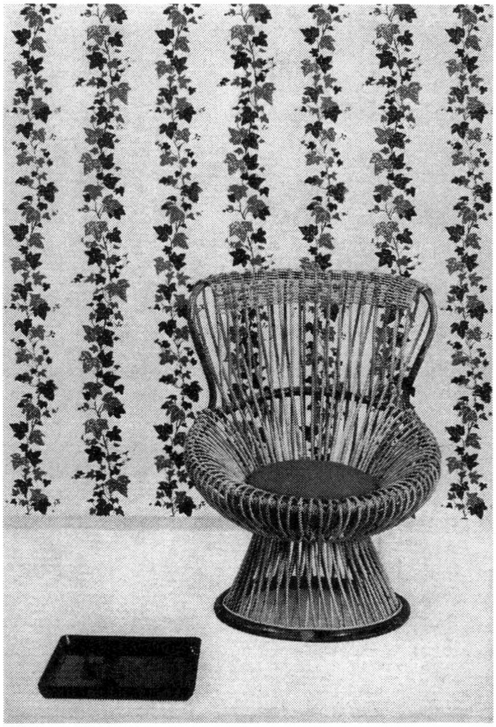

Rattan chair by Franco Albini and wallpaper by Rath & Doodeheefver in the exhibition *My Room*, Stedelijk Museum, Amsterdam, 1953.

his management, the new department of Graphic and Typographic Design pushed aside the modernist principles to make room for a more personal-artistic approach to the subject. This fitted in well with the working methods of W. J. Rozendaal, who as a teacher in the Drawing and Painting department had already introduced students to highly imaginative book illustrations. In Beljon's enthusiastic, rambling collection of essays on art and design, *Ontwerpen en verwerpen* (Design and Reject), written in 1959, he accentuated the subject's artistic capacity.[3] The fact that Kho Liang Ie, who had trained as an analytical designer and had worked for a few years for the Goed Wonen Foundation, also decided to participate in this project was more significant proof of a change in climate. Moreover, Kho had already broken with Wim Crouwel's studio to continue with his own projects, with a view to approaching them in a more intuitive manner.

However, the text in the *My Room* catalogue promised more than the exhibition itself could actually deliver. In the museum display the ordinary wallpapers by R&D were craftily combined with well-known modernist work by leading designers like Charles Eames, Poul Kjaerholm, Gio Ponti and Ludwig Mies van de Rohe, as well as the Dutchmen Martin Visser, Coen de Vries and, of course, Kho himself. The public and the newspapers were duly disappointed: one reviewer of the weekly newspaper *De Groene Amsterdammer* remarked: 'I didn't see *my room* among those on display.'[4]

But the first steps had been taken, and as the 1960s ran their course even the Goed Wonen Foundation, once so completely certain of being in the right, started to readjust its rigid view of interior design (see chapter Three).[5] The show houses and the interiors depicted in the magazine of the same name became increasingly varied and more luxurious over the years. The subjects no longer revolved around modest terraced houses with through lounges, but gave villas, second homes, students' garrets and singles' apartments coverage too. At the same time it also focused more closely on the political and social aspects of the entire built environment. A lot of copy space was devoted to do-it-yourself (DIY) as well: detailed drawings and clear photographs showed the reader how he could make this simple furniture himself. It also responded to individual needs by featuring 'Swedish' wall-unit systems made of wood, for example, alongside a Dutch variant developed by Pastoe; and on a smaller scale, the metal shelves produced by Pilastro and Tomado. These economies of space created room for a record player and long-playing records, not forgetting the space needed for a television, which by this time was beginning to make its appearance in every living room.

Architect Aldo van Eyck's articulate criticism of modernism was furthermore very influential. Van Eyck, once a member of the Amsterdam functionalist architects group De 8 and a participant in the CIAM congresses (see chapter Three), became convinced by the late 1950s that rigid functionalism had killed off many designers' creativity. He observed that functionalists paid far too little attention to the individual and to what he referred to as the 'human proportion' of things. He expressed his views in the periodical *Forum*, which he and the architects Jaap Bakema and Herman Hertzberger edited from 1959 to 1963. They argued for a more humane type of architecture and drew attention to the need for more emotional experiences and for more consideration of the effect buildings and interiors could have on the behaviour and state of mind of their users. This was clearly shown in Van Eyck's design for the Amsterdam Burgerweeshuis (Civic Orphanage) of 1959. Instead of basing it on an austere, understated functional analysis, and on minimal design, he drew an organic sequence of smaller units, allowing space for the children to create little hideaways of their own. Seating units and tables fastened to walls formed an integral part of the

Two views of Aldo van Eyck's Municipal Orphanage, Amsterdam, as illustrated in *Goed Wonen*, 1960.

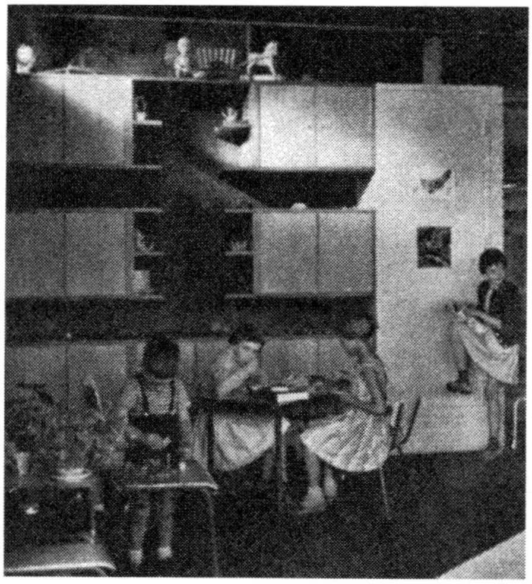

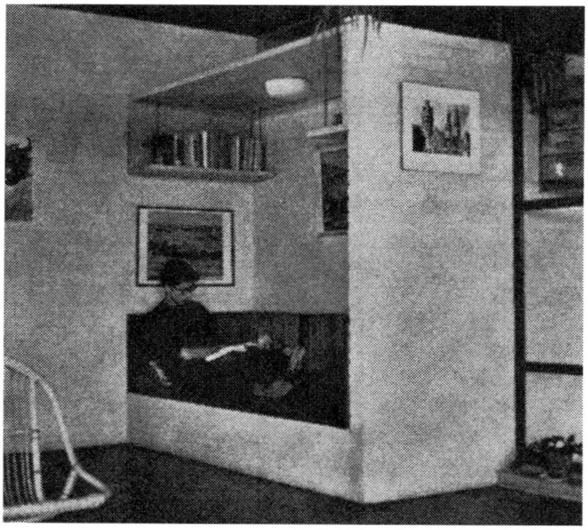

total architectonic landscape. Van Eyck's inspiration had come from primitive settlements in Africa; apparently people living in the neighbourhood had not failed to notice that fact, since the orphanage was popularly referred to as the 'kaffer dorp' (Bantu village). This meant that the Burgerweeshuis was one of the first manifestations of a return to small-scale thinking and individualization that was to characterize interiors in the 1970s.[6]

A growing aversion to concrete flats and endless rows of uniform houses in post-war reconstruction areas stimulated a style of building in the 1970s that made use of traditional brick, slanting roofs, variations in level and winding streets. This was done to achieve a feeling of security and community among its residents by allowing them to survey their habitat. A typical product of this new culture was the residential area closed to through traffic ('het Woonerf'), creating outside space where children could play safely, and where motorized traffic was subject to clear restrictions.[7]

Interior designers, too, no longer believed that functionality and affordability were of the foremost importance, nor was there much faith left in the power of light, yet sober, design to conjure up feelings of well-being, a belief that had held sway shortly after the war. The furnishings and fittings of the house were increasingly considered to be a way of expressing personal taste and identity. Industrious do-it-yourself enthusiasts, who saw their standard residences as unimaginative and monotonously designed, started to convert them.[8] On their Saturdays off they knocked down partition walls, lowered ceilings, removed the sliding doors in their through lounges and hollowed out sunken sitting areas in the garden and sometimes even in their living room. After all that had been completed, they then painted the walls brown, purple or orange, or covered them with jute or wood panelling. Somewhat later, in the 1980s, all those colours and the wood panelling were removed and the walls were then coated in white. For floors, wall-to-wall carpeting became popular, although students and intellectuals preferred rush mats or wooden boards. These were replaced ten to twenty years later by cheap, practical laminate, or by wooden parquet floors, although a few preferred to have modern linoleum in their homes. Bookcases, hobby corners, breakfast bars and bunk beds were knocked together from ready-to-assemble, do-it-yourself kits, which were widely stocked at the home improvement centres opening all over the Netherlands. For those who were not so skilful, but nevertheless still wanted to create a relaxed DIY-style interior, the reasonably priced alternative in the 1970s was the adjustable racks of shelves by the Swedish firm Lundia, manufactured under licence in the Netherlands. These popular pine shelves, which were really designed for warehouses and places of work, were used as practical bookcases and storage units that could be adjusted to suit the client's individual requirements. In 1978 IKEA started its offensive in the Netherlands with the opening of its first furniture and furnishings store in Sliedrecht. In none of the thirty countries in which the Swedish firm had set up shop did these modern, uncomplicated, unpretentious products catch on as well as in the Netherlands.[9]

New furniture by professional designers was scarce in the 1970s and conformed to new ideas that saw an interior as an informal 'living landscape'. Chairs and couches in the living room were replaced by comfortable corner settees, while progressive consumers preferred 'sitting units' that could be joined together and regrouped at liberty. The furniture made by that time was larger and heavier than the light, easily movable Goed Wonen products. A striking example of this new trend was the couch, consisting of separate components, that Jan des Bouvrie designed for the Gelderland furniture factory in 1972. It is a somewhat more informal variation on his still popular 'cube' couch designed three years earlier. Also illustrative is the Levi Chair, upholstered in denim, which Gijs Bakker designed for the firm Castelijn.[10]

From the 1970s onwards consumers not only condemned 'boring' and 'indistinctive' houses and furniture, they had also had enough of products manufactured with the same impersonal mass-produced 'modern' look, and sought refuge en masse in the nostalgia of the good old days or country living. Once again pans and kitchen utensils could be colourfully enamelled

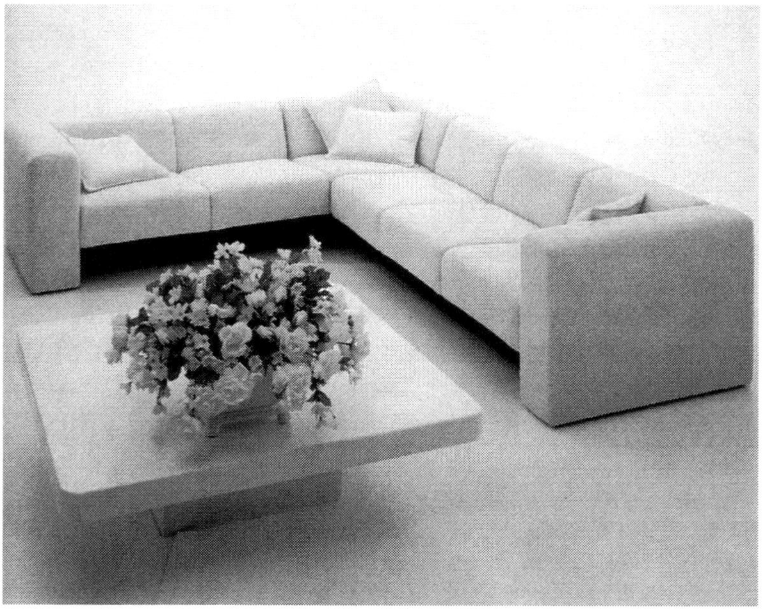

Jan des Bouvrie (furniture factory, Gelderland), cubic couch, 1969.

or even decorated with primitive flower patterns. Country style earthenware with hand-painted flower motifs and similarly decorated French crockery were favourites; old glass storage jars replaced plastic or metal ones; wooden spoons, pudding moulds and coloured bottles were reinstated and used as decorative elements; and self-crocheted lace curtains appeared at windows everywhere. In some progressive milieus Persian carpets once again adorned the floors, and Indian fabrics were draped over couches. Later the throw became a generally accepted means of hiding a couch that was too plain or worn out. This atmosphere was well suited to take on board an antique cupboard or one from grandma's day, either distressed-looking or repainted. An oil lamp, old milk churn or watering can filled with plants would not go amiss either. And, of course, there was the inescapable type-case – hung up flat against the wall, so the square sections where the letters were once kept could be used as a home for tiny decorative ornaments. It took only a few years before such nostalgic and quasi-artistic 'unique' objects were being manufactured in large batches. Many a young family's cosy interior was rounded off to perfection with terracotta pots filled with the plants that had proliferated lavishly not so long before in handmade macramé plant hangers. The new-style pots could be bought from a hardware store, a garden centre or IKEA.

Craft: A Critical Alternative

It was hardly surprising that in this atmosphere of nostalgia, with its predilection for the unique and the personal, handicrafts could continue to flourish as an alternative to the uniformity of mass production.[11] Throughout the years Arts and Crafts maintained its hold as a small but tenacious movement alongside, and sometimes even as a part of, other progressive movements. A few well-known potters and weavers from the pre-war Arts and Crafts movement, among them Bert Nienhuis, Gerrit de Blanken, Thera Hofstede Crull and Kitty van der Mijll Dekker, remained influential long after the war and had themselves trained large numbers of young craftsmen.[12] Pieter Groeneveldt's workshops in Voorschoten, Zaalberg in Leiden, Mobach in Utrecht and De Driehoek in Huizen even managed to carry on for much longer. That also held for the textile workshop Het Paapje in Voorschoten and the hand-weaving workshops De Cneudt in Soest and De Knipscheer in Laren. New potters and textile workshops also opened their doors: Dirk Hubers, Jan Oosterman, Jan de Rooden, Johnny Rolf and the textile printers 't Seghel in The Hague and Harry van Kruiningen in Amsterdam were all making a name for themselves at the time. Still active were various traditional workers in precious metals like Marinus Zwollo, Archibald Dumbar, Chris Steenbergen and Joseph Citroen, and stained-glass artists like Lex Horn, Nico Wijnberg and Berend Hendrix. Finally, the studio glass made by Andries Copier and his pupils Floris Meydam, Willem Heesen and Sybren Valkema continued to be appreciated without interruption.[13]

All these enthusiastic craftsmen were more likely to find modern mass production challenging than discouraging. Moreover, in 1948 the Central Organ of Creative Trade (Centraal Orgaan Scheppend Ambacht, COSA) was set up, although of course it was really intended for the less well known among them. Until well into the 1980s COSA was to craftsmen what the IIV had been to industrial designers. With the financial support of the Ministry of Economic Affairs, COSA's director J.J.E. Salden made every effort 'to strengthen and widen the commercial basis providing professional working craftsmen with a livelihood'.[14] They promoted the craftsmen's work through COSA's mouthpiece *Scheppend Ambacht* (Creative Handicraft), exhibitions and competitions. COSA also mediated when large monumental works were commissioned and ran a gallery of its own in Delft. Moreover, it was not the only organization that supported artistic craftsmen. The Art and Business (Kunst en Bedrijf) foundation, founded in 1950, also acted as a mediator and the Society of Practitioners of Monumental Art (Vereniging van Beoefenaars van Monumentale Kunst, VBMK) supported and advised its members in every conceivable way. Though concerned with craft, their work nonetheless found a platform in industrial design circles. The products made by small independent artists in their studios were recommended in the periodical *Goed Wonen* and could be obtained in progressive furniture and furnishing stores like Metz & Co. and Bas van Pelt. Some workshops were even members of the BKI, until this organization amalgamated with the IIV in 1950. In the 1950s the Stedelijk Museum in Amsterdam focused on both new industrial design as well as this 'artistic' applied art. In 1957 its director Willem Sandberg organized *GKf Hand and Machine*, an exhibition that stressed the unity of free and applied art, and which was later sent in its entirety to the Triennale in Milan. The artistic craftsmen's entry caused some resentment among the other GKf members, due both to its presumed questionable quality and what the industrial designers saw as evidence of a too simplistic attitude towards industry.[15]

The main focus at the Stedelijk Museum in Amsterdam at that time was textiles, leading it to purchase and exhibit both industrial and hand-printed fabrics. Meanwhile, at the Boijmans Van Beuningen Museum in Rotterdam, modern artisanal ceramics were encouraged. It was there that in 1953 the exhibition *Five Contemporary Potters* was organized, featuring work from the older artists Bert Nienhuis, Harm Kamerlingh Onnes, Piet Wiegman and Franz Wildenhain, as well as that of the young artist Dirk Hubers. In 1962 six young Amsterdam ceramic artists exhibited in the Boijmans: Hans de Jong, Jan de Rooden, Johan van Loon, Jan van der Vaart, Sonja Landweer and Johnny Rolf. It was partly due to the platform provided for them on this and numerous other occasions by the curator Bernardine de Neeve that a real revival of artisanal ceramics came about in the Netherlands. 't Kruithuis in 's-Hertogenbosch, the Gemeentemuseum in Arnhem and the Gemeentemuseum in The Hague also extended their collections by adding examples of modern Arts and Crafts, although in The Hague this section was to remain part of the modern art collection until well into the 1970s. Several specialized galleries took it upon themselves to present and promote this new applied art, including Het Kapelhuis in Amersfoort, Nouvelles Images in The Hague, Marzee in Nijmegen and the galleries Ra, Sieraad and Appenzeller in Amsterdam. In 1976 Mieke Spruit-Ledeboer wrote a doctoral thesis at the University of Amsterdam on this modern form of ceramics, which would previously have been an unthinkable topic for scientific research.[16]

How, then, should handicrafts relate to industrial design? Most industrial designers saw little to commend in 'dabbling' with the forms and techniques used by artistic craftsmen. The industrial designers' own search for an object's 'essence' aimed at a total, almost primitive, purging of form that was far more interesting in their view than the relatively non-commercial 'rampant growth of over-strained individualism' observable in the craft world. At least that was the way Karel Sanders sketched it in a brochure published by the IIV in 1955. In retaliation, craft artists argued that their hand-made objects were superior in most respects: they were more spontaneous, more out of the ordinary and 'more aristocratic' than the indistinctive and superficial nature of industrial products.[17]

René Smeets, director of the Academy of Industrial Design in Eindhoven, in a lecture given on the occasion of the COSA's fifteenth anniversary in 1964, found a middle ground to this argument. In his lecture he endorsed handicrafts as a good way of experimenting with shapes and materials. He recognized the imperfect, but organic, individual and multiform character of handicrafts as a constructive quality. He even went as far as saying that he considered the use of traditional methods to be an important and inevitable reaction to the 'alarming over-perfection of technical, dispassionate mass production'. In Smeets's view handicrafts had an educational function and could therefore be deployed to counter the numbing of the senses and the spiritual damage caused by popular culture.[18]

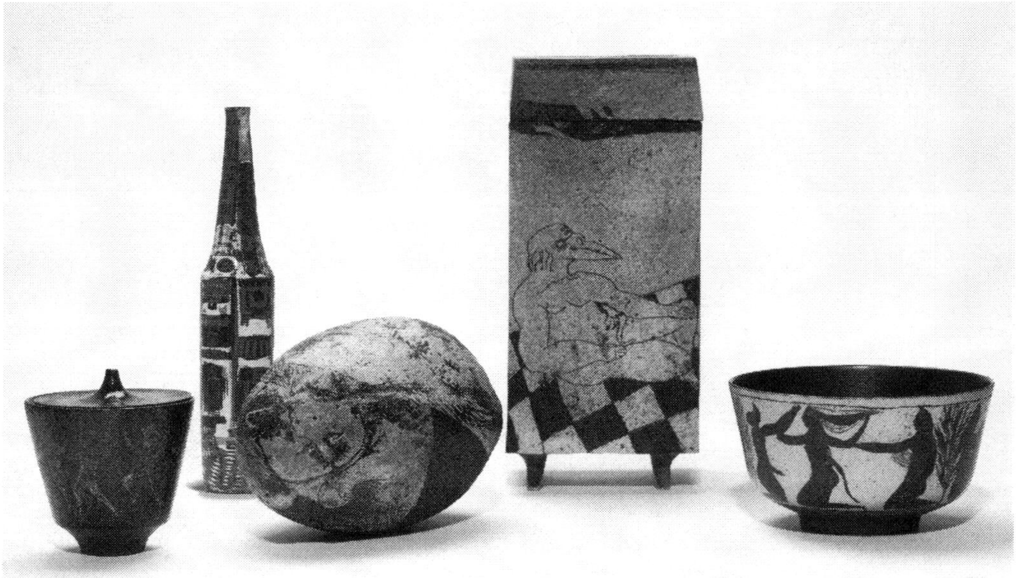

Some examples of the Experimental Department of De Porceleyne Fles in Delft, c. 1960: lidded blue pot by Lies Cosijn and Jet Sielcken; bottle by Jet Sielcken; cat sculpture by Lies Cosijn; lidded square pot by Lies Cosijn; bowl by Lies Cosijn and Adriek Westenenk.

The belief that an artistic, artisanal method of production could be inspiring and fruitful for industry was also shared by the management of the De Porceleyne Fles pottery works in Delft, where between 1955 and 1977 its experimental Unica department worked with this conviction in mind. They had also become acquainted with comparable studios in Scandinavia. The sculptor and ceramic artist Theo Dobbelman, who also taught at the IVKNO, was taken on as the manager of this unusual workshop, where young and ambitious ceramic artists like Lies Cosijn and Jet Sielcken were able to experiment for a few years. Yet those in the department operated a little too experimentally, gradually turning their backs on the factory's normal mass production while pursuing their persistently freer and more autonomous work. A comparable laboratory function underlaid Bernardine de Neeve's initiative to set up the European Ceramic Work Centre in 1973 in Heusden (now in 's-Hertogenbosch).[19] In Leerdam, Andries Copier continued to make unique glass objects in collaboration with the factory's glass-blowers, while Sybren Valkema, Floris Meydam and Willem Heessen were experimenting there with artistic glass. But, like Copier, they also designed industrial glassware. After Copier had left, a Glass Design Centre (Glasvormcentrum) was established in Leerdam in 1968, just a stone's throw from the factory, where the opportunity to continue to experiment with shapes, materials and techniques continued.[20]

Despite all these supportive initiatives, handicrafts continued to be of marginal economic importance. Even if these artistic products perfectly meshed with a social tendency to condemn mass production and uniform products more vociferously, the small-scale nature of handicraft production meant that it could never really compete with industry. Many craftsmen changed course at the end. A number of them trained themselves to become art teachers; others made a conscious decision to abandon the functionality of hand-made products and switch to making unique works of art. From then on the weavers made decorative tapestries and the potters free models.

In the 1970s, in addition to *Scheppend Ambacht* (Creative Craft), the periodical *Bijvoorbeeld* (For Instance), established in 1968, became increasingly vital in reporting the vicissitudes of this specific design sector. Consecutive volumes of this periodical give us an appreciation of the field's increasing complexity, somewhere in between handicraft, industrial design and fine art – what Marjan Unger, editor-in-chief of *Bijvoorbeeld*, named in the 1980s 'vrije vormgeving' (free design). We see in *Bijvoorbeeld* how, after having made decorative tapestries, needlework artists and weavers moved on to making three-dimensional textile sculptures. The weaves made by Ria van Eyk, Loes van der Horst and Margot Rolf became more austere and developed into high-quality abstract-geometric works of art. Potters started to call themselves ceramic artists and began experimenting with flat monumental tile-pictures, or were not interested in making anything except unique pots. Then there were ceramic artists like Helly Oestreicher, who was one of the first in the early 1960s to present her abstract ceramic forms as 'anti-pots'.[21]

In addition, the work of a number of small-scale fashion, shoe and bag designers gradually started to fill the pages of *Bijvoorbeeld*; coverage here contributed to their becoming household names in artistic circles. The publication featured clothes by the Amsterdam designer couple Puck and Hans, Jan Jansen, Lola Pagola, shoes by Freddie Stevens and bags by Maria Hees and Hester van Eeghen. Jewellery-makers in particular were given a lot of column space in *Bijvoorbeeld*, because developments in their field were extremely radical.[22] After Chris Steenbergen and Archibald Dumbar had taken the lead in the 1950s and early '60s with their modern, simple pieces of jewellery, in the late 1960s Gijs Bakker, Emmy van Leersum, Françoise van den Bosch and Hans Appenzeller changed course radically in terms of design, choice of materials and techniques. With their preference for an abstract-geometric design language and

Maria Hees, garden hose bracelet, 1969.

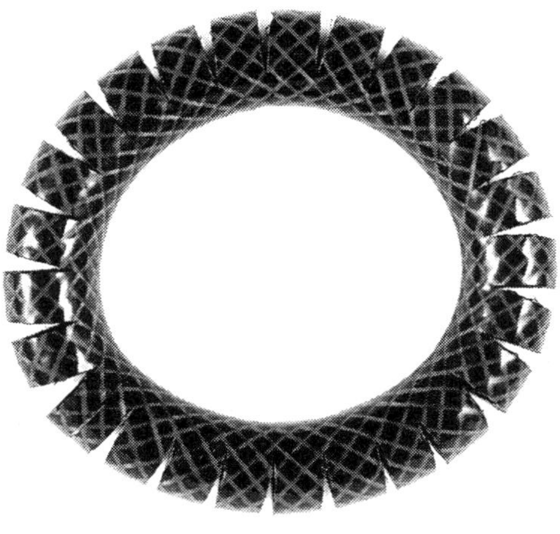

their sometimes almost sculpture-sized 'anti-jewellery', they too were aiming at affiliating with the fine arts. Bakker and Van Leersum showed their first products in Galerie Swart, a fine art gallery in Amsterdam. In 1969 a travelling exhibition of their work and that of Nicolaas van Beek, Françoise van den Bosch and Bernard Laméris went by the name *Objects to Wear*. However, the other side of the picture was that young jewellery designers associated themselves more with industrial design, driven by the need to produce objects in series and to opt for cheaper, more everyday materials like Plexiglas, rubber and aluminium. For instance, in 1969 Maria Hees devised a bangle that could be made simply from garden-hose tubing; Gijs Bakker invented a neck ornament made from a piece of stovepipe; while Marion Herbst had no qualms about combining silver and Perspex to fashion a necklace and matching earrings.

Although, ultimately, almost everything seemed to be possible in the field of artistic crafts, this process did not take place without the mandatory discussions. Should all these new experiments be called 'art', or were applied art and design to remain more suitable terms after all? Was an artist working with clay destined to remain a ceramic artist and an artist working with textiles a textile artist? If not, under which denominator should their work be exhibited, purchased or made public? The warring factions kept passing the buck and accusing one another of being bigoted and ignorant, as well as lacking in technical knowledge and artistic competence. These fierce debates were given new ammunition around 1980 by controversial exhibitions like *West Coast Ceramics* (1979) in the Stedelijk Museum in Amsterdam, *Who Is Afraid of American Pottery* (1983) in the Kruithuis in 's-Hertogenbosch and Rhonda Zwillinger's furniture (1984) in the Groninger Museum.[23] As will become clear below, the last word on the matter had not been spoken.

Design and Political Debate

In the meantime industrial production in the Netherlands was suffering a downward trend. In 1970, after years of growing prosperity, the economy had ended up in a deep depression.[24] Scores of factories set up so hopefully in the post-war years had to shut their gates for good. The worldwide oil crisis in 1972, combined with an alarming report from the Club of Rome think tank in 1973, had even led to Den Uyl's cabinet introducing a number of compulsory car-free Sundays in the Netherlands. By then the seriousness of the situation was beginning to dawn on everyone: there were limits to growth. There was even a sizeable crack in the unquestioning progressive idealism of

Dutch industrial designers. The generation who had brought about the post-war reconstruction of the Netherlands was succeeded by a generation forced to recognize the disadvantages of unbridled economic growth. The gradual rejection of modernist dogmas ran parallel to these changes in society and to the shifting political climate. The emergence of the New Left, far-reaching democratization at schools and universities, the increased influence of the trade union movement, the struggle to achieve redistribution of power and incomes, all these changes had repercussions on the design debate.

During the last years of its turbulent existence, the Institute of Industrial Design (IIV), which from its last reorganization in 1970 until it closed down in 1976 was called the Industrial Design Foundation (SIV), tried to encourage discussion on the new position of industrial design in society. The designer and publicist Simon Mari Pruys was assigned the task of writing a critical memorandum on the subject. His fundamental contribution to the design debate appeared in 1972 under the title *Dingen vormen mensen* (Things Form People).[25] His collection *De nieuwe onzakelijkheid* (New Non-objectivity) had already been published one year earlier, and a compilation of essays previously published in the daily newspaper *NRC-Handelsblad* followed in 1974 under the title *De paradijsbouwers: anti kunstzinnige opmerkingen over de gebouwde omgeving* (The Paradise Builders: Anti-artistic Remarks on the Built-up Environment).

Dingen vormen mensen is the first inventory of design in the Netherlands written as a sociology of culture. Pruys introduces his line of argument by claiming that industrial design, apart from being a profession, is also an ideology. He emphasizes that every activity developed in this framework, and every word written about it, fits into this deliberate or non-deliberate view of man and society. Each new product stands in an area of tension in which a role is played by the interests of the entrepreneur, investor, manager, construction engineer, production engineer, supervisor, operative, retail trader, middleman, consumer or user, designer, government and society. The designer and the manufacturer are plainly not the only people on the playing field.

Pruys very deliberately no longer asked the question: 'What is good design?' He preferred to consider whether good or bad design existed. After all, design can be successful in many ways: for the employer, user, designer or retail trader. And what these parties desire is generally not the same.[26] Even if the debate is restricted to the user, then the question of what is good or bad is equally difficult to answer, if it can be answered at all. It is not just a well-designed object complying with official quality standards that can meet certain requirements: a kitsch product can do the same, for example, by meeting the need for security, familiarity and safety. In any case, demonstrating that ugly objects have a negative influence on our well-being is a very difficult thing to do, and he thought the unsolicited act of ramming this view down people's throats was downright suspect. This is why Pruys was one of the first in the Netherlands to attempt to elevate the concepts of good and bad in terms of design to a higher level, to see what its implications were for society as a whole. He thus observed that with car design it was the car's status and the trade that played the main roles. On social grounds, the escalation of modernization is certainly not a good thing: society and the individual fall 'victim to a form of visual communication which is not only primitive, but which also has undeniably harmful consequences'.[27]

Simon Mari Pruys's work was pioneering in the way it drew attention to the object as a means of communication, or the product as a symbol, or carrier, of a message. He introduced semiotics to the Dutch world of design. Within this framework he discusses Jean Baudrillard's study *Le Système des objets* (The System of Objects, 1968), which has since become a classic text, in which consumption is not looked upon as a material-needs fulfilment but as a systematic processing of symbols. He claims that designers would do better if they were to become far more conscious of this side of their work, and no longer direct their attentions exclusively to the safe area in which aesthetics and functionality lay down the law from top to bottom.

In *Dingen vormen mensen* Pruys argues for a totally different type of designer, namely someone who has outgrown the teething troubles experienced when playing at being an artist and who can approach the real problems of his time in a grown-up fashion. He warns that the competitive conduct displayed by designers and employers will ultimately get them nowhere. This artificial rivalry to achieve ever more originality, sham-innovation and pseudo-progressiveness is merely damaging. On the other hand, showing too much reticence and too much interest in simplicity and naturalness can also lead to problems, as the 'tragedy' of a journal like *Goed Wonen* has shown. This type of propaganda in favour of unpretentiousness in design conjures up associations with moralistic fanaticism.[28]

Pruys's study was the first to document systematically all the pressing social questions connected to the phenomenon of industrial design: planned obsolescence, material shortages, alienation, sustainability and environmental problems, over-consumption and waste. He concluded: 'We must gradually start to acknowledge that both advertising and design, inasmuch as these activities do nothing else but time and again make the chairs we sit on old-fashioned, are the deadly enemy of our civilization.'[29]

The criticism and the dogma of good form, the economic recession in the 1970s, the environmental issues and the first mention of socio-political issues that the industrial design profession had ever run up against caused a great deal of confusion among designers of the period. This was particularly so among the Art Academy students for whom *Dingen vormen mensen* was compulsory teaching material. What, for whom and how were they supposed to design: what were their responsibilities? And last but not least: how on earth were they supposed to find work?

The influence of this sociological approach to design even had consequences in the museological world. In 1980 the Gemeentemuseum in The Hague organized four exhibitions that together were called *Massa Cultuur* (Popular Culture), involving everyday subjects like ready-to-wear clothes ('fashion for everyone'), the ordinary Hague interior ('home culture'), synthetic material in daily life ('the first plastic age') and featuring a low-brow weekly publication from the first half of the century ('*Life*'). The initiative aroused a great deal of debate and in the end attracted only a small following. This almost anthropological approach to design was too far removed from the usual manner of working and thinking in museums in the Netherlands at that time.[30]

The Debate in the Graphic Sector

If there was one field in the Netherlands in the 1970s and '80s where the debate, in the broadest sense of the word, took centre stage, then that was in the world of graphic design.[31] This still relatively new discipline balanced continuously on the unstable borders between commerce, culture and politics. It was also being chased hard at heel by rapid technological developments.

In order to see the situation in its true perspective, it is worth briefly surveying post-war developments in the graphic sector. Straight after the war Dick Elffers played a principal role in the professionalization of graphic designers and in gaining wide recognition for their work.[32] Elffers's own roots went back before the war to the lessons given by pioneers such as Jac. Jongert and Piet Zwart at the Rotterdam Art Academy and Paul Schuitema's design office, where he worked briefly. After the war his work became far more personal, adopting a new design language, as seen in a series of capricious colourful posters (his 'street paintings'), book covers, brochures and corporate photography books. His placards for the Holland Festival in the 1950s and '60s received wide acclaim. The work of kindred spirits like Otto Treumann, Jan Bons and Willem Sandberg had a comparable effect and influence on the subject. During the 1950s these four designers were together able to create a freer climate for the graphic designer's work.

In 1969 the GKf and the VRI, the two professional associations for graphic designers that had originally been so different in their outlook, amalgamated to form a new organization called Dutch Graphic Designers (Grafische Vormgevers Nederland, GVN).[33] This meant the end of a split that had rankled for years on end between Amsterdam's mainly politico-culturally orientated GKf and The Hague's and Rotterdam's more practical and commercially minded VRI. The merger of the two associations was brought about by the charismatic designer Jurriaan Schrofer, who became the first chairman of the new organization. The effort put into it by Elffers and Schrofer was, however, not sufficient to restore a modicum of peace in the traditionally turbulent world of graphic designers, typesetters and printers.

This exciting climate carried on through the 1970s and into the '80s in discussions on modernism. The criticism of the supposed impersonal and dispassionate International Style, as demonstrated especially in the work of Wim Crouwel and other fellow workers at Total Design (TD), began increasingly to determine the character and subject matter of the discussions. An increasing number of colleagues and critics considered Total Design's work to be too commercial, austere and routine, and they were making a case for permitting the introduction of more imaginative and illustrative elements. In 1972, when Wim Crouwel, permanent designer for both the Stedelijk Museum and the Fodor Museum in Amsterdam, became involved in an exhibition and a catalogue about the work of his colleague Jan van Toorn, his appointment caused a great deal of commotion.[34] In contrast to the commercially minded and analytical Crouwel, Van Toorn was an emotional designer, critical of the prevailing social structure. He was, among other things, a permanent designer at the Van Abbemuseum in Eindhoven and regularly produced covers for the national museum journal *Museumjournaal.* The idea that Crouwel should design the catalogue did not please Van Toorn at all. The heated argument between the two designers was fought out in the public domain, a unique episode in the world of Dutch design. In the presence of a few hundred critical spectators, including many designers and Art Academy students, the two leading designers entered into a debate in the Fodor Museum. Crouwel calmly emphasized his standpoint that the graphic designer should only mediate between the client and his public, without wanting to give a personal touch to the task. In Van Toorn's view, however, no such neutrality or objectivity was possible and he heatedly reproached Crouwel for employing a method of work that reduced him to a mere extension of the system. His point was that design could not and should not be a value-free activity. A designer had to relate critically to his client and leave some space for his own viewpoint; ultimately this would serve to improve the quality of the work. The result of the debate was that Crouwel designed the cover of the catalogue in the way he thought best, while Van Toorn decided on the contents, which he very obstinately reduced to just a folded poster.

Despite Crouwel's tenacity, Van Toorn seemed to have touched a sensitive spot. As a member of the board of directors at Total Design, Crouwel attempted to ward off criticism that their production was uniform by taking on several designers, including Jurriaan Schrofer and Anthon Beeke, who worked more intuitively. Earlier Beeke had made designs for the provocative teenage paper *Hitweek*, which from 1965 was published under the editorship of Willem de Ridder.[35] Beeke had lasted only one year at the Amsterdam Art Academy due to his free-and-easy, unconventional views. He felt more at home with the anarchic Fluxus movement in the fine arts, and believed that anything was permitted in the graphic field too. Anthon Beeke worked for years with Swip Stolk, who endorsed similar free, highly imaginative views. Their imagery in the 1960s was influenced by the revaluation of Art Nouveau and Art Deco taking place at the time.[36]

The desire for more freedom in those years was stimulated even more by the circumstances: the prospects for improved technology were greater than ever before. Between the early 1960s and the early '80s, the time-honoured typesetting craft was gradually replaced by present-day computer typesetting. Modern printers using the offset process and photosetters changed graphic design into a process of drawing, cutting and pasting, often involving free composing with the aid of transfer letters and the repro camera. Traditional setting by hand using lead letters had most definitely become a thing of the past.

Wim Crouwel expressed his views on the subject in 1974 in his essay *Ontwerpen en drukken* (On Designing and Printing). His text was the first in a series of publications on graphic design published by the Gerrit Jan Thieme Fund. Crouwel's instructor Dick Elffers followed with *Vorm en tegenvorm* (Form and Anti-Form), after which in 1977 the designer Piet Schreuders dropped a bombshell with the publication *Lay in, Lay out*.[37] The free autodidact Schreuders reacted against the functionalists and could see nothing at all in Crouwel's claims for timelessness and objectivity. Neither did Schreuders have a good word to say about his New Alphabet letter design from 1967, a font suitable for the computer's digital base, remarking wittily that it was 'one of the few letter-types in the world requiring subtitles'.[38] But Schreuders had just as low an opinion of the free, more illustrative course

Dick Elffers, poster advertising the Holland Festival, 1960.

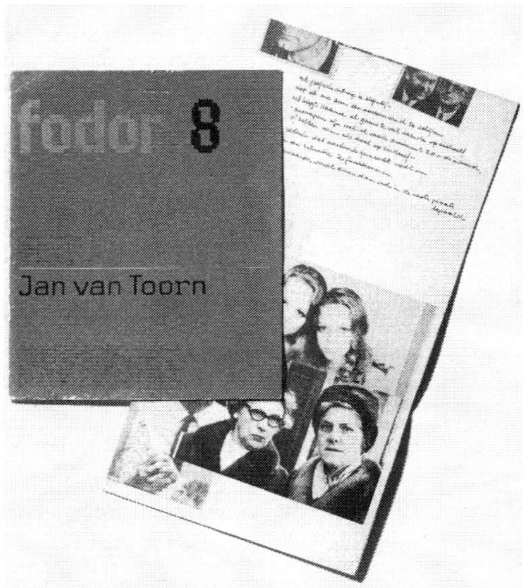

Wim Crouwel (Total Design), cover of *Jan van Toorn* catalogue (Museum Fodor, Amsterdam, 1972), including a folded poster by Jan van Toorn.

advocated by Anthon Beeke and his followers, who wanted to do everything differently and for whom nothing was too far-fetched. That, he stated, was an equally abhorrent trend; they were artists and decorators, who did not work at typography skilfully, but were aiming to elevate it to free design.

This controversy about modernism continued in the late 1970s when a discussion flared up as a result of the modernized telephone directory designed by Total Design.[39] The new order of personal details, and the adoption of the sans serif font Univers, of which only the lower-case letters were used, were capable of stirring up a strikingly wide range of reactions. The debate this time was carried out in the public media. The writer and journalist Renate Rubinstein (under the name Tamar) devoted a few critical articles

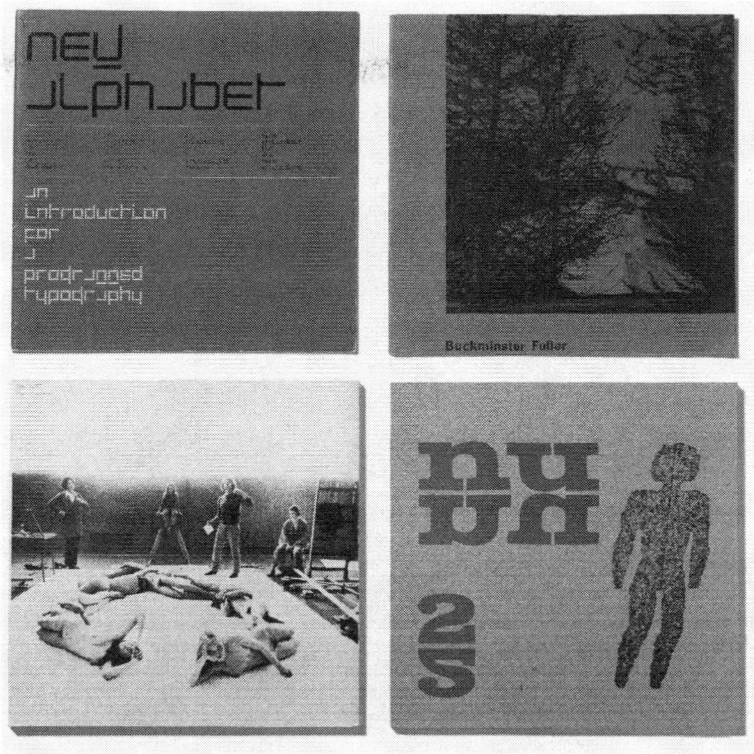

Four 'Square Papers' ('Kwadraatbladen'), (Steendrukkerij de Jong & Co., Hilversum): Nr 23: Wim Crouwel, *New Alphabet* (1967); Nr 9: Buckminster Fuller (1958); Nr. 28: Anthon Beeke, *Alphabet* (1970); Nr 26: Willem Sandberg, *Nu 2* (1968).

to the topic in the weekly paper *Vrij Nederland*, expressing her loathing of the colourless uniformity of the work produced by Total Design. In this discussion she introduced the slogan 'Nieuwe Lelijkheid' (New Ugliness).

A more productive dialogue was carried on almost uninterrupted in the graphic world itself: many distinct designs provoked a response in the form of a counter-design. For years on end the appropriate platforms for these artistic controversies between graphic designers were to be found in the *Kerstnummers* (Christmas editions) of the *Drukkersweekblad* (Printers' Weekly), the *Kwadraatbladen* (Square Papers) published by Steendrukkerij de Jong & Co in Hilversum and the calendar sheets from the printing firm Spruyt and de Erven Van de Geer. Also, since 1925 the annual election of the 'Fifty Best-Looking Books' had provided the opportunity for designers, printers, publishers and other representatives of the graphic industry to engage in a debate with one another about design and typography, but unfortunately this tradition was interrupted in the crucial period between 1971 and 1985.[40]

The *Kwadraatbladen*, which appeared between 1955 and 1974, edited by Pieter Brattinga, aesthetic adviser to the printing firm Steendrukkerij de

Jong & Co., were devoted to various cultural subjects.[41] Each number was designed by a different graphic designer. For the printing firm the *Kwadraatbladen* did not function primarily as a platform for discussion, but chiefly as a public relations tool and an opportunity to carry out experiments. The *Kwadraatbladen* owed their name to their square format (25 × 25 cm) and were designed by such as Willem Sandberg, Harry Sierman, Jan Bons and Otto Treuman. Foreigners like Marc Chagall, R. Buckminster Fuller and Dieter Roth also contributed. Occasionally they responded directly to an earlier published issue. In 1967 Wim Crouwel's New Alphabet provoked further reaction from Gerard Unger, as well as from Schreuders. In a hand-written text Unger made a case for making existing fonts more suitable for computer use, making reference to the fact that Wim Crouwel's progressive alphabet was barely legible. After this Pieter Brattinga asked Anthon Beeke to present a new font in a *Kwadraatblad*. His provocative response was an alphabet composed of naked women which he had carefully laid out in the shape of Baskerville letters on the floor of a large gymnasium, and then photographed.

The various counter-cultural amateur magazines and papers set up at the time by young artists, primarily art students, were redolent with comparable rebellious acts. The mimeograph, the photocopier and even a small offset press all came within the financial reach of many designers in that period. The influential magazine *Hitweek* was renamed *Aloha* in 1969 and continued until 1974.[42] In this form it focused a little less on pop music and more on graphic design and the new underground strip-cartoon culture. In imitation, *Tante Leny presenteert* (Aunt Leny Presents) also made room for a

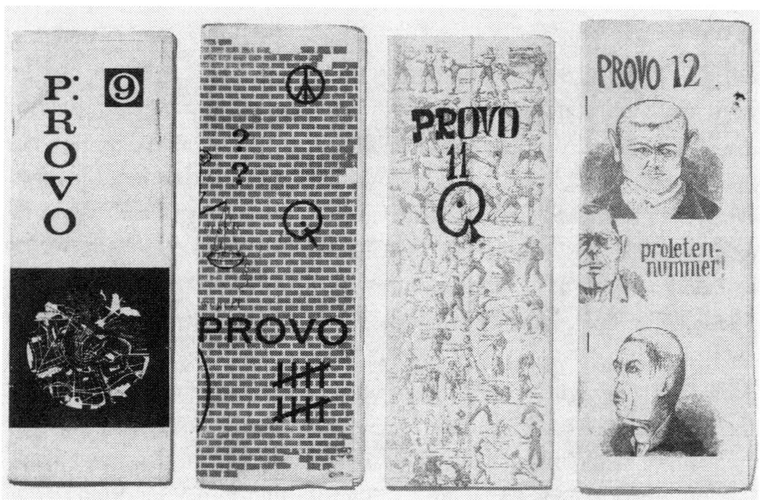

Issues 9 to 12 of *Provo*, magazine of the 1960s Provo movement, Amsterdam, 1966.

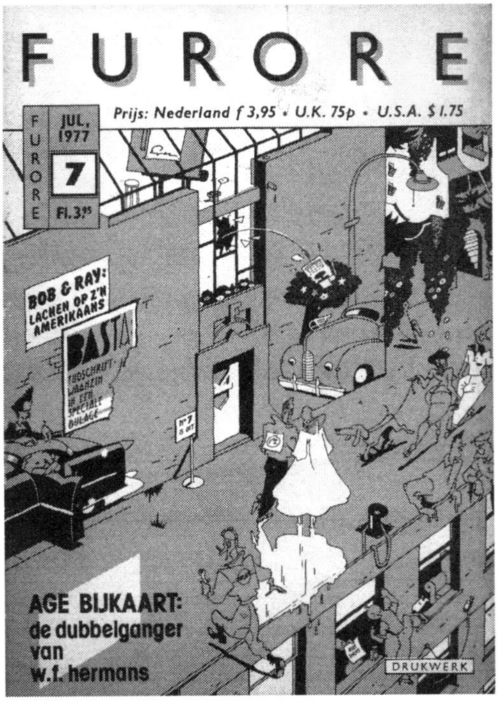

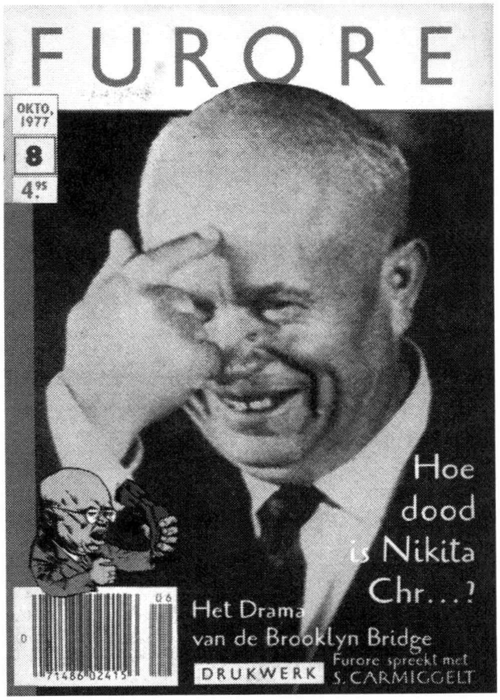

Piet Schreuder, two covers of his magazine *Furore*, 1977 (issues 7 and 8).

new generation of strip cartoonists between 1971 and 1976, including Joost Swarte and Piet Geradts. *De Wolkenkrabber* (Skyscraper), *Furore* and the *Poezenkrant* (Pussy Paper), which Evert Schreuders published on his own with only small print runs, were alienating and recalcitrant. *Utopia,* started in 1976 by a few architecture students from Delft Technical University, was equally saturated with this carnival atmosphere so typical of the 1970s: it was given the subtitle *tweemaandelijks tijdschrift voor wetenschappelijk amusement* (Bi-monthly Paper for Scholarly Amusement). Two editors of *Utopia*, Hans Kamphuis and Jan Pesman, later started up the design periodical *Items*, while their fellow editors on *Utopia*, Peter de Winter and Hans Oldewarris, founded in 1983 the publishing house 010 (the telephone code number for Rotterdam), which is still the most important Dutch publishing house for books on architecture and design.[43]

Students from the art academies in Amsterdam, Rotterdam and Enschede set up design collectives in the early 1980s that turned out to have staying power. Hard Werken (Hard Work) in Rotterdam, Wild Plakken (Illegal Bill Posting) in Amsterdam and De Enschedese School (The Enschede School) in Enschede worked in an eclectic, unconventional and experimental style. Their sources of inspiration were to be found in the

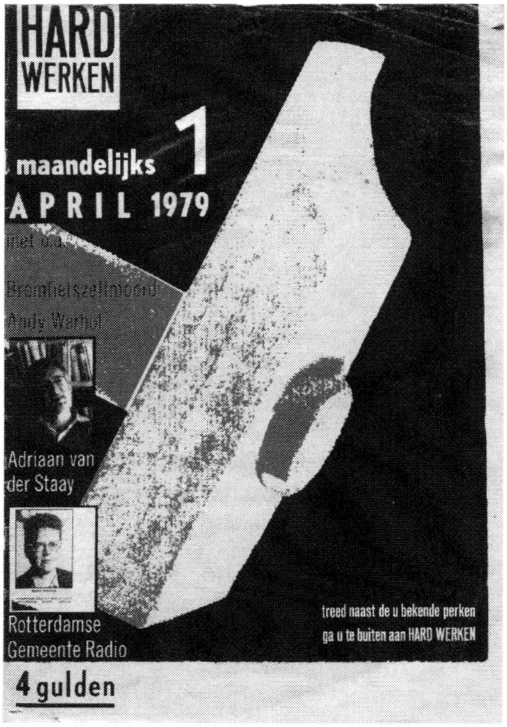

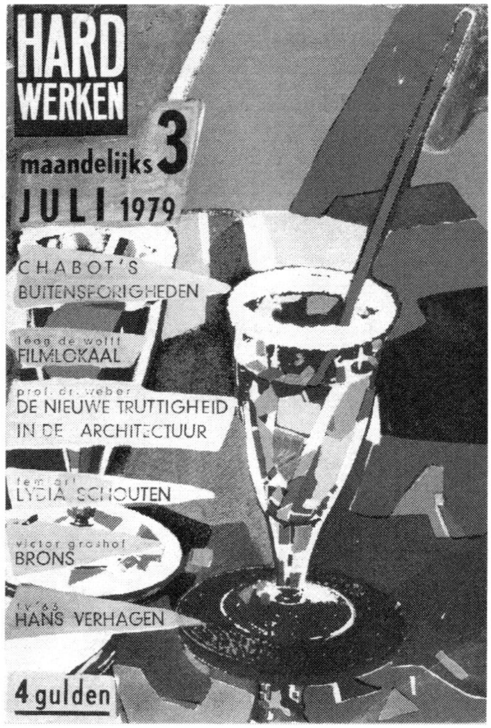

Two covers of *Hard Werken*, 1979 (issues 1 and 3).

world of films, strip cartoons and popular culture, but also in the history of avant-garde graphics.[44] Hard Werken was not interested in spreading an explicit ideological message and, with designers like Gerard Hadders, Rick Vermeulen and Henk Elenga, the commissions soon started to roll in from trade and industry and the government. On the other hand, Wild Plakken, with designers like Lies Ros and Rob Schröder, both pupils of Jan van Toorn, did stand for a definite political point of view. Initially their clients were chiefly the women's movement and the students' movement, the Dutch Communist Party and the squatters' movement. Later their commissions mainly came from cultural organizations, such as theatre groups, museums and the Netherlands Opera, and from the PTT. When it came to accepting new commissions during this period, the originally Amsterdam-based designer Gielijn Escher (grandson of Jacob Jongert) was also guided by his personal preferences and ideals. But for the rest, his carefully designed and colourful posters are quite unlike anything else at the time.

It is beyond dispute that the graphic sector in the Netherlands gained a great deal from this free and open climate in the 1970s and '80s. Furthermore, there was no lack of broad-minded clients prepared to commission work

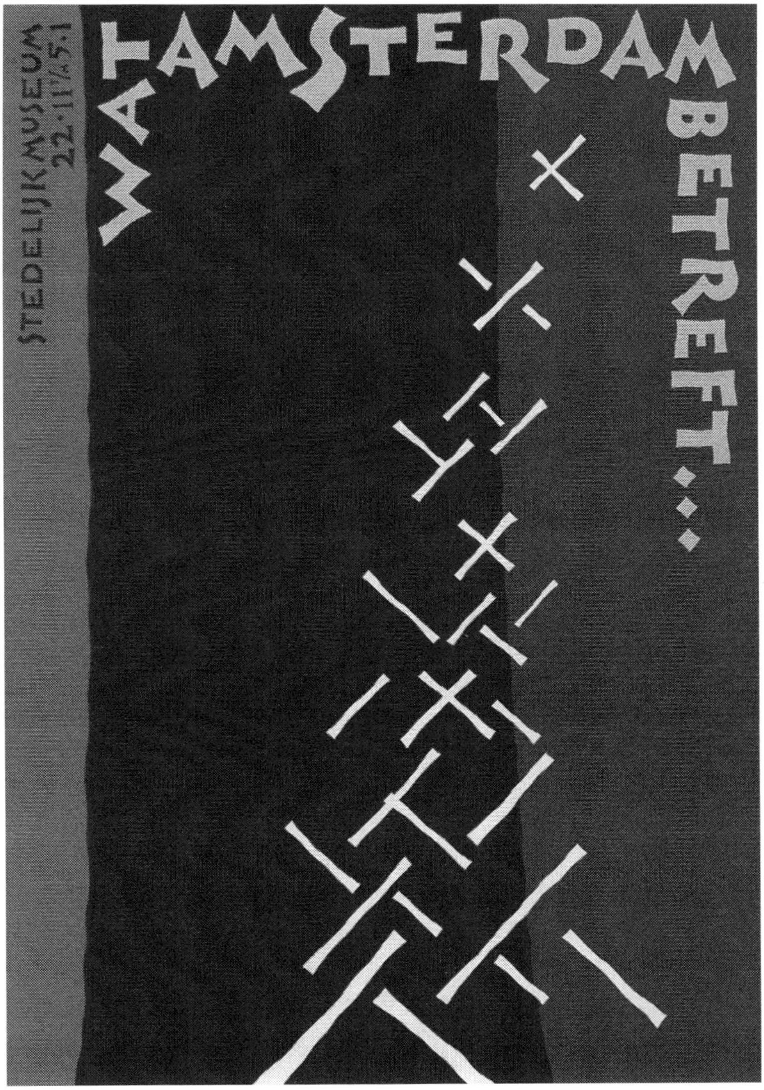

Gielijn Escher, *Concerning Amsterdam* poster, 1985.

from these designers. It is partly due to this support that for decades the most internationally esteemed areas of Dutch design have been graphic design – advertising – and book design, typography and commercial printing.

Culture or Economy

In contrast, Dutch product design was less successful during the 1970s and '80s. Two government-aided travelling exhibitions were organized in the early 1980s to generate more interest in modern consumer products among the business community, potential buyers or partners, as well as among the cultural scene and ordinary consumers. Quite unintentionally, these two presentations stimulated one of the design debate's crucial discussion points: is design a cultural or artistic phenomenon, or should it be considered primarily as an economic activity?

In the exhibition *Design from the Netherlands/Design aus den Niederlanden*, mounted in 1981, design was presented without hesitation as a cultural affair.[45] Organized by the Fine Arts Abroad Agency (Bureau Beeldende Kunst Buitenland), a sub-department of the Ministry of Culture, Recreation and Social Welfare, the exhibition highlighted the work of the creative designer, the designer-artist, whose products could enrich the life of the consumer, not only in a practical way, but especially artistically. The exhibition was mounted with a view to it travelling abroad, which it did for six years, calling at Stuttgart, Groningen, Brussels, Düsseldorf, Jerusalem, Helsinki, Stockholm, Budapest and Berlin.

As guest curator, Gijs Bakker was responsible for choosing the participants. By the 1970s he was not only involved in jewellery, but had also developed an interest in product design. Bakker worked for various Dutch factories, including the furniture factory Castelijn. He also taught at the Arnhem Art Academy right up to the end of the decade. Considering his own fascination with the design of jewellery and furniture, it was only natural that Bakker should select products for the exhibition based on what he considered to be an explicit and original visual concept. Whether the products he selected were a commercial success, or reasonably priced, or satisfactory in terms of the technicalities of the production process, did not matter, nor did he pay much attention to whether the objects had been produced industrially, or by craftsmen, in small series or piece for piece. Some well-known names were to be found among the twenty designers selected, including Friso Kramer with his Mehes office system for Ahrend, Aldo van den Nieuwelaar with light systems for Artimeta and Frans de la Haye with a bicycle for the firm Union, an experimental prototype that had been constructed with the aid of steel cables. Also represented were a few large design offices including Philips's CIDC, the Premsela Vonk studio and Kho Liang Ie Associates. It was remarkable that so much attention was focused on fashion and textiles, even though fashion in the Netherlands had yet to spring to life and thrive. Among the select were the fashion designer Frans Molenaar, the textile designer Ulf Moritz, the 'jumper-knitter' Marijke de Ley and the artisanal shoemaker Charles Bergmans. The later internationally successful children's clothing atelier Oilily (then still called Olly) was also part of the

Noudi Spönhoff and Loek Kemming for Mieke Teunen Design Vertrieb, poster, *Design in the Netherlands*, 1982, showing desk lamp by Herman Hermsen, 1979; hanging lamp by Vormgevers associatie, 1980; Tethrahedron lamp by Frans van Nieuwenborg/Martijn Wegman, 1977; DK Clock by Vormgevers associatie/Paul Schudel, 1980; tube lamp by Bruno Ninaber van Eyben, 1977; case by Maria Hees, 1980; neck watch by Bruno Ninaber van Eyben, 1976.

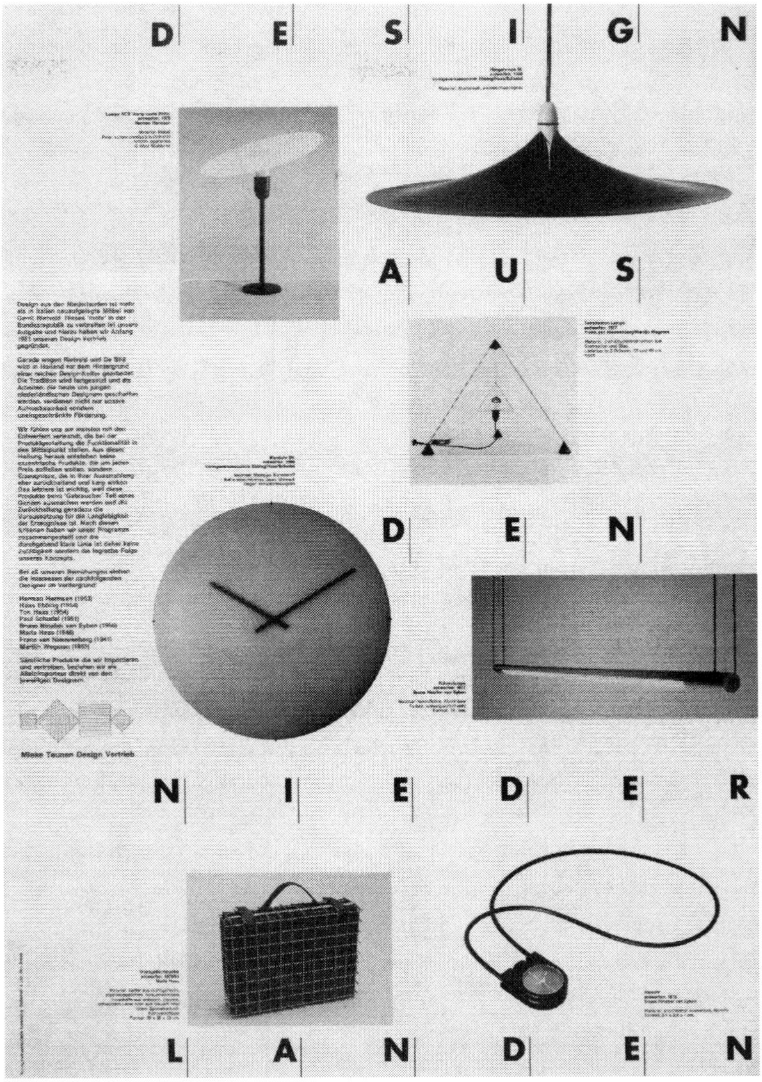

chosen group. The most surprising entries, however, were from the young designers who had chosen to make their products on their own, such as Bruno Ninaber van Eyben, Maria Hees, Henk Lampe and Hans Ebbing, Ton Haas and Paul Schudel (the Designers' Association). Reinder van Tijen stood out from the rest, presenting appliances intended for use in developing countries, made from scrap iron and waste, including a cement mixer from an oil drum and a pair of bellows from an old car tyre. The only ceramics in the exhibition came from Jan van der Vaart, who had been working on a collection of practical vases produced using traditional methods. Gijs Bakker was himself represented at the exhibition by a few chairs that he had designed for Castelijn. But he also showed his surprising umbrella lamp, one of the first Dutch objects with a hint of irony that was later to become so characteristic of Dutch Design.[46]

In 1983 the second travelling exhibition, titled *Ontwerpen voor de Industrie* (Designs for Industry), started its tour at the Bonnefanten Museum in Maastricht. Here industrial design was primarily presented as an economic activity. Visitors to the exhibition could become acquainted with a collection of high-quality products that a team had developed with the support of a problem-solving industrial designer. The Ministry of Economic Affairs supported this exhibition, which went on show in eight Dutch municipalities.

The compilation was in the hands of Wim Crouwel, the graphic designer Gertjan Leuvelink and the product designer from the firm Océ van der Grinten, Louis Lucker. The selection criteria employed may be found in the book published to accompany the exhibition, *Ontwerpen in Nederland 1* (Design in the Netherlands 1); the number one in the title suggested that a sequel to the project might have been on its way, but in actual fact it never materialized.[47] The objects exhibited in Maastricht were totally different from the products selected by Gijs Bakker. In the first place it was strictly stipulated that for this exhibition the objects on show, without exception, had to have been industrially manufactured. The other conditions were enumerated in a long list, including usefulness, safety, longevity, ergonomic adaptation and environmental friendliness. All were given a higher priority than the aesthetic quality of the design.

In Crouwel's introduction this view of the design discipline is put forward straight away. He advocates drawing up a series of systematically ordered conditions to be met each time a new product is required. This should be followed by an analytical approach to the design process whereby, as a matter of course, some thought should be given to the environment and the social and cultural circumstances. According to Crouwel, designers were often too critical in dismissing commercial principles used in large-scale industrial enterprises, and in their rejection of styling and corporate identity. Moreover, many young designers mistakenly considered good design to be an elitist phenomenon. Crouwel did not consider Italian or Scandinavian products to be a good example: the Italians attached far too much importance to design, whereas the Scandinavians had got bogged down in their modern tradition, which was far too rooted in Arts and Crafts. The English and the Germans provided better role models because,

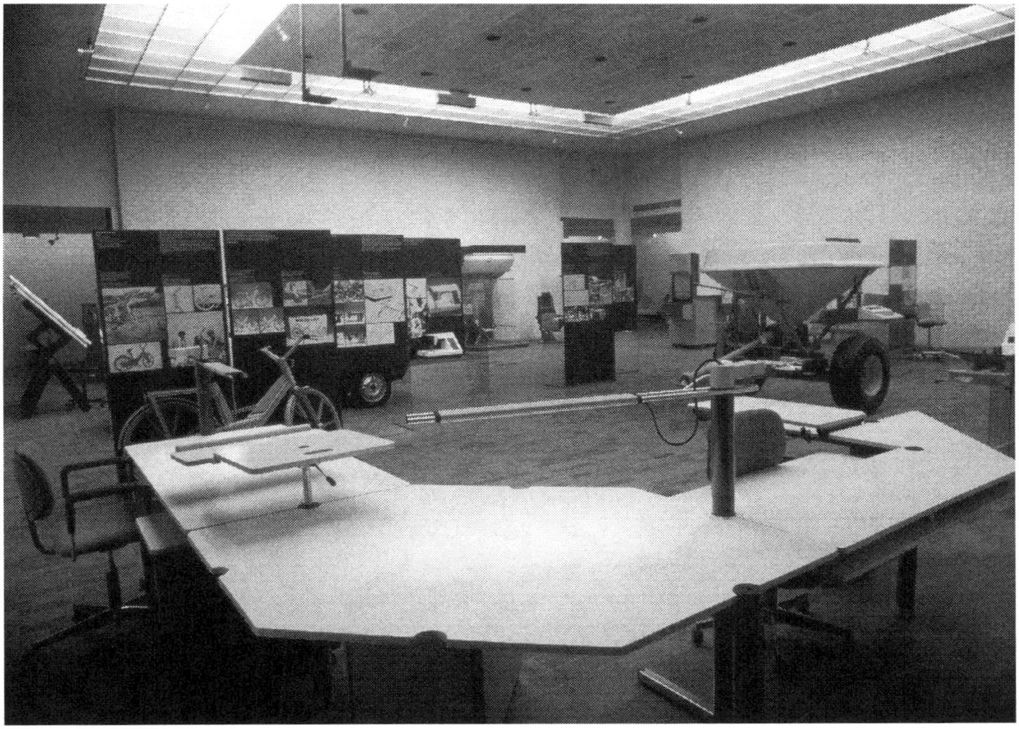

A view of the *Designs for Industry* exhibition (Schiedam, 1984).

according to Crouwel, their industrial design had acquired a name for being a 'relational and systematic design process planned with optimal economic efficiency'.[48]

At the exhibition in Maastricht itself about one hundred products illustrated this view. The visitors' interest in the discipline was to be stirred by means of office furniture, pans, clocks, toasters, calculators, automatic coffee-makers and wall spotlights. They could even see a crane, a compressed air dryer and medical measuring equipment. Only three products had already been represented in Gijs Bakker's touring exhibition: the Ahrend office system by Friso Kramer, the 'Auronde' bed for Auping by Frans de la Haye, and the 'Lagos' couch by Kho Liang Ie Associates, produced by Artifort in Maastricht.

Two different trends in Dutch design thus started to take shape. The Maastricht presentation *Ontwerpen voor de Industrie* crystallized the view that had already been formulated and institutionalized shortly after the war, seeing industrial design as an instrumental and interdisciplinary activity. It was this view of the discipline that was mainly taught at Delft Technical University, where Wim Crouwel had been a professor since

1972.[49] It was expected of the highly trained, well-informed and cultured designers who trained in Delft that on their own, or as part of a company design team, they would contribute to ensuring that trade and industry thrived and the country's economic prosperity increased. At the same time they were expected to look after the interests of the consumer. In this vision the designer was no more than a fairly anonymous cog in the wheel of trade and industry, devoid of glamour and without too many artistic pretensions, but still carrying a great amount of responsibility.

Ontwerpen in Nederland was a mouthpiece for the designers who opposed this somewhat technocratic view. Quite different questions were raised here, such as the cultural importance and artistic potential of a product, the broader cultural and social responsibility shouldered by the designer, and the question of whether there might be another characteristic style apart from the modernist idiom.

While the first group of classical industrial designers actually represented the great majority of designers in the Netherlands who continued to work in a steady way, the second, smaller, but more critical group of designers has attracted national and international attention over the last twenty years. This is the group to which the label Dutch Design has been attached, the designers who have featured in the press and who for a larger public have fixed the image of Dutch design. Without doubt, they have given the discipline an important new stimulus. It is seldom, however, that mention is made that by far the largest part of the Netherlands is, and was, designed by a much larger group of anonymous colleagues.

Designer-Makers

At the exhibition *Ontwerpen in Nederland* a few so-called designer-makers were represented. At the time this was a comparatively new phenomenon in Dutch design culture. Their motive for taking the production and distribution of their designs into their own hands was that initially there was a shortage of commissions coming from industry, although artistic and ideological motives also played a role.

Bruno Ninaber van Eyben was quick off the mark in following this unpredictable path. After having trained as a jewellery-maker at the Art Academy in Maastricht, he worked for a while for the silver factory Van Kempen en Begeer as assistant to the head designer, Gustav Beran. This factory was too narrow in scope for his innovative ideas, so in 1971 he opened his own design and production studio. Ninaber van Eyben's well-known early designs comprise a few simple bracelets and a necklace-watch in elementary forms made from Perspex, rubber and stainless steel, which for the jewellery of the time were innovatory materials. These he made in small series and distributed under his own name. This system, however, meant that he was not in a position to win attractive orders from trade and industry. His inspiration stemmed from the need to make a completely new type of product, one that you were more likely to identify with and not just a standard industrial product. The innovatory quality of his work lay not only in its clever design, but also in the material he chose and the experimental technology he applied. Its quality was quickly recognized. In 1979 Ninaber van Eyben was awarded the Kho Liang Ie prize, a distinction conferred for the first time that year, for his then still modest œuvre.[50] Gijs Bakker was on the jury whose report stressed the innovative character of Ninaber's products and his unusual working methods. As a teacher at the Art Academy in Arnhem, Bakker stimulated his students to adopt the artist-cum-entrepreneur's attitude. He preferred them not to become dependent on industry, with its presumed concomitant constraints, because in his view that was likely to become suffocating. In the late 1970s the Arnhem course produced so many students adhering to this attitude that they were referred to as the Arnhem School.[51]

Items, a journal launched in 1982, offered a platform for young designers and those starting up in business to discuss their own new products and ideas in print. After one year the original concept was broadened, enabling the editorial staff to cover design history, exhibitions and all sorts of news items related to the field. *Items* was involved in mounting the exhibition *Furniture from the Netherlands, 1980–1983* in the Bouwcentrum in Rotterdam. Here a great deal of the idiosyncratic work by young designer-makers was on show for the first time. But it was here, too, that a few Dutch manufacturers proved they were open to modernization and change. The Dutch Design Centre (see chapter Four), for example, used this occasion to show the newest products from the various associated furniture companies.[52]

A year later the Fodor Museum in Amsterdam organized the exhibition *Ontwerpen in de marge* (Fringe Design), where about forty designer-makers were able to show their work.[53] As in 1972, when the modernization of graphic design was the central theme, Fodor organized a panel discussion specially for the occasion – and, as before, feelings were running high. The graphic designer Rob Schröder, known through the designers' collective Wild Plakken (Illegal Bill Posting), was one of the panellists, as was Frank Oosterhof of the Enschede School. The Dutch Labour Party politician Felix Rottenberg came up with proposals whereby the government would be able to support these new initiatives, whereas fellow party member Arie van der

Zwan questioned the wisdom of granting that type of government support too nonchalantly. This evening session in the Fodor Museum was not an isolated incident and was followed by other debates and panel discussions on the theme. It was obvious that the designers' community in the Netherlands had been forced to sit up and take notice. Frederike Huygen, editor of *Items* and design curator of the Boijmans Van Beuningen Museum, wrote an article in which the designer-makers were put on the design map as a real movement or new trend.[54] Almost all the designers featured took the view that producing their own ideas themselves, in small series, provided them with more creative scope than just sitting around waiting for industry to show some interest. Moreover, it was patently obvious that these potential clients were not keen on the prospect of innovatory design, nor were they eager to experiment with materials and techniques.

The Design Association (Vormgevingsassociatie) was one of the most high-profile groups using these new working methods in practice. Founded in 1978 by Hans Ebbing, Ton Haas, Paul Schudel and the graphic designers Loek Kemming and Noudi Spönhoff, they were encouraged by their teacher at the Arnhem Art Academy, Gijs Bakker, to produce their own designs under their own name if commissions from industry should fail to be forthcoming. Developing a product entirely on their own and experimenting with forms and techniques, unhindered by the commercial limitations of a factory, did indeed stimulate their creativity. The Minimalist globule clock (DK klok) by Paul Schudel is one of the group's most well-known designs. Made from a slightly bulging plate of sandblasted glass, behind which just two hands indicate the time, the clock crystallized the idea of 'visible immateriality'. It was brought onto the market under the brand name Designum. Kemming and Spönhoff took care of the packaging and the publicity.[55]

The architect Mart van Schijndel became one of the most successful designer-makers without deliberately setting his sights on it.[56] He first exhibited his Delta vase (1981), made from three rectangular pieces of glass, as a one-off in Hans Appenzeller's gallery in Amsterdam. The public reacted so enthusiastically that he and his wife had to glue together about a thousand vases in the following months in order to meet the demand. He signed each one. When demand for the vase continued, its production was contracted out and to date tens of thousands of Delta vases have been sold.

An almost equal success was experienced by the architect Rob Eckhardt when he turned his hand to furniture design.[57] His furniture, sold under his own name, caught the eye of the progressive public and attracted a following that could appreciate these original, postmodern designs. Popular pieces

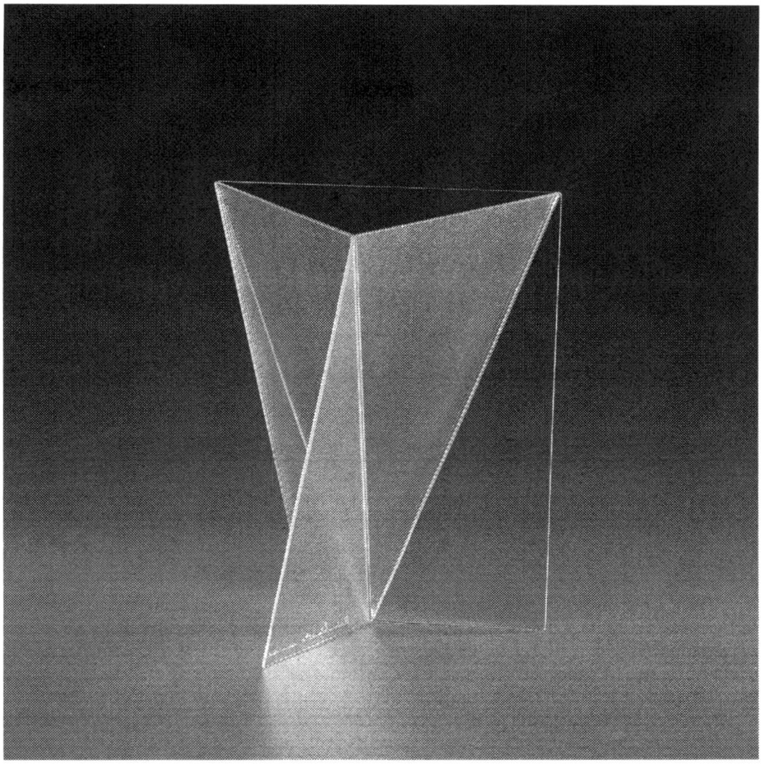

Mart van Schijndel, Delta vase, 1981.

include his asymmetric chair, titled 'Groeten uit Holland' (greetings from Holland), and an equally asymmetric chaise longue given the name Karel Doorman (a renowned Second World War Dutch admiral). From 1982 Eckhardt sold his own products, and those of a few selected avant-garde designers, from the design shop he opened in the centre of Amsterdam.

Many other designers are also worth mentioning, such as Maria Hees for her alternative bags, Jeroen Vinken for his original carpets and curtains, the small-scale studio Bon Bon for their silk screen-printed curtaining and Peggy Bannenberg for her jewellery.[58] Producing such items on a small scale under their own name, during the 1980s these designers struggled out of the straightjacket of functionalism that had dominated the industry for so long. Although most of them still maintained a predilection for simplicity and a geometric style language, they no longer gave functionalism and affordability the highest priority. For them expressiveness, daring and creativity took the place of the purely commercial pursuit of a neutral, 'good form'.

The emergence of this new generation of designers ran parallel with Italian and postmodernist influences gaining an increasing hold on Dutch design culture. Although the relationship between Italy and the Netherlands has not been properly studied up to now, strangely enough it appears that interest in Italy as *the* design Mecca seems not to have been a notable factor. During the years of post-war reconstruction Italy underwent a period of economic growth and industrialization that was many times greater than that experienced in the Netherlands. Italian firms like Kartell, Pirelli, Olivetti, Artemide, Cassina and Fiat had thrown open their doors to allow experimentation more enthusiastically and in doing so had capitalized on the country's rich artisanal tradition, which was then still in existence. In contrast to the Netherlands, the Italian consumers were soon showing enthusiasm for the phenomenon of design.

Of course, the landmark products of Italian design also filtered through to the Netherlands. There was much interest in the Vespa scooter, the Fiat 500 and, particularly, in Olivetti typewriters. Gio Ponti, Marco Zanuso (who in 1956 arranged the Italian section of the industrial design exhibition in the Stedelijk Museum), Marco Bellini and Ettore Sottsass were prominent names in designers' circles.[59]

The product designer Andries van Onck and the graphic designer Bob Noorda were among the handful of Dutch designers who risked venturing to the south.[60] They were not to regret it. Van Onck, after completing an industrial design course in The Hague, studied for five years at the Hochschule in Ulm. From there he headed straight to Milan in 1959, becoming Ettore Sottsass's assistant at Olivetti; together they designed one of the first computers. Bob Noorda made a career for himself as a graphic designer in Italy, too. When he arrived the discipline was still in its infancy. He took care of the commercial graphics for, among others, Pirelli, the COOP and Agip. Working with the Italian designer Massimo Vignelli in the 1960s, he designed the signage for the underground in Milan, which was later adapted for use in New York and São Paulo.

The design authorities in the Netherlands thought that Germany was more important than Italy. The more modest design ideals of this country were said to suit the Netherlands better, while the innovations in Italy were regularly dismissed as mannerist and formalist. Supposedly, fear of too much competition also played a role in the attitude they took: the foundation of the EEC in 1958 had made it easier to import goods from Italy, as a result of which some sectors of Dutch industry, including ceramics, were actually under threat.

An exception, as seen earlier, was Kho Liang Ie, who did familiarize himself properly with the Italian market. The effect this had can be seen in his furniture designs for Artifort. Kho became a personal friend of Ettore

Sottsass, who in 1971 commended him at the opening of an exhibition devoted to Kho's work.

It was not until the early 1980s that Italy began to exert real influence on the design debate and on design in the Netherlands due to its new pluralist style. *Items* had already drawn attention to postmodern Italian products with their provocative design language and their surprisingly decorative and expressive presentation. After the Memphis group exhibition in 1984 in 's-Hertogenbosch, the editors of the *KIO-bulletin*, the designers' journal, also made a serious attempt to fathom out this new and confusing idiom: one without any familiar aesthetic connotations, but full of ironic references to popular culture, kitsch and classical antiquity.[61] From then on the glossy monthly magazine *Avenue*, with its much wider readership, regularly report on the new Italian trends.

In this same period the highly imaginative work of the Amsterdam-based Czech designer Borek Šípek became well known.[62] Like the Italians, Šípek was inspired by various styles and cultures. Sensory experiences and emotions were more important to him than functionality or affordability. His glasswork in particular was soon widely known in the Netherlands. In 1989 Šípek was awarded the Kho Liang Ie prize for his work. This was indicative of the changing views on design culture in the Netherlands that now placed Šípek among the elite.

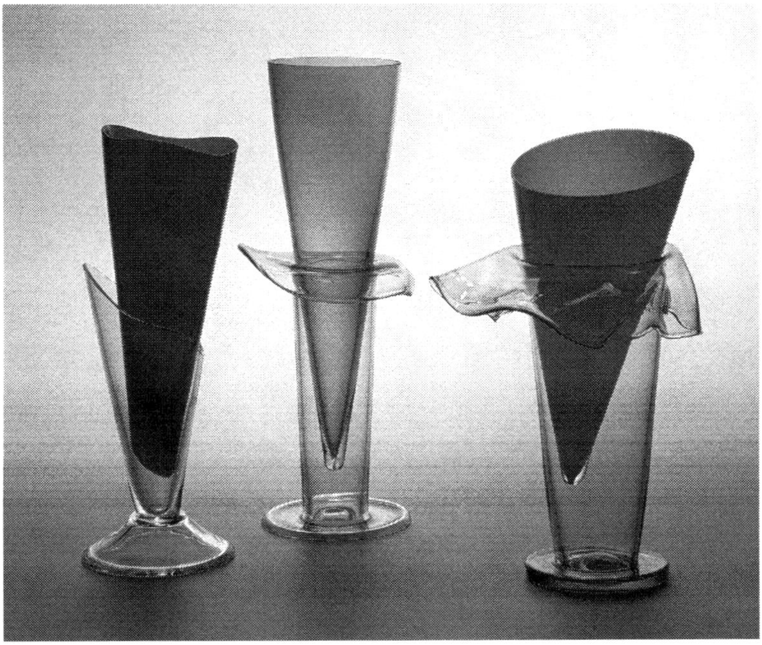

Borek Šípek (Anthologie Quartett, Germany), three glasses from the series 'Veno Pro Xenophona', 1983.

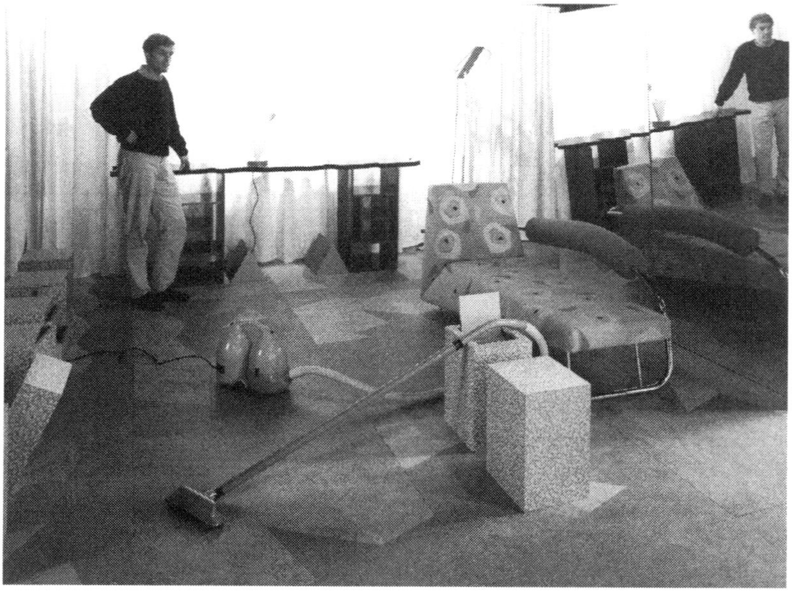

Ed Annink in a post-modern interior, designed by Annink for the Centraal Museum, Utrecht, 1984.

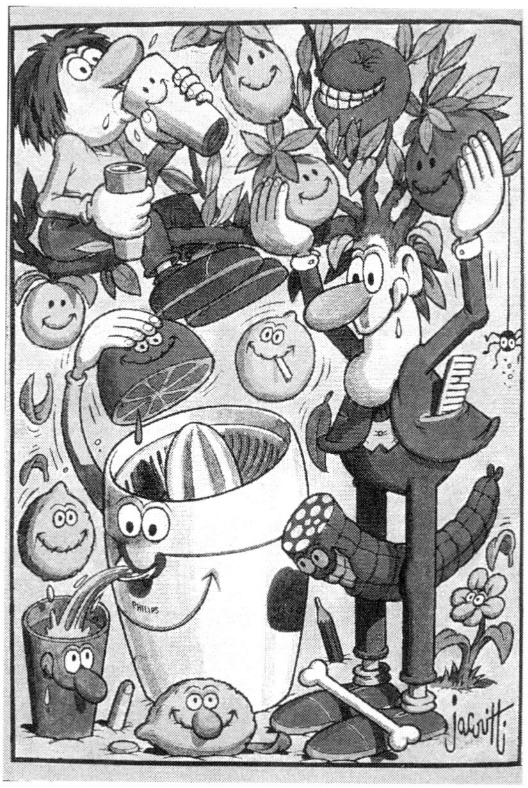

Cartoon used as a promotion when the Philips Alessi line was introduced in 1994.

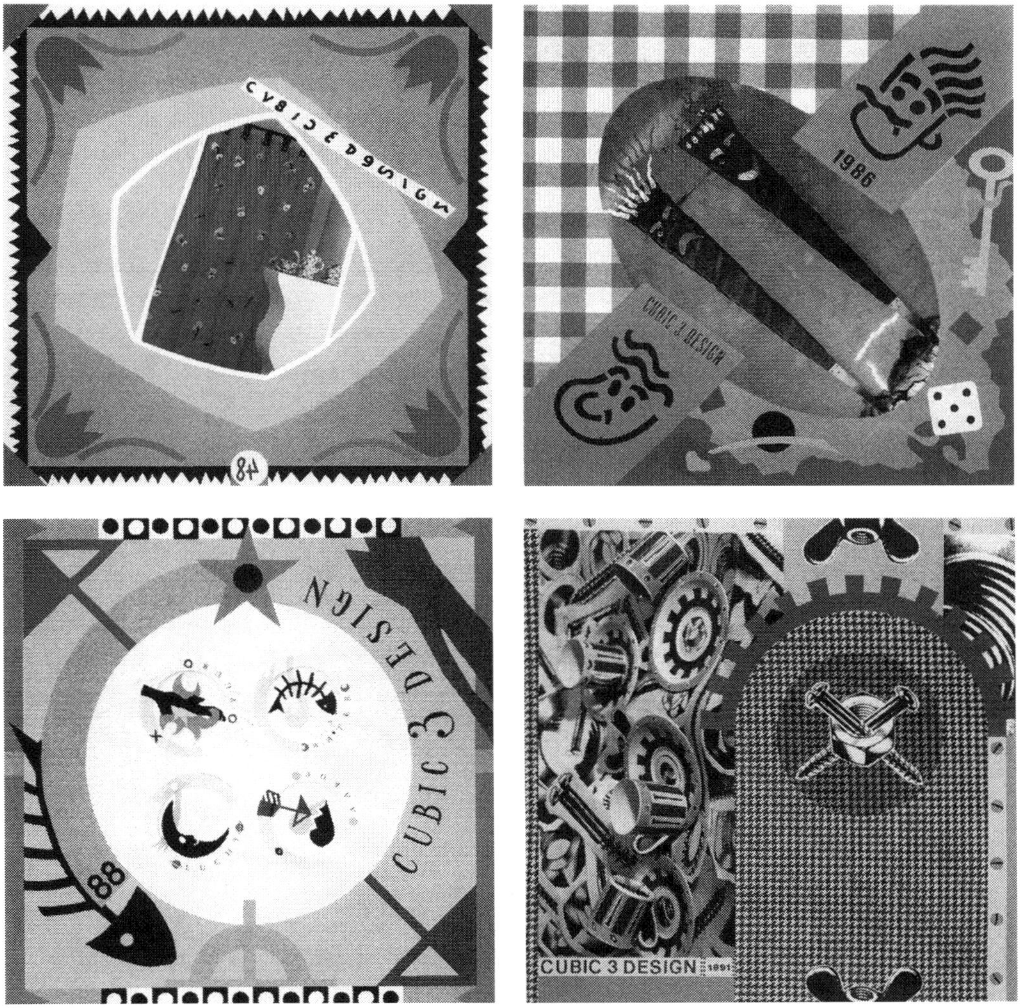

Handkerchiefs printed on the occasion of the tenth anniversary of Cubic 3 Design (Gerwin van Vulpen and Ton Hoogerwerf) (1991) from *Man* (1991).

Shortly afterwards, in imitation of the Italians, a few Dutch designers also started to think that the product's message was more important than a beautiful or functional form. Ed Annink and Rob Eckhardt were notable among this group, but there were also Ton Hoogerwerf and Gerwin van Vulpen (with their Cubic3 Design bureau in The Hague), Peer de Bruyn, Bob Verheijden and the young Marcel Wanders.[63] Ed Annink's early designs included a lectern in the shape of a butterfly (1982), which symbolized the transitory quality of the spoken word. In 1984 he was commissioned by the Centraal Museum in Utrecht to select the fabrics and furnishings for a postmodern period room.[64] A prominent place was reserved in it for the 'Groeten uit Holland' chair by Rob

Eckhardt. The chair was upholstered in a material printed with a brightly coloured tulip pattern produced by Cubic3 Design. Perhaps the most obvious signs of the influence of Italian postmodern design were to be found in Cubic3's work, with their predilection for colourful decoration, bizarre, imaginative shapes and historical citations. This influence later filtered through in a weaker, more diluted, form to many other Dutch products, including Pastoe furniture, Philips's electric appliances and the HEMA's household goods.

Not everyone was equally enthusiastic about the rise of this new 'unprincipled' design being produced by the Italians, which in fact made the Moral Modernism crisis complete. Hein van Haaren, for example, who in 1984 held the influential position of aesthetic adviser to the PTT, understood the need for something new, but thought that the postmodern alternative was elitist and pretentious:

> Who are these objects made for? Certainly not for the popular-culture-man who cannot recognize the baroque profusion of symbols, nor the irony and ambiguity of the abstract language being used. Neither is it for those who out of fascination follow developments in art critically. For them all the information carried by neo-design is old hat . . . Functionalism deserves a more profound response than the coquettish design of the Milanese.[65]

Free Design: Design as Art

During the 1980s the borders between art, craft and even fashion became blurred due to the influence of Italian postmodernism and designer-makers who were attracting a great deal of attention. The distinction between the disciplines had already become indefinite in the preceding two decades (see above). Whereas earlier many potters and weavers working in the old tradition had moved into the field of art, during the 1980s industrial designers also began to make small-scale series, or even unique, objects. A good illustration of this is the renowned furniture designer Martin Visser who, after his Spectrum period from 1955 to 1974, had principally occupied himself with collecting art. From 1979 to 1983 he was a curator in the Modern Art department at the Boijmans Van Beuningen Museum. Around 1986 he returned to his old trade and with his partner, the textile designer Joke van der Heijden, he made a collection of monumental chairs and tables. Although these items were most certainly usable, they were principally well-thought-out autonomous objects and studies in form, colour and material, rather than products for everyday use.[66]

Jan van der Vaart, five vases, 1953–90.

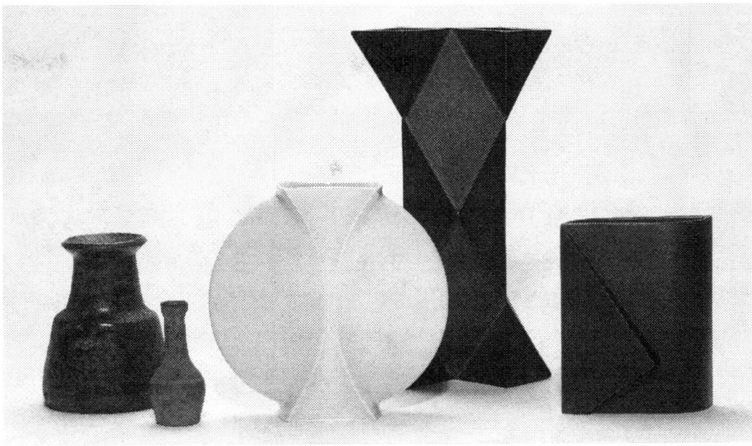

However, a movement in the opposite direction also became visible. In 1986 the Amsterdam design gallery Binnen organized an exhibition of the work of fifteen potters who showed great industrial potential.[67] With this exhibition Jan van der Vaart, Vincent de Rijk, Pauline Wiertz and others hoped to attract the attention of the producers. Indeed, even completely autonomous artists started to set their sights on the world of design. After some unique pieces of furniture had been presented for the first time in a traditional free art setting in 1987 at the Documenta in Kassel, the Boijmans Van Beuningen Museum organized the exhibition *Het meubel verbeeld* (Furniture as Art) in 1988 with work by such internationally famous artists as Sol Lewitt, Donald Judd and John Armleder, complemented by furniture by the Dutch artists Carel Visser, Paul Beckman and Frank Mandersloot. In 1990 Het Kapelhuis in Amersfoort followed with *Gebruiksbeelden* (Useful Sculptures) and in 1992 the Commanderie van St Jan Museum in Nijmegen did likewise with their exhibition *Meubelsculptuur* (Furniture Sculptures), where work by artists such as Thijs van Kimmenade, Wilma Sommers and Frank Bezemer could be admired.[68]

This new category of art objects did not continue to be a purely museological affair. To the great dismay of tourists – and the disgust of many residents – in 1991 the centre of Amsterdam suddenly saw the arrival of a large number of new cast-iron lamp-posts, litter bins, benches, bicycle racks and bollards intended to prevent illegal parking, all designed in an unorthodox manner that had little to do with the Dutch tradition. The totem-pole elements from which they were made appeared to be arbitrary shapes piled on top of one another, then spray-painted in an arresting bluish-green and adorned with gold accents. They were the work of the

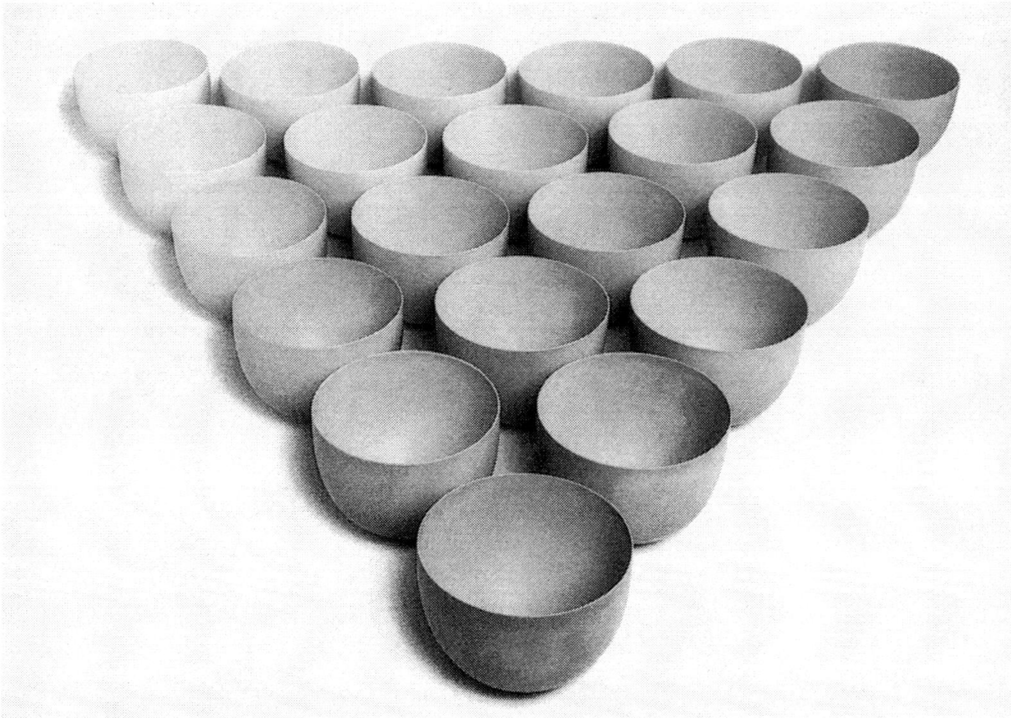

Geert Lap, installation of twenty-one stoneware bowls in different colours, 1988.

sculptors Alexander Schabracq and Tom Postma, who had not been guided by the motto 'less is more', but by 'less is a bore'. The vehement discussion that this work aroused was welcomed by the artists: they had never had such a frenzied reaction to their autonomous sculpture![69]

In the early 1990s the Rotterdam sculptor Joep van Lieshout made his entry into design too. His straightforward tables, cupboards and bath elements in lively colours, made from rough polyester and cheap blockboard, had an equally puzzling effect. Was this art? Was it design? Whatever the answer may have been, his work was a great success both at home and abroad and commissions from private individuals and cultural institutions poured in. In addition, Van Lieshout worked on various projects for the Office for Metropolitan Architecture (OMA) run by Rem Koolhaas, including, in 1994, designing two bars and the toilet areas for the Congress Centre in Lille, France. From 1995 these projects went under the designer's name Atelier van Lieshout (AVL). Meanwhile, the artist had become the leader of a workshop employing twenty employees.[70]

The final new development in the 1990s was the emergence of the Netherlands as a fashion nation, marked at the same time by the fusion of

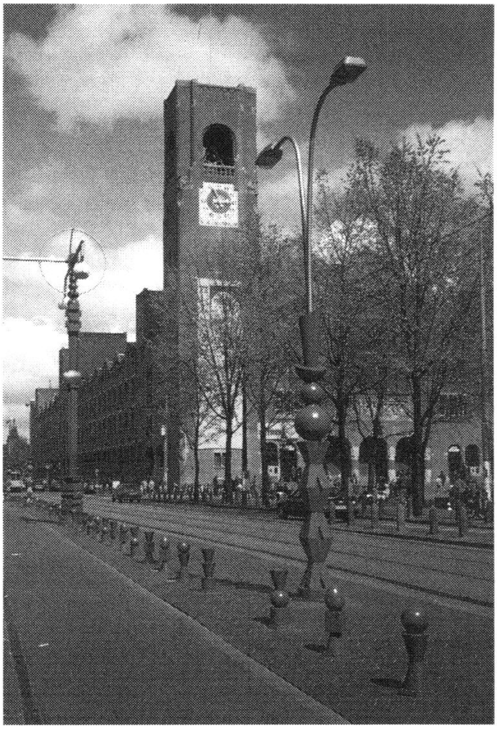

Alexander Schabracq and Tom Postma, street furniture on the Damrak in Amsterdam, 1991.

fashion with art and design.[71] In this area the Arnhem Art Academy again played a pioneering role. Alexander van Slobbe, for instance, was trained there in the 1980s. His garments, under the brand name Orson & Bodil, appeared to be simple, almost abstract clothes; nevertheless, they could also be seen as pioneering art experiments, textile constructions with more consideration than usual paid to the material used and the way it fell into pleats.[72] The six graduates of the Arnhem Academy in 1992 – Marcel Verheijen, Lucas Ossendrijver, Viktor Horsting, Rolf Snoeren (Viktor & Rolf), Saskia van Drimmelen and Pascal Gatzen – showed a promising collection. They joined forces the following year, calling themselves Le Cri Néerlandais, and had great success with their group presentations in Paris and Milan. For these shows they were given every conceivable form of support by a group of authoritative figures from the Dutch fashion and design world: the photography for this successful promotion campaign, for example, was by Rineke Dijkstra. In 1994 the Netherlands was represented at *prêt-à-porter* shows in Paris for the first time by seven fashion

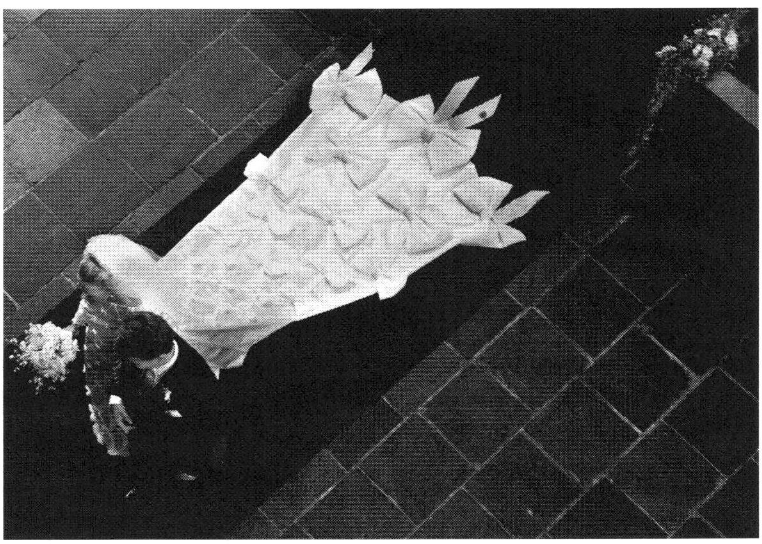

Prince Johan Friso and Mabel Wisse Smit's wedding, Delft, 24 April 2004. Wedding dress designed by Viktor & Rolf.

designers – Alexander van Slobbe plus the six from Le Cri.[73] Meanwhile, Van Slobbe has become an internationally respected, commercially successful designer. Viktor & Rolf, with their numerous high-profile presentations, have also earned themselves a prominent position, operating on the borderline between fashion, design and art.[74]

The new cross-border tendencies in design were snatched up eagerly in official art circles. In 1985 the Netherlandish Office for Fine Arts (Rijksdienst Beeldende Kunst) spent a major part of the government's acquisition budget on the procurement of many of these exciting products, which could be called neither pure art nor pure design. The large museums also took the step of extending their, still young, collections of industrial design with these types of one-off art objects or, at the very most, objects produced in small series.[75] At the end of the 1990s, the Boijmans Museum even went as far as purchasing a few fashion garments on the grounds that they belonged to the new definition of contemporary design, which had extended to cover a larger area than before. In addition to the existing galleries of applied art, new ones were opened that operated in the specific area between art and design: Binnen and Frozen Fountain (Amsterdam), Vivid (Rotterdam), Puntgaaf (Groningen), Intermezzo (Dordrecht) and Galery Yksi (Eindhoven).

Critical articles about these new developments were regularly published in the periodicals *Bijvoorbeeld* and *Items*, and also from the late 1980s in *Industrieel Ontwerpen* (Industrial Design) and since 1990 in *Vormberichten*. The discussions printed here, however, were often typically ambivalent and irresolute: if beauty, functionality and affordability were no longer the criteria, what could objects still be judged by?

Conceptual Design

In 1992 work by young designers that seemed to be taking off in yet another direction was on show at the Gallery Marzee in Nijmegen.[76] The revolutionary factor in their designs did not lie primarily in their pioneering form, but more in their new ideas about the function and meaning of design. The way it was explained in the catalogue was that they were furniture designers who also wanted to be 'meaning-givers': 'Their pieces of furniture are not solely a chair, a table or a cupboard. They are designed ideas and experiences reflecting everyday surroundings and furniture art itself.' Not long afterwards this trend became known as 'conceptual design', that is, the idea behind it was more important than the design. Among the exhibitors in Gallery Marzee were Tejo Remy, Jurgen Bey, Jan Konings and Marcel Wanders, designers who were to become the stars of the Droog Design label.

This Droog Design collection was soon after to be considered the international business card of new avant-garde Dutch design, but its creation would have been unthinkable if certain developments had not already taken place in the 1980s. After all, the penultimate decade of the twentieth century had witnessed resistance to the impersonal nature of modernism, the struggle for more expression and emotion, the intermingling of design and art, and even a certain degree of conceptualization. However, the new element was the young designers' belief that the message conveyed in their work should refer especially to the product, the design and the technique itself, in order to stimulate the design debate.

In 1992, when Renny Ramakers, the editor-in-chief of *Industrieel Ontwerpen* (Designing for Industry), saw Jurgen Bey's and Jan Konings's bookcase made of rough-hewn wood, paper and textile in Gallery Marzee, she immediately felt that something new was happening. The two young designers had not been aiming at making something marvellous or a representative bookcase, but were merely in search of a simple solution for storing books in an orderly fashion. The bookcase's harmonica-style structure allows it to expand as the number of books increases. Equally pioneering were two objects shown by Tejo Remy at Gallery Marzee. One was the chest of drawers titled *You Cannot Lay Down your Memory*, a collection of old drawers, originating from a great many different cupboards and desks, tied together by a large strap. The other object was a chair made from a large pile of old blankets and clothes, which were kept in place by strong tape. Renny Ramakers placed these two objects and a cupboard by Piet Hein Eek, made of assorted pieces of scrap wood, on her magazine's stand at the furniture show in Kortrijk. The objects, with their mandatory hint of irony, elevated waste as an issue worthy of debate and propagated the artistic recycling of materials.

Following the overwhelming success of the small display, Ramakers decided to organize a presentation of these, and a few comparable objects, under the name Droog Design at the Salone del Mobile in Milan in 1993. The show was realized in close association with Gijs Bakker, who since 1987 had taught at the Eindhoven Design Academy. The birth of a new movement in design was then a reality.[77] A year later the Droog Design Foundation was established with financial support from the Ministry. The name Droog (literally 'Dry') refers to the supposed character of the products: subdued, straightforward and austere. Droog Design is not a designers' association or organization, nor is it a style. The aims of Droog, with its radical rejection of aesthetics in favour of an emphasis on the process of conceptualization and design, correspond to the aspirations of

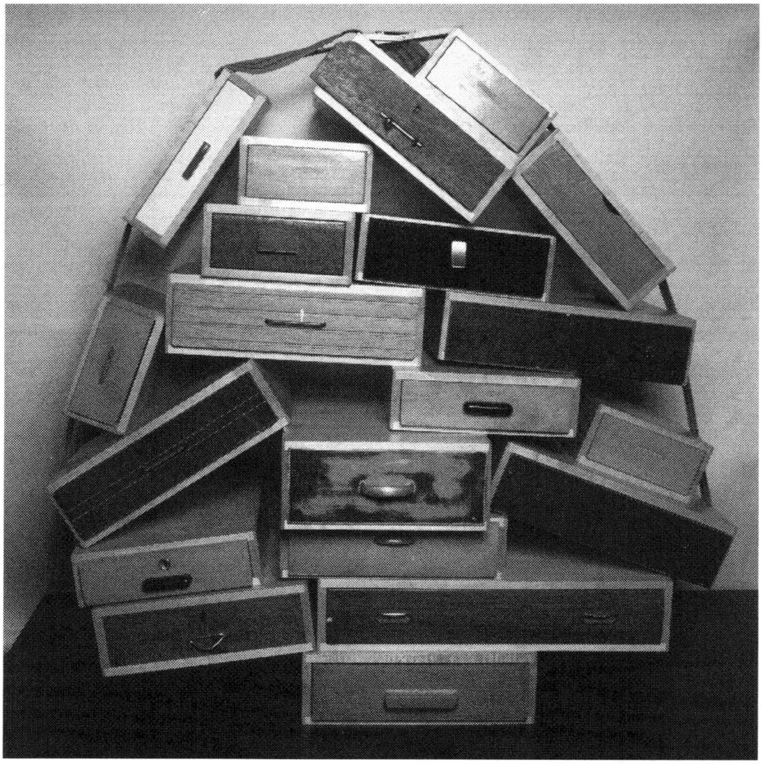

Tejo Remy (Droog Design), 'You Can't Lay Down Your Memory', chest of drawers, 1991.

Rem Koolhaas, who broke into the world of international architecture in the same period. Droog Design is formally a foundation run by Ramakers and Bakker, who select new products for the label on the basis of an explicit design mentality. Products are regularly chosen from young art academy students' finals projects, after which Droog supports their production and distribution and promotes them worldwide. Creation, innovation and debate have of late been at the centre of Droog's policy.

Over the almost fifteen years that Droog has been in operation, the accent has been on constantly changing themes. Initially their central focus was on 'recycling' old products, or parts of them, in a new design: new products like Eibert Draisma's automatic coffee-maker and desk lamp assembled from waste material, Tejo Remy's hanging lamp made from twelve traditional milk bottles and Rody Graumans's chandelier made from a bundle of 85 light bulbs on the end of a lead.[78] Jurgen Bey went a step further and fitted reflecting foil around all sides of an old chandelier, so that by day one sees a taut-lined modern lamp, but in the evening the traditional lavishly designed chandelier emerges. Some products were specially designed with a view to

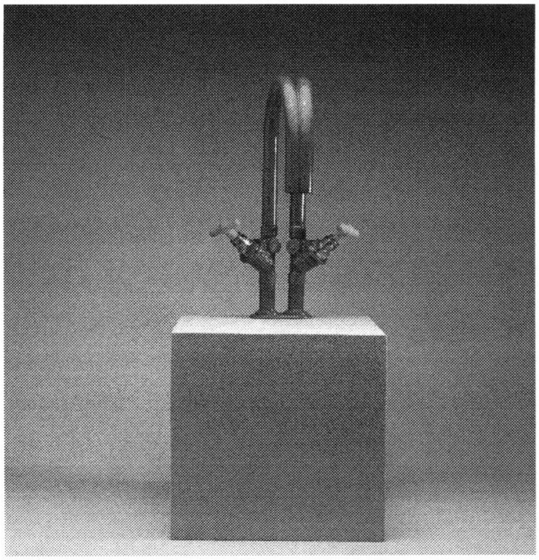

Dick van Hoff (Droog Design), tap, 1995.

being reused: Djoke de Jong designed a shower curtain with the pieces of a jacket printed on it, in the same way as a paper pattern.

The following theme in Droog's history was 'simplicity', an express quest for the most elementary, archetypal form of a product, or for designs that seemed to embody their function most concisely. At the same time, however, they are objects that radiate a sort of controversial, anti-design sentiment, as illustrated by Dick van Hoff's products and Richard Hutten's furniture.[79] In the case of Van Hoff, a water tap consists merely of two water pipes bent towards one another, and two traditional twisting-headed brass taps, thought to be a primitive form of a tap. In fact it is a reaction to the over-designed products that he detests, reacting against the type of product that mystifies the poor unsuspecting user, such as those taps that only start when you put your hands underneath them. Richard Hutten made chairs and tables under his motto 'no sign of design' that were also reduced to their most elementary form. Ineke Hans, a designer who has no connection with Droog, did something comparable, with her simple, pictogram-related furniture made from recycled plastic.[80]

Experiments followed with new materials or the unexpected uses of old ones. Marcel Wanders knotted rope to make a chair, which he then impregnated with epoxy resin to make it strong enough to sit on. He did something similar with a piece of traditional needlepoint lace, which was transformed into an elementary stool by using the same method. Hella Jongerius made washbasins and vases from rubber, but also embroidered on ceramics.[81] Saar Oosterhof devised springy floor tiles made of polyurethane and Arian Brekveld a hanging lamp made of soft PVC. In this way materials, techniques and functions were reconsidered, and then deployed in new ways. Sometimes this involved engaging the help of others, for instance, Delft Technical University's Faculty of Space and Aviation Technology, the Rosenthal porcelain factory (Germany) and Tichelaar's Pottery (Makkum), to provide the premises, money and wherewithal to realize revolutionary ideas.

Droog reached its peak in conceptual design in 2000 with the Droog *do create* collection. This comprised a series of objects that had to be 'finished off' or decorated by the user. A metal cube by Marijn van der Poll had to be

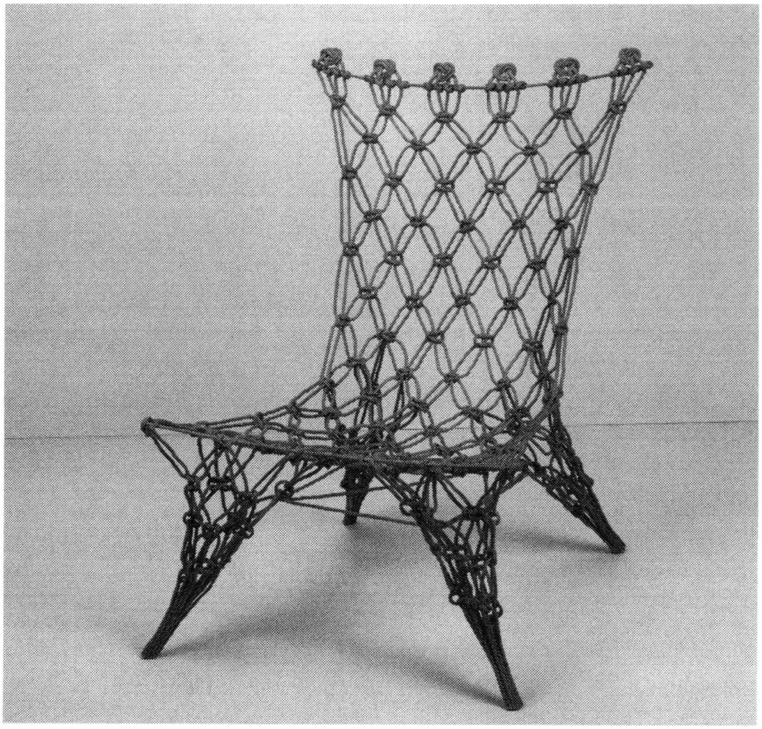

Marcel Wanders (Droog Design), chair knotted of carbon fibre and steeped in epoxy, 1996.

forcefully modelled into the shape of a chair with the aid of an enclosed mallet. A vase could be dropped and smashed into pieces, but its latex inner-wall kept it all in one piece. A chair by Jurgen Bey, with one leg much too short, had to be kept in balance by a pile of books. The whole collection was presented with the aid of a provocative publicity campaign run by the firm KesselsKramer. It was obvious, though, that these types of products would not reach the general consumer market.[82]

Most Droog products were very successful among those in the cultural circuit. They principally featured in art magazine articles, and their major purchasers were museums. Tejo Remy's chest of drawers was included in the *International Design Yearbook* (1994), compiled by Ron Arad, and was acquired in that same year by the Boijmans Van Beuningen Museum. In 1997 the Centraal Museum in Utrecht obtained all the products – more than a hundred – that had appeared under the Droog label up to that date. Furniture by Droog and by other comparable high-profile Dutch designers was used by the firm Opera Ontwerpers (Opera Designers) for the Museum of Modern Art's restaurant in New York, furnishing the Dutch Garden Café with the support of the Design Institute, the Ministry of Education, Cultural

KesselsKramer/
Droog design, presentation of a *Do-create* design for a chair by Marijn van der Poll, 2000.

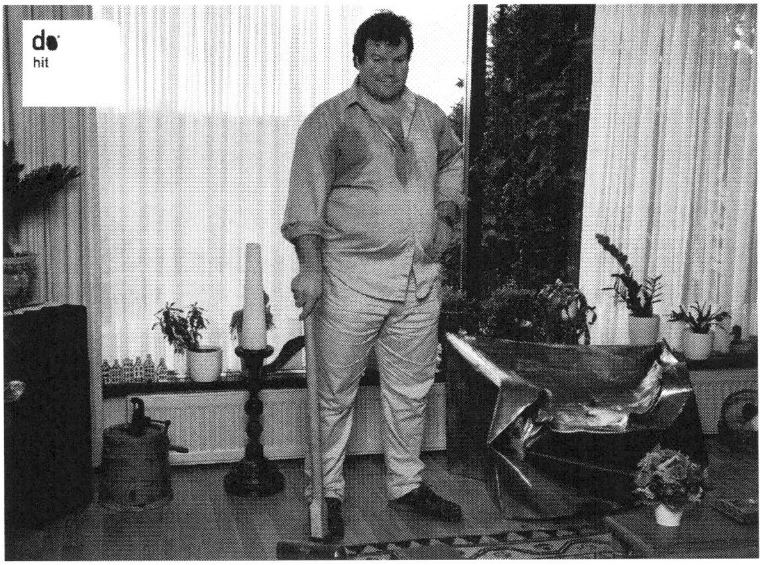

Affairs and Science (OCW), and the Dutch Consulate. The chairs and tables were by Piet Hein Eek, the lamps by Rody Graumans and the bar and serving trolleys by Joep van Lieshout. Also part of this effective propaganda campaign for avant-garde Dutch design was a presentation of Dutch posters.

Nonetheless, commercial results were not forthcoming, with the exception of a few incidental successes. Conceptual ideas in Dutch design, and Droog's conceptual ideas in particular, were on the receiving end of criticism.[83] Their idiosyncrasy and irony were regularly dismissed for

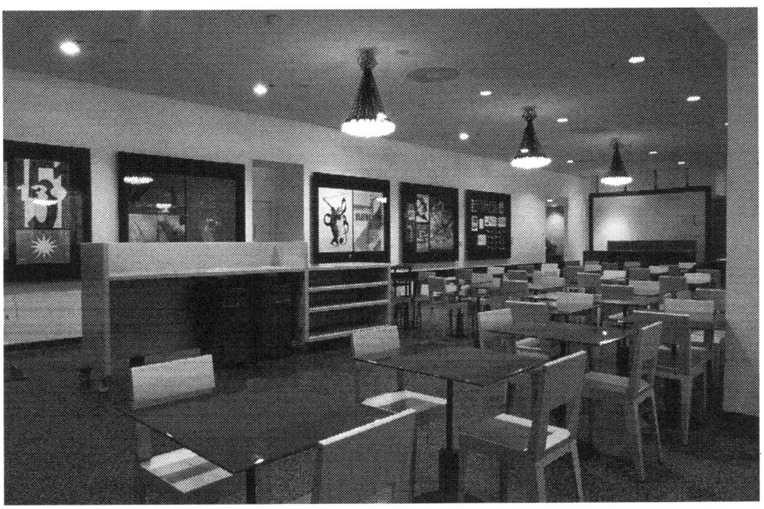

Dutch Garden Café at the Museum of Modern Art in New York with furniture by Piet Hein Eek, a bar by Joep van Lieshout and lamps by Rody Graumans, 1995.

being too non-committal, while this deliberate reflection on their own functioning was regarded as a dead end leading nowhere or a diversionary tactic. A number of critics were more inclined to regard their dogged search for the simplest form, or rather the product's alleged primitive form, as a sign of intellectual poverty rather than a durable idea. Moreover, they questioned whether the consciousness-raising the conceptual designers had hoped to incite was actually taking place. Were the Dutch en masse starting to think differently about design? Were styling and fashion judged more critically? Were people leading more frugal lives and were more products being recycled? Were consumers becoming more conscious of the materials used, and the technology employed, to turn out the products surrounding them in everyday life? Droog was repeatedly criticized for being a media machine. The designers were reproached for shirking their duty to society by wallowing in their non-committal stance, which was still acceptable in art circles.[84] At present, the feeling of euphoria that was induced by Dutch Design does not seem to have run its course completely, although in several places a re-evaluation of industrial design can be seen to be taking place. For the time being, however, it is too soon to draw any conclusions about this new development.

On a global level, Dutch designs not produced under the Droog label were also capable of success, as was shown at the exhibition *The Foreign Affairs of Dutch Design*, which was mounted by the Dutch Designers' Union (Bond van Nederlandse Ontwerpers, BNO) and the Premsela Foundation (Stichting Premsela) and travelled the world from 2004 to 2006.[85] In addition to the typical Droog products by Hella Jongerius, Marcel Wanders and Jurgen Bey, they also displayed the internationally successful Bugaboo children's prams by Max Barenbrug and Eduard Zanen, as well as Gerard Unger's letter designs, Jan Jansen's shoes, Alexander van Slobbe's fashion wear, Dick Bruna's children's books, KesselsKramer's publicity campaigns for Diesel jeans, Paul Mijksenaar's signposting and Marlies Dekkers's lingerie. The collection could have been enlarged to include many other hits too, since it would not have been amiss to include the clever Maxi Cosi by Huibert Groenendijk; Van Berkel's slicing machines, which could be found in butchers' shops all over Europe and had recently been redesigned by Well Design; the extremely popular Senseo automatic coffee-maker by Waac's Design for Philips and Sara Lee; and last, but not least, the Philips Compact Disc and its associated CD jewel case designed by Peter Doodson.[86]

Despite the decline in production of consumer goods, these and a few thousand other Dutch designers were still able to achieve much. The TNO

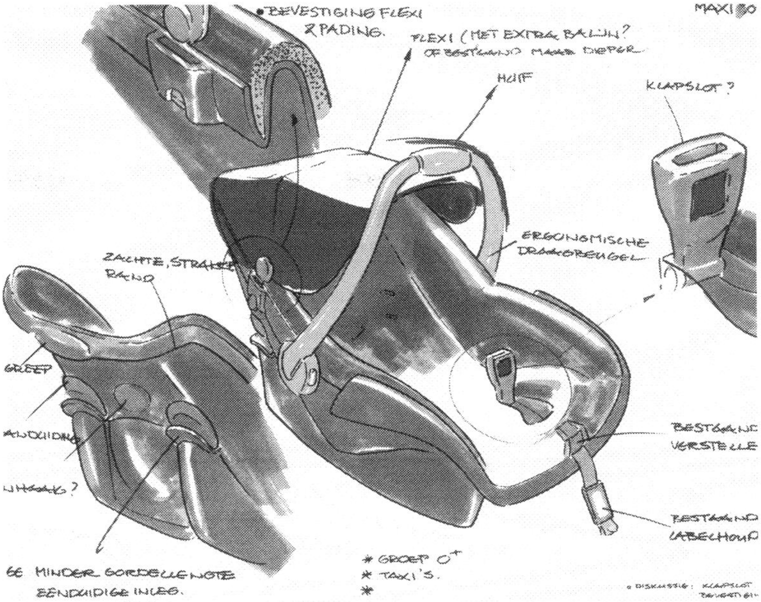

Huibert Groenendijk, designs for the Maxi Cosi Plus car seat, 1998.

report of 2005 quoted in the Introduction showed evidence of this: design is an important facet of the creative economy of the country, even though the production of many of these new designs increasingly takes place in low-wage countries in Eastern Europe and Asia, rather than in the Netherlands.

Two factors explain how, out of all these designs, those of conceptual Dutch design have been able to attract so much publicity despite their lack of commercial success. First, this has been down to the exceptional interest shown for these objects by those in design education. At the Academy of Industrial Design in Eindhoven, for example, which since 1997 has been known as the Design Academy, they have been focusing on conceptual design since the early 1990s. The original departments of product design, packaging and textiles (see chapter Four) were replaced by less commercial-sounding branches of study like 'Man and Leisure', 'Man and Identity' and 'Man and Living'. Since then creative invention and controversial points of view have been more appreciated in this 'house of concepts' than expertise in the technological side of production or analytical insight.[87] Job Smeets, Joris Laarmans, Maarten Baas and the twins Joep and Jeroen Verhoeven are among those who graduated from there not so long ago and achieved rapid worldwide fame.[88] An increasing amount of interest in this new outlook on design has been shown by other academies, too: in 1995, for example, a postgraduate Free Design course was opened at the Rietveld Academy in Amsterdam.

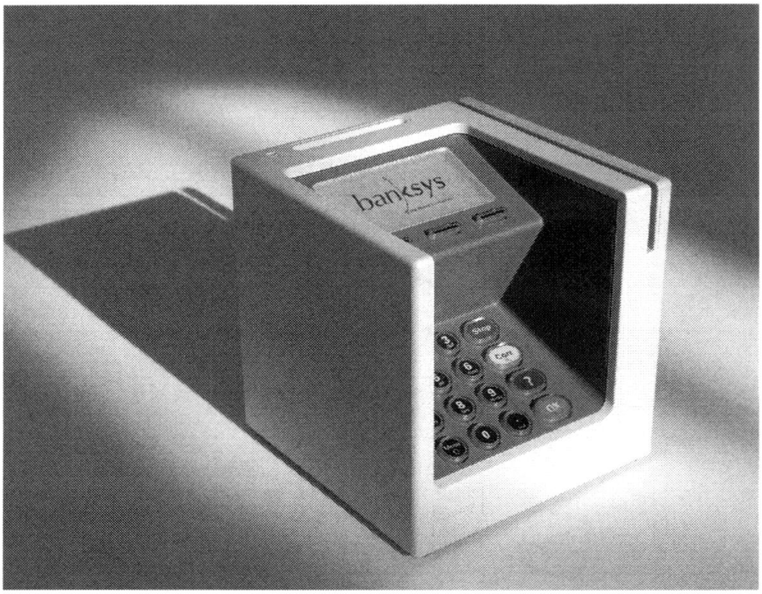

Well Design, 'Xenta' countertop pay terminal for Banksys Worldline Brand, 2005.

A second factor that was important for the remarkable success of Dutch design is related, particularly in the eyes of foreigners, to the unique subsidy culture in the Netherlands. As far back as the 1980s the Dutch government viewed design very much as a cultural asset, and in accordance with their convictions they embarked upon an incentives policy, believing that modern design was a good way of bringing people into contact with art and culture. This constituted a shift in government design policy from stimulating contacts with trade and industry to supporting individual designers and subsidizing exhibitions and publications aimed at attracting an audience. The Dutch Art and Design Fund can financially support a whole range of schemes thought up by designers and artists: with the aid of a starters' grant an office or studio can be set up, the equipment required financed by the Facilities Fund (Voorzieningenfonds) and the rent for a place to work reimbursed from yet another source. There are also separate grants available for taking part in international shows and exhibitions, and government-financed studios and workshops in various large capitals are available for a circumscribed period of time. In this way, with a little luck and talent, a young designer is granted a number of relatively carefree years in which he has the opportunity to build a name for himself.[89]

Activities and initiatives organized by institutions, foundations and publishing houses – mainly exhibitions and publications – are supported by the Mondriaan Stichting. Grants awarded by the Ministry of Foreign

Affairs (known as HGIS-Cultuurmiddelen), local funds and various municipal and provincial financial arrangements complete the picture. All this has led to the paradoxical situation that the more defiant, critical and non-conformist the designers or artists were inclined to be, the more strange, colourful, funny or extreme their designs became, the greater their cultural standing seemed to become, and the more financial aid they were able to attract. This investment in high-profile, pioneering design was seen by the government as a demonstration of social-democratic commitment to visual culture in the public domain.

Institutes and Prizes

Finally, the vicissitudes of the National Institute of Design and the way the large national design prizes are run are two subjects that have continued to preoccupy those in the design world. Both have provoked vehement discussion without ever arriving at viable conclusions.

After the war there were more than four national design organizations in the Netherlands (the exact number depends on the way such an organization is defined). After the IIV closed in 1976, the Industrial Design Foundation (Stichting Industrieel Ontwerpen, ION), the Dutch Form Foundation and the National Design Institute tried to function as such, but for totally different reasons they did not succeed. Since 2002 the Premsela Stichting, however, seems to have been successful. The ION, founded in 1984, had a typically pragmatic attitude and considered its main tasks to be the 'improvement of industrial design', as well as mediation between designer and trade and industry, and the search for a potential market. It was a platform for both the industrial designer and the business world. From 1985 onwards the foundation published the periodical *Industrieel ontwerpen* (Designing for Industry) and since 1987 it has conferred the annual recognition for Good Design for Industry on products from various sectors. The Dutch Form Foundation was initiated in 1989 to fill the space left open for theory, debate and the more artistic side of the discipline.

The national Design Institute (Vormgevingsinstituut), established in 1993, took a different direction with culture at its core. This organization chose to promote discourse in the field of design and focused on new media. Unfortunately, most of the discussions about the discipline initiated by the Institute were not ultimately about the subject, but about the way the institution itself functioned.

Regardless, interesting initiatives have been taken, such as the successful conferences about new media titled *Doors of Perception*, the initiatives of the *Young Designers & Industry*, the contacts with developing countries in the *Crossroads* project and its activities in the field of ecological design.[90] A great drawback, however, was that they appealed to only a very few Dutch designers and firms. Most of them pulled out of the organization. Their objection, in a nutshell, amounted to the fact that the Institute's activities were no longer about them. So, after the Institute had been in existence for ten years, the curtain came down for the last time.[91] It no longer had sufficient support. The Culture Council (Raad voor Cultuur), which had functioned since 1995 as the Dutch government's statutory advisory body on formulating cultural policy, published a pejorative report about the Vormgevingsinstituut and in 2000 it was closed down. The ION, which was still in existence, amalgamated with the Good Industrial Design Foundation (Stichting Goed Industrieel Ontwerp) in 2001 to form the Designlink Foundation.

In 2000 the Temporary Design Advisory Committee (Tijdelijke Adviescommissie Vormgeving) convinced the government that nonetheless a new Design Institute should be founded. This is why the Premsela Foundation (Premsela Stichting) was established in 2002. Its aim is to bridge the gap between designers, society and trade and industry, thus bringing culture and commerce closer together. The foundation has been in existence only for a short while, but it is beginning to look as if it has struck the right note and will be successful in improving the Dutch design climate and in integrating the fragmented world of design culture.[92]

There has also been a great deal of change in the employers' organizations and in professional associations in recent years. Since 1996 the Professional Association of Dutch Designers (beroepsvereniging Nederlandse Ontwerpers, BNO), incorporating the Industrial Designers Circle (KIO), has built up a membership of more than two thousand. Its first chairman was Wim Crouwel, the *éminence grise* of Dutch design.[93]

Finally, the specialist journals were confronted with an equally turbulent ending to the millennium, if they actually managed to reach it at all. In 1996 *Bijvoorbeeld* was discontinued. *Items* and *Industrieel Ontwerpen* amalgamated in 1993 to carry on jointly under the former title. *Vormberichten*, the BNO's mouthpiece, managed to survive, but had to submit to several editorial reorganizations. Since 1993 the Delft Faculty of Industrial Design has published the periodical *Product. Vaktijdschrift voor productontwikkeling* (Product. Specialist Journal for Product Development). Other new magazines are the luxuriously glossy *Frame*, mainly focusing on interiors, and *Identity Matters* (IM), a periodical devoted entirely to corporate identity, packaging and publicity. Finally, since 2005 the Premsela Foundation has published a collection of old and new texts on design biannually under the name *Morf*.

The second bone of contention was design prizes. Design competitions and the presentation of prizes and awards, some of which have already been mentioned, gave rise to heated debates. For many years there were different prizes for the various sectors in design. The Gouden Noot (Golden Nut) trophy, for example, has been awarded since 1958 for the best-designed packaging. Designers or organizations that had been important in the field of colour could win the Sikkens prize (Sikkens was an independent Dutch paint firm, and is now part of Akzo-Nobel).[94] Each year there is a presentation of the Fifty Best-Looking Books. However, the prize that has perhaps generated the most publicity since 1988 is the HEMA design competition, run by the popular chainstore selling clothes and household goods to be found in every Dutch town and open to designers still in training. The winning design must meet the HEMA criteria of being simple, nice and affordable; the prizewinning design then goes into production and is sold in their stores.

The Rotterdam Design prize was initiated in 1993 by the Rotterdam Art Foundation (Rotterdamse Kunst Stichting).[95] During the last decade of the twentieth century this was the prize that caused the most controversy. This competition is open to everyone and, once the entries have been received, a national nomination committee makes a first selection that is presented at the Design prize exhibition. The next step is that an international jury makes its final choice. The winner receives a large sum of money. Every year the entries cover the most divergent subjects, varying from books, jewellery, vases and lamps to a light buoy, a postman's carrier tricycle and even small buildings like a heat exchange unit. People send in traditional designs, functional products and examples of controversial conceptual design, although in recent years the balance is starting to tip in favour of the last category. This great diversity in itself invariably leads to many discussions. How can such different entries actually be compared with one another fairly? Is it not just an attempt to compare apples to pears?

Jeroen Bruijn and Tijl Akkermans of Thonik design studio, winners of the Rotterdam Design Contest 2007, with Lideweij Edelkoort of the Design Academy in Eindhoven on the right and the politician Jan Marynissen on the left.

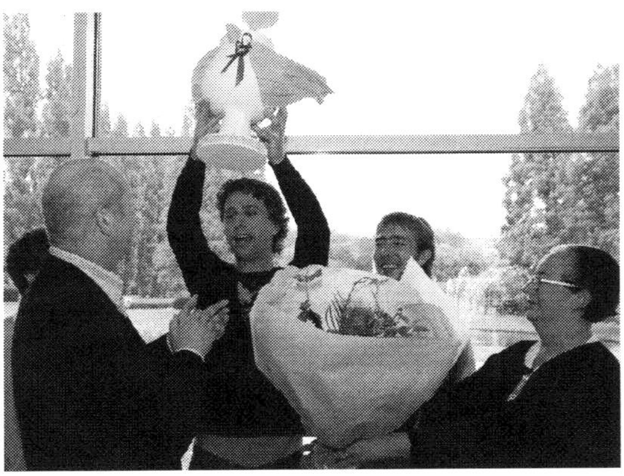

Despite all these objections, the nomination and jury reports still manage to give a good idea of the aspects that the jury in that specific year thought were important and what the criteria were for their judgement of the designs. Rather than giving an overview of the 'best' Dutch products, the successive selections have represented the way attitudes towards design culture in the Netherlands have shifted or, as the design critic Gert Staal stated in 2003: 'the selection procedure, no matter how painstakingly performed, was mainly a vehicle for supporting and presenting innovations in the discipline'.[96] This was in fact the idea from the very start, as the presentation of the first prize illustrates. The 1993 nomination report records new standards, in addition to the frequently used, traditional Dutch quality criteria of simplicity, sobriety and unpretentiousness. The selected entries were now also described as bold, daring and revolutionary, making it clear that Dutch design and the way it was judged were undeniably changing.

In 2003 many people were discontented with the disproportionately large amount of interest shown towards conceptual design, so much so that the Designlink Foundation, together with the BNO, took the initiative to organize a new, more pragmatic and commercial design competition. In this new annually organized Dutch Design competition, entries can be submitted in sixteen different categories. Unlike the Rotterdam Design prize, the organization remains in close contact with trade and industry, once again arousing discussions and debates.[97]

Conclusion

In 2007 the Dutch government promoted a new history syllabus programme, with Gerrit Rietveld's slatted chair as one of the central logos, under the heading 'Revolution in Design', reinforcing the chair's iconic status to school pupils. Yet whichever way you approach it, choosing what is most characteristic of Dutch design – or indeed its most characteristic design object – remains debatable.

This book has attempted to illustrate how design has been characterized in Dutch culture, as well as how designers and manufacturers have responded to social and economic conditions. While the book cannot offer an encyclopaedic survey or a comprehensive quantitative study, it has tried to highlight how design has been subject to certain interpretations and how, throughout the twentieth century, it has been subject to under-exposure as well as over-exposure. If a firm failed to submit an entry to the Paris Exposition Universelle (1900), like many important Dutch firms at that time, no interest was generated in its products and consequently nobody wrote about them. The reverse also holds true: many publications in the 1920s and '30s overrated Berlage's role in the modernization of design. As we have also seen, the choice of illustrations in VANK publications was controlled, while the Goed Wonen Foundation's and the IIV's policies in the mid-twentieth-century were in fact determined by a very small group of fellow-believers; later, the interpretation of subsidy schemes run by the government in the 1980s and '90s may have been biased. This holds true to this day: we will need time and a certain amount of objectivity to examine critically how appropriate the enormous amount of recent interest has been in Dutch design.

What does remain undisputed, however, is that nowadays in all layers and all sectors of Dutch society there has been a steadily increasing interest in design. The outer appearance of products, indeed the entire visual and material culture in the broadest sense of the word, plays a far greater role in the everyday lives of the Dutch today than it did a hundred years ago. Design has become an expression of identity, with more attention paid to the decor of personal and public life than ever before. The public pay more attention too to the designers themselves, familiar as they are not only with international celebrities such as Philippe Starck, Ettore Sottsass and Karl Lagerfeld, but also Dutch names such as Benno Premsela, Wim Crouwel, Marcel Wanders and the fashion designers Viktor & Rolf, especially after the haute-couture team designed the wedding dress of one of the Dutch princesses a few years ago. Names from the past like Willem Gispen and, of course, Gerrit Rietveld have steadily filtered through to the public's collective consciousness.

As regards which themes, as opposed to which names, have been prevalent in the history of Dutch design, I will hazard a few conclusions drawn from the preceding chapters. It is striking how persistent certain topics in design culture have proven to be. Contemporary designers sometimes forget that the issues preoccupying them today were at the forefront of discussion over a hundred years ago. One example is the role of handicraft. Throughout the whole of the last century designers have displayed an almost unremitting, sometimes romantically tinged, devotion to handiwork and small-scale enterprise, which they felt allowed room for personal expression and the production of unique objects. Even in the 1950s and '60s, when industrialization was welcomed with open arms, designers continued to cherish handicraft. Another example relates to how the design of everyday objects and appliances should relate to something approaching 'art'. Throughout the entire century there proved to be designers whose main ambition was to create artistic artefacts, who felt that the design of a good, beautiful or handy household object was not enough of a challenge.

We have also noted that artisan or artistic design has always attracted far more publicity than anonymous mass-produced articles, no matter how functional, commercially successful or beautiful these may have been. This has led to a rift between the two branches of design culture. A handful of artist-designers have become household names, whilst the great multitude of draughtsmen and designers who worked for large firms like Daalderop, Inventum and the scores of smaller furniture factories in the first quarter of the century, or those at Tiger, Gazelle, Brabantia and even Philips at the end of the century, have remained anonymous – not to mention the hundreds of unidentified advertisement designers. Who is familiar with the names of the designers of HEMA products – despite the fact that HEMA items can be found in every Dutch household? This rift has also widened in recent years as globalization has progressed. Industrial production in the Netherlands reached its peak in the 1960s and has been on a steady decline ever since. During the last few years a growing number of established firms such as Philips have decided to move their manufacture to Eastern Europe or Asia. Designers still carry on doing their work, but increasingly they are working for colossal companies in countries a long way off, or with artisans in those faraway countries. Globalization did not narrow the gap between these long-distance designers and the group of designers mainly addressing the cultural circuit at home, with their still often socially committed messages.

One persistent issue perhaps best typifies Dutch design: the designer's social and ethical responsibility. Whether this has become a main concern as a result of Calvinism, or whether it is related to the democratic principles and middle-class values that have been typical of Dutch society for the last few centuries, goes beyond the scope of this book. It is, however, true to say that the focus of this presupposed moral or social responsibility has shifted. Many designers at the start of the century chiefly addressed themselves to the task of educating the general public and spreading 'good taste', after which they concentrated on the 'social question'. In the 1970s democratization occupied centre stage; today globalization and the environment are the most important issues. But the one connecting thread from the past up to the present day is designers' faith in their work as a means of contributing to a solution to these successive problems.

Designers have worked with a whole range of aesthetic principles but above all it is simplicity, austerity and affordability that have gained support. In the Netherlands you can generally rely on an ordinary, serviceable and preferably cheap product being more widely appreciated than an expensive, eye-catching model whose material, shape and colour have been meticulously chosen. Too much interest in design is soon suspect in a country where, even in the past, rich clients have been few and far between, and where the tenor is: 'it's better to be just normal than to try to stand out from the crowd'.

Yet it is often forgotten that there has always been a niche market for the more lavish and luxurious product. At the beginning of the twentieth century, for instance, Carel Lion Cachet's and Theo Colenbrander's expensive and extremely imaginative designs were very highly rated. Later the Amsterdam School's exceedingly inventive design was held in great esteem. In the 1950s the public admired liners and large, streamlined American cars, even while they lived in plain Goed Wonen interiors. Today the most extravagant and expensive one-off products by the newest generation of designers find their way into the homes of interested buyers – objects such as Maarten Baas's scorched chairs and Piet Hein Eek's scrap-wood furniture.

A final theme to note is how just as much interest has always been shown in the means of selling new products as in the products themselves. Graphic design, typography and letter designs have always been important in the Netherlands. Effectively communicating a message is probably still rated higher in Dutch design than a demonstrable commercial success. This might be explained by the fact that the Netherlands has continued to be a nation in which trade is more important than production. Good and effectively designed advertisements or packaging are essential to increasing turnover. From advertising posters from the early 1900s, print work for the Amsterdam Municipal Council in the 1920s, revolutionary phototypography in the 1930s, stirring cultural posters in the 1950s and '60s, almost the entire collection of post-war design carried out for the PTT, to house-style campaigns in the 1970s and avant-garde theatre posters in the '80s, this tradition continues to the several prestigious Dutch advertising and publicity studios working internationally today.

The medium is the message: The Dutch have always been able to see the truth of this maxim, which may explain why irony and conceptuality play such a prominent role in contemporary Dutch design culture. This has probably been strengthened by the fact that no one style has been dominant since the 1980s, and freedom and independence have been more valued than ever before.

References

Introduction

1 Paul Rutten et al., *Vormgeving in de Creatieve Economie* [TNO-rapport 33553] (Delft, 2005).

2 *Rapport der Rijks-Commissie tot het instellen van een onderzoek naar de toestand der Nederlandsche Kunst-Nijverheid* ('s-Gravenhage, 1878).

3 J. Bouman, P. Schuitema and P. Zwart, *Rapport inzake de richtlijnen en mogelijkheden eener technisch-kunstzinnig verantwoorde industrieele productie van gebruiksvoorwerpen op basis eener sociaal economisch verantwoorde productie* (The Hague, 1944).

4 See, for instance, Henri Baudet, *Een vertrouwde wereld: 100 jaar Innovatie in Nederland* (Amsterdam, 1986), where technique and innovation are starting points for essays on the history of design. The book that accompanied the television Teleac-course, Reyer Kras, *Nederlands Fabrikaat: Industriële Vormgeving* (Utrecht and Bussum, 1997), treats the history of industrial design in the Netherlands from several interesting and unexpected angles. Timo de Rijk studied the history of the design of electrical appliances from the background of design policy related to new philosophies on consumption and marketing: Timo de Rijk, *Het Elektrische Huis: Vormgeving en acceptatie van elektrische huishoudelijke apparaten in Nederland* (Rotterdam, 1998). Anthropological and sociological sides of the history of design got full attention in articles by Irene Cieraad in Jaap Huisman et al., *100 jaar Wonen in Nederland* (Rotterdam, 2000). Important surveys of the twentieth century include: Ellinoor Bergvelt, ed., *Industry and Design in the Netherlands, 1850–1950*, exh. cat., Stedelijk Museum, Amsterdam (1985); Gert Staal and Hester Wolters, eds, *Holland in Vorm: Dutch Design, 1945–1987* (The Hague, 1987); M. Titus Eliëns, Marjan Groot and Frans Leidelmeijer, *Dutch Decorative Arts, 1880–1940* (Bussum, 1997); Frederike Huygen, *Visies op Vormgeving: Het Nederlandse ontwerpen in teksten Deel I: 1874–1940* (Amsterdam, 2007: part II forthcoming, Autumn 2008). Studies dealing with specific aspects of design include: Ellinoor Bergvelt, Frans van Burkom and Karin Gaillard, eds, *From Neorenaissance to Postmodernism: A Hundred and Twenty-five Years of Dutch Interiors, 1870–1995* (Rotterdam, 1996); Jan Middendorp, *Dutch Type* (Rotterdam, 2004); Marjan Groot, *Vrouwen in de vormgeving, 1880–1940* (Rotterdam, 2007). A recent study, indicating a marked preference for architecture and graphic design, is Aaron Betsky and Adam Eeuwens, *False Flat: Why Dutch Design Is so Good* (London and New York, 2004).

1 New Art, Old Craft, 1875–1915

1 On Dutch culture around 1900 in general, see Jan Bank and Maarten van Buuren, *1900: Hoogtij van burgerlijke cultuur* (The Hague, 2000). Recent general surveys of applied arts and design in this period with extended bibliographies are: Ellinoor Bergvelt, Frans van Burkom and Karin Gaillard, eds, *From Neorenaissance to Postmodernism: A Hundred and Twenty-five Years of Dutch Interiors, 1870–1995* (Rotterdam, 1996); Titus. M. Eliëns, Marjan Groot and Frans Leidelmeijer, *Dutch Decorative Arts, 1880–1940* (Bussum, 1997); Jan Jaap Hey, *Vernieuwing en bezinning: Nederlandse beeldende kunst en kunstnijverheid ca 1885–1935 uit de collectie van het Drents Museum*, exh. cat., Drents Museum, Assen (Zwolle, 2004). The first ground-breaking overview of this period in Dutch decorative arts was: L. Gans, *Nieuwe Kunst: De Nederlandse bijdrage aan de Art Nouveau* (Utrecht, 1966). Most of the artists, designers, workshops, firms and

factories mentioned in this chapter are also dealt with in *Industry and Design in the Netherlands, 1850–1950*, exh. cat., Stedelijk Museum, Amsterdam (1985), or in Timo de Rijk, ed., *Designers in Nederland: Een eeuw productvormgeving* (Amsterdam and Gent, 2003). Specially for female designers, see Marjan Groot, *Vrouwen in de vormgeving, 1880–1940* (Rotterdam, 2007). An interesting overview of primary sources can be found in Frederike Huygen, *Visies op Vormgeving: Het Nederlandse ontwerpen in teksten Deel, 1: 1874–1940* (Amsterdam, 2007).

2 *Verslag der Centrale Commissie tot inrichting van de afdeelingen van Nederland en zijne Koloniën en tot behartiging van de belangen der inzenders in die afdeelingen op de Wereldtentoonstelling te Parijs in 1900* (Haarlem, 1902).

3 Jos Hilkhuijsen, *Delftse Art Nouveau: Onderwijs en ontwerp van Adolf le Compte (1850–1921), Karel Sluyterman (1863–1931) en Bram Gips (1861–1943)*, exh. cat., Drents Museum, Assen (Zwolle, 2001), pp. 71–3. On Art Nouveau in Belgium and 'Congo-style', see Claire Leblanc, *Art Nouveau and Design: Sierkunst van 1830 tot Expo 58* (Tielt, 2005), pp. 94–6.

4 Robert Fock, *Maastrichtse serviezen, 1917–1937* (Zwolle, 2007), pp. 12–18; Marie Rose Bogaers, *Drukdecors op Maastrichts aardewerk, 1850–1900* (Lochem, 1992).

5 G.P.J. Verbong, *Technische innovaties in de katoendrukkerij en -ververij in Nederland, 1835–1920* [NEHA series III], (Amsterdam, 1988); *Katoendruk in Nederland*, exh. cat., Textielmuseum, Tilburg, and Gemeentemuseum, Helmond (1989).

6 O. Gerdeil, 'La Hollande à l'Exposition Universelle', *L'Art décoratif*, III (November 1900), pp. 72–83.

7 *Verslag der Centrale Commissie*, pp. 166–7.

8 J.M.W. van Voorst tot Voorst, 'Nederland op de Wereldtentoonstelling van 1851 te Londen', *Nederlandse kunstnijverheid en interieurkunst*, Nederlands Kunsthistorisch Jaarboek 30 (Haarlem, 1980), pp. 475–92; Titus M. Eliëns, *Kunst Nijverheid Kunstnijverheid: De nationale nijverheidstentoonstellingen als spiegel van de Nederlandse kunstnijverheid in de negentiende eeuw* (Zutphen, 1990), pp. 59–62. For a complete survey of the interior arts and furniture industry in the Netherlands in the second half of the nineteenth century, see J.M.W. van Voorst tot Voorst, *Tussen Biedermeier en Berlage: Meubel en interieur in Nederland, 1835–1895* (Amsterdam, 1992).

9 H. W. Lintsen, ed., *Geschiedenis van de Techniek in Nederland: De wording van een moderne samenleving, 1800–1890*, Techniek en samenleving VI (Zutphen, 1995). An excellent publication on the modernization of the Dutch infrastructure in the nineteenth century is Auke van der Woud, *Een Nieuwe Wereld: Het ontstaan van het moderne Nederland* (Amsterdam, 2006).

10 Mienke Simon Thomas, 'Het ornament, het verleden en de natuur, drie hoofdthema's in het denken over vormgeving in Nederland 1870–1890', in *That Special Touch: Vormgeving tussen kunst en massaprodukt*, Nederlands Kunsthistorisch Jaarboek 39 (Haarlem, 1989), pp. 27–60.

11 Eliëns, *Kunst Nijverheid Kunstnijverheid*, pp. 97–113.

12 J. R. de Kruyff, *De Nederlandsche kunstnijverheid in verband met den Internationalen Wedstrijd bij gelegenheid van de in 1877 te Amsterdam te houden tentoonstelling van Kunst toegepast op Nijverheid uitgeschreven door de afdeeling Amsterdam der Vereeniging tot bevordering van Fabrieks- en Handwerksnijverheid in Nederland* (Amsterdam, 1879), p. 3.

13 *Rapport der Rijks-Commissie tot het instellen van een onderzoek naar de toestand der Nederlandsche Kunst-Nijverheid* ('s-Gravenhage, 1878).

14 H. H. Pijzel-Dommisse, 'Het Museum en de School voor Kunstnijverheid in de periode 1977–1926', in *Paviljoen Welgelegen, 1789–1989: Van buitenplaats van de bankier Hope tot zetel van de provincie Noord Holland* (Haarlem, 1989), pp. 151–72; Huygen, *Visies op Vormgeving*, pp. 14–22.

15 F. W. van Eeden, 'De Internationale Tentoonstelling te Londen in 1862', *Tijdschrift uitgegeven door de Nederlandsche Maatschappij ter bevordering van Nijverheid*, 26 (1863). See from the same author in this magazine in 1864 the series 'Versiering en kunststijl in de nijverheid'.

16 Jan de Maeyer, 'P.J.H. Cuypers in internationaal, comparatief perspectief: de Nederlandse Viollet-le-Duc?', in *P.J.H. Cuypers (1827–1921): Het complete werk*, ed. Hetty Berends (Rotterdam, 2007), pp. 43–51; A.J.C. van Leeuwen, *P.J.H. Cuypers, Architect, 1827–1927* (Zwolle, 2007).

17 For the start of design education in the nineteenth century, see Van Voorst tot Voort, *Tussen Biedermeier en Berlage*, pp. 81–100; Adi Martis, 'Voor de Kunst en voor de Nijverheid: Het ontstaan van het Kunstnijverheidsonderwijs in Nederland', dissertation, University of Amsterdam, 1990.

18 H. L. Boersma, 'Meer dan een Verslag', *De Tijdspiegel*, II (1879), pp. 129–47.

19 Mienke Simon Thomas, *De Leer van het Ornament: Versieren volgens voorschrift, 1850–1930* (Amsterdam, 1996).

20 Marty Bax, *Het Web der schepping: Theosofie en kunst in Nederland van Lauweriks tot Mondriaan* (Nijmegen, 2006).

21 M. de Bois, *Chris Lebeau, 1878-1945*, exh. cat., Drents Museum, Assen (Zwolle, 1983); L. F. Jintes and J. T. Pol-Tyszkiewicz, *Chris van der Hoef, 1875–1933*, exh. cat., Rijkmuseum Het Koninklijk Penningkabinet, Leiden, and Drents Museum, Assen (1994).

22 Simon Thomas, *De Leer van het Ornament*, p. 79.

23 J. R. ter Molen, ed., *Frans Zwollo sr, 1872–1945, en zijn tijd / Frans Zwollo sr, 1872–1945, und seine Zeit*, exh. cat., Museum Boijmans Van Beuningen, Rotterdam (1982).

24 Titus Eliëns et al., *Delfts aardewerk: geschiedenis van een nationaal product, III: De Porceleyne Fles* (Zwolle, 2003).
25 Yvonne Brentjens, *Rozenburg: Plateel uit Haagse kringen (1883–1917)*, exh. cat., Gemeentemuseum, The Hague (Zwolle, 2007); Titus Eliëns, *T.A.C. Colenbrander (1841-1930): Ontwerper van de Haagse Plateelbakkerij Rozenburg*, exh. cat., Gemeentemuseum, The Hague (Zwolle, 1999).
26 H.E. van Gelder, *Pottenbakkerskunst* [De toegepaste kunsten in Nederland] (Rotterdam, 1923).
27 Marijke E. Spliethoff, *Feestelijke geschenken voor de jonge koningin, 1898–1913* (Amsterdam and The Hague, 1998), pp. 110, 113.
28 Adri van der Meulen and Paul Smeele, *De pottenbakkers van Friesland, 1750–1950: het ambacht, de mensen, het aardewerk* (Leiden, 2004).
29 Mariannne Heslenfeld, *De collectie Holland: art nouveau-keramiek van de NV Faience en Tegelfabriek 'Holland', 1894–1918*, exh. cat., Museum Het Princessehof, Leeuwarden (2007); Eliëns et al., *Delfts aardewerk*; F. D. Doornberg et al., *Purmerends Jugenstil Aardewerk, 1895–1907* (Purmerend, 1995); Hans Vogels, *N.V. Plateelbakkerij Zuid-Holland*, exh. cat., Museum Het Catharina Gasthuis, Gouda (Zwolle, 1994); Jan Daniël van Dam, *Amstelhoek, 1897–1910*, exh. cat., Museum Het Princessehof, Leeuwarden (1986). For these firms, see also Eugène Langendijk and Mienke Simon Thomas, *Dutch Art Nouveau and Art Deco Ceramics, 1880–1940*, Museum Boijmans Van Beuningen, Rotterdam (2001).
30 Mienke Simon Thomas, 'K.V.T.', in *Industry and Design in the Netherlands*, pp. 94–7.
31 Mechteld de Bois, ed., *C.A. Lion Cachet, 1864–1945*, exh. cat., Drents Museum, Assen (Zwolle, 1994), pp. 28–51; Yvonne Brentjens, *G. W. Dijsselhof (1866–1924): Dwalen door het Paradijs*, exh. cat., Gemeentemuseum, The Hague (Zwolle, 2002), pp. 81–3.
32 Cor de Wit, *Chris en Agathe Wegerif, dragers van de nieuwe kunst in Apeldoorn* (Apeldoorn, 1994); Marjan Groot, 'Een dilemma voor de Nieuwe Kunst: Modern eclecticisme in de meubelkunst van Chris Wegerif voor de firma Arts and Crafts, 1901–1906', *Jong Holland*, 2 (1998), pp. 36–51; Joop Joosten, 'Henry van de Velde en Nederland, 1892–1902: De Belgische art nouveau en de Nederlandse Nieuwe Kunst', *Cahiers Henry van de Velde*, XII/XIII (1974), pp. 28–32.
33 T. M. Eliëns, *H. P. Berlage (1856–1934): Ontwerpen voor het interieur*, exh. cat., Gemeentemuseum, The Hague (Zwolle, 1998), pp. 23–4.
34 Annelies Krekel-Aalberse and Willem Voorthuysen, *Zeist, zilver, werken* (Zwolle, 2004); S.A.C. Begeer et al., *Mensen en zilver, bijna twee eeuwen werken voor Van Kempen en Begeer*, Zonnehof, Amersfoort, and Museum Boijmans van Beuningen, Rotterdam (1975). On Dutch silversmiths around 1900 in general, see Annelies Krekel-Aalberse, *Art Nouveau and Art Deco Silver* (London, 1989), pp. 137–90; Annelies Krekel-Aalberse, *Zilver / Silver / Silber: 1880–1940: Art Nouveau / Art Deco*, exh. cat., Gemeentemuseum, The Hague (Stuttgart, 2001).
35 A. Krekel-Aalberse and E. Raasen-Kruimel, *Jan Eisenloeffel, 1876–1957*, exh. cat., Singermuseum, Laren, and Drents Museum, Assen (Zwolle, 1996).
36 *Eerste Internationale Tentoonstelling van Moderne Decoratieve Kunst te Turijn 1902: Verslag van de Nederlandsche Afdeeling* (Haarlem, n.d.), pp. 14–15; M. Boot, 'Olanda', in *Torino 1902: le arti decorative internazionale del nuevo secolo*, exh. cat. (Turin, 1994), pp. 488–529.
37 Mario Benders, 'Van Vlissingen', in *Industry and Design in the Netherlands*, pp. 90–93.
38 Karin Gaillard, 'Sober Honesty, Comfortable Simplicity' in Bergvelt, Van Burkom and Gaillard, eds, *From Neorenaissance to Postmodernism*, pp. 58–83; Eliëns, *H. P. Berlage (1856–1934)*.
39 L. Tibbe et al., *Jac. van den Bosch, 1868–1948*, exh. cat., Drents Museum, Assen (Zwolle, 1987).
40 Hey, *Vernieuwing en bezinning*, pp. 178–9.
41 Enrico Thovez, 'The International Exhibition of Modern Decorative Art at Turin: The Dutch Section', *The Studio*, XXVI (1902), pp. 204–13. Only one side of the screen was saved in the collection of the Gemeentemuseum in The Hague. Brentjens, *G. W. Dijsselhof*, pp. 88–94.
42 For a recent publication on this subject, see Lieske Tibbe, 'Gemeenschapskunst: de samenleving in symbolen', in M. Bax and C. Blotkamp, *In't diepst van mijn gedachten . . . Symbolisme in Nederland ca 1890–1930*, exh. cat., Drents Museum, Assen (Zwolle, 2004).
43 Thovez, 'The International Exhibition'.
44 P.J.W.J. van der Burgh, 'De Nederlandse inzending op de eerste internationale tentoonstelling voor moderne decoratieve kunst te Turijn', *Elsevier's Geïllustreerd Maandschrift*, XIII/25 (1903), pp. 3–20.
45 Manfred Bock, *Anfänge einer neuen Architektur: Berlages Beitrag zur architektonischen Kultur der Niederlande im ausgehenden 19. Jahrhundert* (The Hague and Wiesbaden, 1983).
46 Eliëns, *H. P. Berlage (1856–1934)*, pp. 64–77; E. P. Tibbe, *R. N. Roland Holst, arbeid en schoonheid vereend: opvattingen over gemeenschapskunst* (Amsterdam, 1994).
47 M. Boot, 'Carel Henny en zijn huis: een demonstratie van "goed wonen" rond de eeuwwisseling' in *H. P. Berlage, 1846–1934; een bouwmeester en zijn tijd*, Het Nederlands Kunsthistorisch Jaarboek 25 (Bussum, 1975), pp. 91–131; Gaillard, 'Sober Honesty'.
48 Harm Ellens, *Onze Disch*, [De Toegepaste Kunsten in Nederland] (Rotterdam, 1926), p. 8.
49 Jo de Jong, *De Nieuwe Richting in de kunstnijverheid in Nederland: Schets eener geschiedenis der Nederlandsche*

kunstnijverheidsbeweging (Rotterdam, 1929), p. 8.
50 Auke van der Woud, *Waarheid en Karakter: Het debat over de bouwkunst, 1840–1900* (Rotterdam, 1999), pp. 293–407.
51 Adi Martis, 'Some Organizations and their Activities', in *Industry and Design in the Netherlands*, pp. 22–9; Van Voorst tot Voorst, *Tussen Biedermeier en Berlage*, pp. 113–17; Yvonne Brentjens, *K.P.C. de Bazel (1869–1923): Ontwerpen voor het interieur*, exh. cat., Gemeentemuseum, The Hague (Zwolle, 2006), p. 113–15.
52 Huygen, *Visies op vormgeving*, I, pp. 462–6; Martis, 'Some Organizations', pp. 25–6.
53 M. Knol, *Klaas van Leeuwen, 1868–1935*, Drents Museum, Assen (Zwolle, 1988).
54 L. Tibbe et al., *Jac. van den Bosch, 1868–1948*, exh. cat., Drents Museum, Assen (Zwolle, 1987).
55 Eliëns, 'Groot en Leidelmeijer', in *Dutch Decorative Arts*, pp. 245–6.
56 Jan Middendorp, *Dutch Type* (Rotterdam, 2004), pp. 40–47.
57 Huygen, *Visies op Vormgeving*, I, pp. 64–5, 126; Hennie van der Zande, 'Van alle markten thuis: Het veelzijdige oeuvre van de kunstenaar Hendrik (Herman) Hana 1874–1952', endpaper, Free University Amsterdam, 2004.
58 C. van Adrichem, *Willem Penaat, meubelontwerper en organisator (1875–1957)* (Rotterdam, 1988).
59 The publication of Tak in *De Kroniek* of 1905 was reprinted in De Jong, *De nieuwe richting* (1929), pp. 42–6, and again recently in Huygen, *Visies op vormgeving*, I, pp. 202–5.
60 *Voorlopig orgaan (VANK)* (March 1911), p. 11.
61 Van der Zande, '*Van alle mark ten thuis*', pp. 56–64; Lien Heyting, *De wereld in een dorp: schilders, schrijvers en wereldverbeteraars in Laren en Blaricum, 1880–1920* (Amsterdam,1994), pp. 189–95.
62 K. Sluyterman, *Gedenkboek uitgegeven naar aanleiding van het vijfentwintig jarig bestaan der Vereeniging Art et Industriae, 1884–1909* (The Hague, 1910), p. 6; K. Sluyterman, 'Hendrik Petrus Berlage Nz.', *Elseviers's Geillustreerd Maandschrift*, XV/29 (1905), pp. 3–21.
63 R. L. Miellet, *Honderd jaar grootwinkelbedrijf in Nederland* (Zwolle, 1993).

2 Design as Art, 1915–40

1 For a general overview of the decorative arts and design in this period, see chapter One, n. 1.
2 Hein A. M. Klemann, 'Ontwikkeling door isolement', in *Wankel evenwicht: Neutraal Nederland en de Eerste Wereldoorlog* (Soesterberg, 2007), pp. 271–309.
3 The lectures of Eisenloeffel and Zwart are published in *Jubileum-Orgaan Nederlandsche Vereeniging voor Ambachts- en Nijverheidskunst (VANK) 1929*, pp. 14–22 and 22–8, reprinted in Frederike Huygen, *Visies op vormgeving. Het Nederlandse ontwerpen in teksten: Deel 1: 1874–1940* (Bussum, 2007), pp. 345–61. A. Krekel-Aalberse and E. Raasen-Kruimel, *Jan Eisenloeffel, 1876–1957*, exh. cat., Singermuseum, Laren, and Drents Museum, Assen (Zwolle, 1996); Kees Broos, *Piet Zwart, 1885–1977*, exh. cat., Gemeentemuseum, The Hague (Amsterdam, 1982). See also Yvonne Brentjens, *Piet Zwart: Vormingenieur*, exh. cat., Gemeentemuseum, The Hague (Zwolle, 2008).
4 Huygen, *Visies op vormgeving*, I, pp. 249–51.
5 For Zwart's stay at the Bauhaus, see his archive at the Rijksbureau voor Kunsthistorische Documentatie (Netherlands Institute for Art History), The Hague.
6 Karin Orchard and Isabel Schulz, eds, *Kurt Schwitters en de Avant-garde*, exh. cat., Museum Boijmans Van Beuningen, Rotterdam (2007), pp. 139–55, 239; Flip Bool et al., *Piet Zwart (1885–1977)*, Monografieën van Nederlandse fotografen 5 (Amsterdam, 1997); Broos, *Piet Zwart*, pp. 34–53.
7 Annette van der Kley-Blekxtoon, *Kristalunie Maastricht* (Lochem, 2003); Broos, *Piet Zwart*, pp. 32–3.
8 Corrie van Adrichem, *Willem Penaat: Meubelontwerper en organisator, 1875–1957* (Rotterdam, 1988); Mienke Simon Thomas, *Corn. van der Sluys: Binnenhuisarchitect, organisator en publicist, 1881–1944* (Rotterdam, 1988).
9 Mienke Simon Thomas, 'Cornelis van der Sluys: Nijverheidskunst in Den Haag, 1906–1916', *Jaarboek Geschiedkundige Vereniging Die Haghe* (1989), pp. 147–74.
10 T. Landré, 'De Hollandsche nijverheidskunst op de Brusselsche tentoonstelling', *Onze Kunst*, IX/2 (1910), pp. 83–98, 118–33.
11 Ellinoor Bergvelt et al., *80 jaar wonen in het Stedelijk*, exh. cat., Stedelijk Museum, Amsterdam (1981), pp. 12–13.
12 W. F. Gouwe, 'Reclame', in *Nederlandsche Ambachts- en Nijverheidskunst: Jaarboek 1930: Werk* (Rotterdam, 1930), pp. 137–58; Wilbert Schreurs, *Geschiedenis van de reclame in Nederland* (Utrecht, 1989), pp. 45–86; Huygen, *Visies op vormgeving*, I, pp. 163–97.
13 *De Nieuwe Courant* [The Hague], 26 October 1935.
14 Gouwe, 'Reclame', p. 147.
15 Jo de Jong, *De Nieuwe Richting in de kunstnijverheid in Nederland: Schets eener geschiedenis der Nederlandsche kunstnijverheidsbeweging* (Rotterdam, 1929), p. 14.
16 J. M. van der Mey, 'Moderne meubelkunst', *Nederlandsche Ambachts- en Nijverheidskunst: Jaarboek 1919* (Rotterdam, 1919), pp. 50–55.
17 P. M. Cochius, 'Nijverheid en Kunst', *Nederlandsche Ambachts- en Nijverheidskunst: Jaarboek 1919* (Rotterdam, 1919), pp. 34–7.
18 Articles in the *Nederlandsche Ambachts- en Nijverheids-*

kunst: Jaarboek 1923–24 are: W. van der Pluym, 'De kunstenaar en zijn zending in het leven'; A.H.L. Bähler, 'Liefde 'n grondslag voor de kunst'; H. C. Verkruysen, 'Kunst, vorm van wijsheid'; M.H.J. Schoenmaekers, 'Religieuze schoonheid'; R. N. Roland Holst, 'Enkele gedachten over de harmonie van constructieve en versierende vormen'.

19 W. H. Gispen, 'Kunst als noodzaak en als spel', in *Nederlandsche Ambachts- en Nijverheidskunst: Jaarboek 1925–26* (Rotterdam, 1926), pp. 38–59.

20 Hetty Berends, ed., *Gispen in Rotterdam: Nieuwe verbeelding van het moderne* (Rotterdam, 2006); A. Koch, *W. H. Gispen, serieproducten, 1923–1960* (Rotterdam, 2005); A. Koch, *W. H. Gispen, a Pioneer of Dutch Design* (Rotterdam, 1998); B. Laan and A. Koch, ed., *Collectie Gispen: Meubels, lampen en archivalia in het Nai, 1916–1980* (Rotterdam, 1996); A. Koch, *Industrieel ontwerper W. H. Gispen (1890–1981): Een modern eclecticus* (Rotterdam, 1988); Jane Beckett, 'W. H. Gispen and the Development of Tubular Steel Furniture in the Netherlands', in B. Campbell-Cole and T. Benton, eds, *Tubular Steel Furniture* (London, 1979), pp. 28–45; Christopher Wilk, ed., *1914–1939 Modernism: Designing a New World*, exh. cat., Victoria and Albert Museum, London (2006), p. 213, cat. nos 137a and 137b.

21 W. H. Gispen, 'Techniek en kunst', *Wendingen*, IX/2 (1928), pp. 2–18. See also Yvonne Brentjens, '"De Woning is nieuwer dan de Mensch": De receptie van het stalen buismeubel in Nederland, 1927–1938', in Titus Eliëns and Marlite Halbertsma, eds, *Volmaakt verchroomd d3 en het avant-gardemeubel in Nederland* (Rotterdam, 2007), pp. 56–73.

22 Hildelies Balk, *De Kunstpaus H. P. Bremmer, 1871–1956* (Bussum, 2006); Titus Eliëns, 'Een verzameling rond de kunsten van het vuur', in *Jaarboek Haags Gemeentemuseum: Jubileumnummer 95/96* (The Hague, 1997), pp. 115–45.

23 Robert Fock, *Maastrichtse serviezen, 1917–1937* (Zwolle, 2007); Arno Weltens, *Maastrichts aardewerk: constructivistische decors uit het interbellum* (Zwolle, 2006).

24 Manfred Bock et al., *Michel de Klerk: bouwmeester en tekenaar van de Amsterdamse School, 1884–1923* (Rotterdam, 1997); Maristella Casciato, *The Amsterdam School* (Rotterdam, 1996); Frans van Burkom, 'Desperate Dreaminess: In Class with the Amsterdam School', in Ellinoor Bergvelt, Frans van Burkom and Karin Gaillard, eds, *From Neorenaissance to Postmodernism: A Hundred and Twenty-five Years of Dutch Interiors, 1870–1995* (Rotterdam, 1996), pp. 134–59; Ellinoor Bergvelt, 'The Decorative Arts in Amsterdam, 1890–1930', in *Designing Modernity: The Arts of Reform and Persuasion, 1885–1945*, exh. cat., Wolfsonian, Miami Beach (1995), pp. 79–110; Frans van Burkom, *Michel de Klerk: Bouw- en meubelkunstenaar, 1884–1923* (Rotterdam, 1990); Ellinoor Bergvelt, Frans van Burkom et al., *Amsterdamse School, 1910–1930*, exh. cat., Stedelijk Museum, Amsterdam (1975).

25 H. Boterenbrood and J. Prang, *Van der Mey en het Scheepvaarthuis* (The Hague, 1989).

26 Michel de Klerk, 'De invloed van Dr Berlage of de ontwikkeling der Nederlandsche Bouwkunst', *Bouwkundig Weekblad*, XXXIV (1916), pp. 331–2.

27 Van Burkom, 'Desperate Dreaminess', pp. 141–51.

28 Bernhard Kohlenbach, *Pieter Lodewijk Kramer, 1881–1961: Architect van de Amsterdamse School* (Naarden, 1994).

29 E. J. Lagerwey-Polak, *Hildo Krop: beeldhouwer* (The Hague, 1992).

30 Ingeborg de Roode and Marjan Groot, *Amsterdamse School textiel, 1915–1930*, exh. cat., Textielmuseum, Tilburg (1999).

31 Mienke Simon Thomas, Elly Adriaansz. and Sandra van Dijk, *Jaap Gidding: Art Deco in Nederland*, exh. cat., Museum Boijmans Van Beuningen, Rotterdam (2006); Susan Day, *Art Deco and Modernist Carpets* (San Francisco, 2002), pp. 72–5.

32 Riet Neerincx, ed., *T.A.C. Colenbrander (1841–1930): Plateelbakkerij 'RAM' te Arnhem (1921–1935)*, exh. cat., Gemeentemuseum, Arnhem (1986).

33 Jean Paul Baeten, *Ontwerp het onmogelijke: de wereld van architect Hendrik Wijdeveld*, exh. cat., Nederlands Architectuurinstituut, Rotterdam (2006).

34 Martijn F. Le Coultre, *Wendingen, 1918–1932: Architectuur en vormgeving* (Blaricum, 2001); Hans Oldewarris, *The Covers of Wendingen* (Rotterdam, 1995).

35 Carel Blotkamp et al., *De Stijl: The Formative Years, 1917–1921* (Utrecht, 1982); Carel Blotkamp et al., *De vervolgjaren van De Stijl, 1922–1932* (Amsterdam and Antwerp, 1996); Marijke Kuper, 'Space Dissolved in Colour, De Stijl' in *From Neorenaissance to Postmodernism*, pp. 160–83; Nancy Troy, *The Stijl Environment* (Cambridge, MA, 1983); Cees Boekraad, et al., *Het Nieuwe Bouwen: De Nieuwe Beelding in de architectuur: Neo-Plasticism in Architecture De Stijl*, exh. cat., Gemeentemuseum, The Hague (1983).

36 Marijke Kuper, *Gerrit Th. Rietveld: L'oeuvre complet* (Utrecht, 1993).

37 Huygen, *Visies op vormgeving*, I, pp. 264–71.

38 *Verslag betreffende de Nederlandsche inzending op de in 1925 in Parijs gehouden internationale tentoonstelling van moderne decoratieve en industrieele kunst* (1925); Michiel Nijhoff. '"De druk zal in goed leesbare letters plaatsvinden": Drie catalogi bij de Nederlandse inzending voor de wereldtentoonstelling van 1925 in Parijs', in *Vormgeven aan veelzijdigheid: opstellen aangeboden aan Wim Crouwel ter gelegenheid van zijn afscheid als directeur van Museum Boymans-van Beuningen* (Rotterdam, 1993), pp. 124–33.

39 Timo de Rijk, *De Haagse Stijl: Art Deco in Nederland* (Rotterdam, 2004).
40 Renny Ramakers, *Tussen kunstnijverheid en industriële vormgeving: De Nederlansche Bond voor Kunst in Industrie* (Utrecht, 1985).
41 Huygen, *Visies op Vormgeving*, I, pp. 215–18. On Vlaanderen, see also Schreurs, *Geschiedenis reclame*, pp. 77–8; Saskia de Bodt and Jeroen Kapelle, *Prentenboeken: Ideologie en Illustratie, 1890–1950* (Amsterdam and Gent, 2003), p. 257; Caroline Boot and Sanny de Zoete, *Artistiek damast van Brabantse bodem 1900–1960, ontwerpen van Chris Lebeau, André Vlaanderen Jaap Gidding en tijdgenoten*, exh. cat., Textielmuseum, Tilburg (2005), pp. 64–8.
42 Ramakers, *Tussen kunstnijverheid en industriële vormgeving*, pp. 42–55.
43 Thimo te Duits, *Glasfabriek Leerdam: 1915–1934. De kunstnijverheidscollectie van de Glasfabriek Leerdam, 1915–1934 Museum Boijmans Van Beuningen*, exh. cat., Drents Museum, Assen (Zwolle, 1998); A. van der Kley-Blekxtoon, *Leerdam glas, 1878–2003: De glasfabriek Leerdam* (Lochem, 2004); T. Eliëns and J.J.M. Meihuizen, *K.P.C. de Bazel – de glazen van de architect*, exh. cat., Glasmuseum, Leerdam (2005); Ed van Hinte and Timo de Rijk, eds, *Wereldverbeteraars: 100 Jaar idealen in glas*, exh. cat., Glasmuseum, Leerdam (2006).
44 Reino Liefkes, *Andries Copier: Glass Designer, Glass Artist* (Zwolle, 2002).
45 Boot and De Zoete, *Artistiek damast*, pp. 16–59; M. de Bois, *Chris Lebeau, 1878–1945*, exh. cat., Drents Museum, Assen (1983), pp. 84–104.
46 Karin Gaillard, *Labor Omnia Vincit: een idealistische meubelfabriek, 1910–1935*, exh. cat., Gemeentemuseum, Arnhem (1991).
47 Sjoerd van Faassen, ed., *W. L. & J. Brusse's Uitgeversmaatschappij, 1903–1965* (Rotterdam, 1993).
48 Petra Timmer, *Metz & Co.: De creatieve jaren* (Rotterdam, 1995).
49 Petra Dupuits, *Metz est venu*, exh. cat., Stedelijk Museum, Amsterdam (1992).
50 E. Stapersma and H. Beukers, 'Het Paapje viert 60 jaar na oprichting zijn eerste lustrum in Twente', *Handwerken zonder grenzen*, no.2 (1990) pp. 5–8; Eliëns, Groot and Leidelmeijer, *Kunstnijverheid in Nederland, 1880–1940* (Bussum, 1997), pp. 234–5.
51 Hans Vogels, *N.V. Koninklijke Plateelbakkerij Zuid-Holland*, exh. cat., Stedelijke Musea, Gouda (1994); Nicolette Sluijter-Seijffert and Hans Vogels, eds, *Van decor naar design: Kunstenaars in de Goudse aardewerkindustrie, 1898–1941*, exh. cat., Museum het Catharina Gasthuis, Gouda (Zwolle, 2001), pp. 31–66.
52 Peter van Dam and Philip van Praag, *Stefan Schlesinger 1896–1944: Atelier voor reclame* (Abcoude, 1997).
53 Arnold Witte and Esther Cleven, *Design is geen vrijblijvende zaak: Organisatie, imago en context van de ptt-vormgeving tussen 1906 en 2002* (Rotterdam, 2006); Gerard Forde, *Design in the Public Service: the Dutch PTT, 1920–1990*, exh. cat, Design Museum, London (1991); P. H. Hefting, *Royal PTT Nederland NV: Art and Design Past and Present, a Guide*, exh. cat., Design Museum, London (1990).
54 Jan Rudolph de Lorm, *Cornelis de Lorm – ontwerper*, exh. cat., Drents Museum, Assen, and Nederlands Postmuseum, The Hague (1987), p. 35; Simon Thomas, *Corn. Van der Sluys*.
55 J. F. van Royen, *Driewerf leelijk* [Rijksmuseum Meermanno-Westreenianum/Museum van het Boek] (Leiden, 1994); derived from J. F. van Royen, 'De typographie van 's Rijks drukwerk', *De witte mier* (1912).

3 Good Design, 1925–65

1 Wouter van Stiphout, 'Stories from behind the Scenes of Dutch Moral Modernism', in *Mart Stams Trousers: Stories from behind the Scenes of Dutch Moral Modernism* (Rotterdam, 1999), pp. 20–45.
2 Wim Beeren et al., eds, *Het Nieuwe Bouwen in Rotterdam*, exh. cat., Museum Boijmans Van Beuningen, Rotterdam (1982); Marlite Halbertsma and Patricia van Ulzen, eds, *Interbellum Rotterdam: Kunst en cultuur, 1918–1940* (Rotterdam, 2001).
3 E. Hoogenboezem, *Jac. Jongert, 1883–1942: Graficus tussen kunst en reclame*, exh. cat., Gemeentemuseum, The Hague (1982). The Museum Boijmans Van Beuningen is preparing an exhibition and catalogue on this designer for 2009.
4 Jacob Jongert, 'Herinneringen van een boerenjongen' (Doorn, 1941) [unpublished autobiography].
5 Adi Martis, 'Voor de Kunst en voor de Nijverheid: Het ontstaan van het Kunstnijverheidsonderwijs in Nederland', dissertation, University of Amsterdam, 1990, p. 277, n. 92.
6 Interview with Beatrijs Esscher-Jongert, daughter of the designer, 15 December 2006.
7 Hetty Berends, ed., *Gispen in Rotterdam: Nieuwe verbeelding van het moderne* (Rotterdam, 2006); A. Koch, *W. H. Gispen, serieproducten, 1923–1960* (Rotterdam, 2005); A. Koch, *W. H. Gispen: A Pioneer of Dutch Design* (Rotterdam, 1998); A. Koch, *Industrieel ontwerper W .H. Gispen (1890–1981): Een modern eclecticus* (Rotterdam, 1988).
8 Joris Molenaar, ed., *Van Nelle: Monument in Progress* (Rotterdam, 2005), pp. 93–4; Halbertsma and Van Ulzen, *Interbellum Rotterdam*, p. 209; Herman van Bergeijk and Otakar Mácel, eds, *'We vragen de kunste-*

naars kind te zijn van zijn eigen tijd': teksten van Mart Stam (Nijmegen, 1999).

9 On the Dutch architects in Siberia, see Cor de Wit, *Johan Niegeman, 1902–1979: Bauhaus, Sowjetunie, Amsterdam* (Amsterdam, 1979), pp. 53–5; Hélène Damen et al., *Lotte Stam-Beese* (Rotterdam, 1993).

10 W. de Wagt, *Van Loghem, 1881–1940: Landhuizen, stadswoningen en bouwprojecten* (Haarlem, 1995). A facsimile of Van Loghem's publication, with an introduction by U. Barbieri, is avaialable as *Ir. J. B. van Loghem: Bouwen, Bauen, Bâtir, Building – Holland* (Nijmegen, 1980).

11 Barbieri, *J. B. van Loghem*, p. 16.

12 I. Pey and T. Boersma, eds, *Michiel Brinkman, 1873–1925* (Rotterdam, 1995); Beeren et al., *Nieuwe Bouwen in Rotterdam*, pp. 31–2.

13 Ed Taverne et al., *J.J.P. Oud: poëtisch functionalist, 1890–1963: Compleet werk* (Rotterdam, 2001); Christopher Wilk, *Modernism: Designing a New World*, exh. cat., Victoria and Albert Museum, London (2006), cat. nos 88, 89, 98, 103, 118; Hans Esser, 'J.J.P. Oud', in Carel Blotkamp et al., *De Stijl: The Formative Years, 1917–1921* (Utrecht, 1986), pp. 138–54.

14 Reprinted in Frederike Huygen, *Visies op Vormgeving*, 1: *Het Nederlandse ontwerpen in taksten Deel 1*, 1874–1940 (Amsterdam, 2007) p. 328; Koos Bosma, ed., *Het Nieuwe Bouwen in Amsterdam*, exh. cat., Stedelijk Museum, Amsterdam (1985).

15 Beatrice Bernini and Timo de Rijk, *Het Nieuwe Wonen in Nederland, 1924–1936* (Rotterdam, 1990); Karin Gaillard, Mienke Simon Thomas and Petra Timmer, 'Nieuwe Bouwen and the interior', in *Het Nieuwe Bouwen in Amsterdam 1*, pp. 112–41. On terminology, see Frederike Huygen, 'Some Terms Defined: Objective, New, Modern, and Functional', in *1928: Beauty,Lucidity, Logica and Ingenuity*, exh. cat., Museum Boijmans Van Beuningen, Rotterdam (1993), pp. 42–63.

16 Taverne et al., *J.J.P. Oud*, pp. 290–304; Wilk, *Modernism*, pp. 182–5.

17 Huygen, *Visies op Vormgeving*, 1, pp. 372–7; Wilk, *Modernism*, pp. 226–47; O. Máçel, *2100 Metal Tubular Chairs* (Rotterdam, 2006); Titus Eliëns and Marlite Halbertsma, eds, *Volmaakt verchroomd: d3 en het avantgardemeubel in Nederland* (Rotterdam, 2007).

18 Peter Fuhring, 'Doelmatig wonen in Nederland: De efficiënt georganiseerde huishouding en de keukenvormgeving 1920–1938', *Nederlandse kunstnijverheid en interieurkunst*, Kunsthistorisch Jaarboek 31 (Haarlem, 1981), pp. 575–85; Mayke Groffen and Sjouk Hoitsma, *Het geluk van de huisvrouw*, exh. cat., Historisch Museum, Rotterdam (2004), pp. 109–14; Irene Cieraad, 'Het Huishouden tussen droom en daad: Over de toekomst van de keuken', in *Schoon genoeg: Huisvrouwen en huishoudtechnologie in Nederland, 1898–1998*, ed. Ruth Oldenziel and Carolien Bouw (Nijmegen, 1998), pp. 31–54; Wilk, *Modernism*, p. 180.

19 Marijke Kuper, *Gerrit Th. Rietveld: L'oeuvre complet*, exh. cat., Centraal Museum, Utrecht (1993). Marijke Kuper, 'Gerrit Rieveld', in Blotkamp et al., *De Stijl: The Formative Years*, pp. 263–86.

20 Wilk, *Modernism*, p. 55; Ida van Zijl, *Rietveld in Utrecht* (Utrecht, 2001); Bertus Mulder, *Rietveld Schröder Huis* (Bussum, 1997); Corrie Nagtegaal, *Tr. Schröder-Schräder: Bewoonster van het Rietveld Schröderhuis* (Utrecht, 1987).

21 Auke van der Woud, *Het Nieuwe Bouwen: International: CIAM: Housing Town Planning*, exh. cat., Museum Kröller-Müller, Otterlo (1983).

22 Wilk, *Modernism*, p. 237; Petra Timmer, *Metz & Co.: De creatieve jaren* (Rotterdam, 1995), pp. 77–149.

23 Elly Adriaans et al., *Brinkman en Van der Vlugt: Huis Sonneveld: Modern Wonen in 1933* (Rotterdam, 2001); 'The Rotterdam Museumpark Villas', *Wiederhall*, 20 (2001).

24 Eliëns and Halbertsma, *Volmaakt verchroomd d3*.

25 Yvonne Brentjens, '"De Woning is nieuwer dan de Mensch": De receptie van het stalen buismeubel in Nederland, 1927–1938', in Eliëns and Halbertsma, *Volmaakt verchroomd*, pp. 56–73; Mienke Simon Thomas, 'The Functional Interior', in Beeren et al., *Het Nieuwe Bouwen Rotterdam*, pp. 120–28.

26 J. B. van Loghem, 'Richtlijnen', *De 8 en Opbouw*, 6 (1935), p. 1.

27 Mienke Simon Thomas, *Corn. van der Sluys: Binnenhuisarchitect, organisator en publicist, 1881–1944* (Rotterdam, 1988), pp. 35–7; Timmer, *Metz & Co.: De creatieve jaren* (Rotterdam, *1995*), (1995), pp. 70–73; Guus Vreeburg and Hadewych Martens, *UMS Pastoe: Een Nederlandse Meubelfabriek, 1913–1983*, exh. cat., Centraal Museum, Utrecht (1983); Timo de Rijk, *De Haagse Stijl: Art Deco in Nederland* (Rotterdam, 2004).

28 Eveline Holsappel, *Ida Falkenberg-Liefrinck (1901): de rotan stoel als opmaat voor een betere woninginrichting* (Rotterdam, 2000).

29 Ed van Hinte and Timo de Rijk, eds, *Wereldverbeteraars: 100 Jaar idealen in glas*, exh. cat., Glasmuseum, Leerdam (2006); A. van der Kley-Blekxtoon, *Leerdam glas 1878–2003: De glasfabriek Leerdam* (Lochem, 2004); Reino Liefkes, *Andries Copier: Glass Designer, Glass Artist* (Zwolle, 2002); Thimo te Duits, *Glasfabriek Leerdam, 1915–1934: De kunstnijverheidscollectie van de Glasfabriek Leerdam, 1915–1934, Museum Boijmans Van Beuningen*, exh. cat. Drents Museum, Assen (Zwolle, 1998).

30 Helen Boterenbrood, *Weverij De Ploeg* (Rotterdam, 1989).

31 Annelies Krekel-Aalberse and Willem Voorthuysen, *Zeist, zilver, werken* (Zwolle, 2004); Annelies Krekel-Aalberse, *Carel J. A. Begeer, 1883–1956*, Drents Museum,

Assen (Zwolle, 2001); S.A.C. Begeer et al., *Mensen en zilver, bijna twee eeuwen werken voor Van Kempen en Begeer*, Zonnehof, Amersfoort, and Museum Boijmans van Beuningen, Rotterdam (1975).

32 Krekel-Aalberse, *Carel J. A. Begee*, p. 100.

33 Timo de Rijk, *Het elektrische huis: Vormgeving en acceptatie van elektrische huishoudelijke apparaten in Nederland* (Rotterdam, 1998), pp. 167–76; Huygen, *Visies op Vormgeving*, I: pp. 258–60.

34 Huygen, *Visies op Vormgeving*, pp. 162–97, 319–22; Peter van Dam and Philip van Praag, *Fré Cohen, 1903–1943: Leven en werk van een bewogen kunstenares* (Abcoude, 1993); Dick Maan, *Paul Schuitema: Beeldend organisator* (Rotterdam, 2006); Alston W. Purvis, *Dutch Graphic Design, 1918–1945* (New York, 1992); Kees Broos and Paul Hefting, *Grafische Vormgeving in Nederland: Een eeuw* (Naarden, 1995), pp. 76–91; Kees Broos, *Piet Zwart, 1885–1977*, exh. cat., Gemeentemuseum, The Hague (1982).

35 Anna Müller-Härlin, '"Reclameontwerpen. Dat klinkt zo makkelijk". . . Karin Orchard and Isabel Schulz: Kurt Schwitters en zijn vrienden als typografen', in *Kurt Schwitters en de avant-garde*, ed. Karin Orchard and Isabel Schulz, exh. cat., Museum Boijmans Van Beuningen, Rotterdam (2007), pp. 138–55; Sjarel Ex, 'De blik naar het oosten: De Stijl in Duitsland en Oost-Europa', in *De vervolgjaren van De Stijl*, ed. Carel Blotkamp (Amsterdam and Antwerp, 1996), pp. 67–112.

36 Jan Tschichold, *Die Neue Typographie* (1928); Kees Broos, 'Typografie', in *Berlijn–Amsterdam, 1920–1940: Wisselwerkingen*, ed. Kathinka Dittrich (Amsterdam, 1982), pp. 237–45; Flip Bool and Ingeborg Leijerzapf, 'Fotografie', in ibid., pp. 246–50.

37 Piet Zwart, 'Nederlandsche ambachts- en nijverheidskunst', *Het Vaderland*, 31 May 1928; reprinted in Huygen, *Visie op Vormgeving*, I: pp. 25–7.

38 Broos, *Piet Zwart*, p. 81.

39 Dick Maan, *De Maniakken: Het ontstaan en ontwikkeling van de grafische vormgeving aan de Haagse academie in de jaren dertig* (Eindhoven, 1982).

40 Jan van Adrichem et al., *Rebel, mijn hart: Kunstenaars, 1940–1945*, exh. cat., Nieuwe Kerk, Amsterdam (Zwolle, 1995); J. W. Mulder, *Kunst in crisis en bezetting: Een onderzoek naar de houding van Nederlandse kunstenaars in de periode 1930–1945* (Utrecht, 1978).

41 Marcel Brouwer and Joep Haffmans, *Cris Agterberg: Beeldhouwer en sierkunstenaar* (Vianen, 2001), pp. 180–81.

42 Koch, *Gispen in Rotterdam*, pp. 149–54.

43 C.J.M. Schuyt and E. Taverne, *1950 Welvaart in zwart wit* (The Hague, 2000).

44 De Rijk, *Het elektrische huis*, p. 261.

45 Piet Zwart, 'Uit de keuken van de keuken', *De Ingenieur*, LXVII/35 (1955), pp. 410–14; Broos, *Piet Zwart*, pp. 84–5; Petra Timmer, 'Total Control: Transparency, Usefulness and Nieuwe Bouwen', in *From Neorenaissance to Postmodernism: A Hundred and Twenty-five Years of Dutch Interiors, 1870–1995*, ed. Ellinoor Bergvelt, Frans van Burkom and Karin Gaillard (Rotterdam, 1996), pp. 212–15.

46 Ellinoor Bergvelt and Hadewych Martens, 'Living as Work: Postwar Reconstruction and Goed Wonen', in *From Neorenaissance to Postmodernism*, pp. 260–83; Wies van Moorsel, *Contact en controle: Over het vrouwbeeld van de stichting Goed Wonen* (Amsterdam, 1992); Ellinoor Bergvelt, ed., *Goed Wonen een Nederlandse wooncultuur, 1946–1968*, special issue of *Wonen TABK* (1979) 4/5.

47 Bergeijk and Máçel, *Teksten van Mart Stam* (1999); Caroline Boot, 'Mart Stam: Kunstnijverheidsonderwijs als aanzet voor een menselijke omgeving. Dessau-Amsterdam', *Wonen TABK*, 11 (1982), pp. 10–21.

48 De Wit, *Johan Niegeman*.

49 Peter Vöge, *Wim den Boon: binnenhuisarchitect, 1912–1968* (Rotterdam, 1989).

50 *Goed Wonen*, p. 88.

51 Ibid.,, p. 38.

52 *Goed Wonen* (1948), pp. 162–3.

53 H. Lindinger, ed., *Ulm Design: The Morality of Objects: Hochschule für Gestaltung Ulm, 1951–1968* (Berlin, 1990); Kho Liang Ie, *Hochschule für Gestaltung Ulm*, exh. cat., Stedelijk Museum, Amsterdam (1965).

54 Frederique Huygen, 'Het calvinisme van de goede vorm', in *Holland in Vorm: Dutch Design, 1945–1987*, ed. Gert Staal and Hester Wolters (Den Haag, 1987), pp. 132–8.

55 Notes critical of this American trend can be found in *Industriële vormgeving in Amerika: Rapport Studiegroep industrie: with a Summary in English* (Rotterdam, 1954). Similar sentiments were published in several articles of this period in the *Maandbericht* [Monthly News] of the Institute of Industrial Design (IIV).

56 Caroline Roodenburg-Schadd, *Expressie en Ordening: Het verzamelbeleid van Willem Sandberg voor het Stedelijk Museum, 1945–1962* (Rotterdam, 2004); Ad Petersen, *Sandberg: Designer and Director of the Stedelijk* (Rotterdam, 2004).

57 Ellinoor Bergvelt et al., *80 jaar wonen in het Stedelijk*, Stedelijk Museum, Amsterdam (1981), pp. 17–25.

58 See the press file on this exhibition in the Gemeentemuseum in The Hague.

59 Timmer, *Metz & Co.*

60 Petra Faber, *My Home Bas van Pelt: Binnenhuisarchitectuur, 1931–1995* (Rotterdam, 1995).

61 Iris Knapen, 'De Amsterdamse Bijenhorfenen het moderne meubel 1945–1961' in *Jong Holland*, 17 (2001), 3, pp. 21–7; Ileen Montijn, *'t Gonst. 125 jaar De Bijenkorf* (Amsterdam, 1995); R. L. Miellet, *Winkelen in Weelde: Warenhuizen in West-Europa, 1860–2000* (Zutphen, 2001), pp. 202–23; R. L. Miellet, *Honderd jaar grootwinkelbedrijf in Nederland* (Zwolle, 1993).

4 Design as Profession, 1945–80

1 For introductions to economic and industrial developments in the Netherlands between 1945 and about 1975, see C.J.M. Schuyt and E. Taverne, *1950 Welvaart in zwart wit* (The Hague, 2000); J. W. Schot et al., *Techniek in Nederland in de twintigste eeuw* (Zutphen, 2003), part VI: *Stad, bouw, industriële ontwikkeling* and part VII: *Techniek en modernisering. Balans van de twintigste eeuw*; J. P. Smits, H. de Jong and B. van Ark, *Three Phases of Dutch Economic Growth and Technological Change, 1815–1997* [Groningen Growth and Development Centre, University of Groningen] (1999); Jan Luiten van Zanden, *Een klein land in de 20ste eeuw: Economische geschiedenis van Nederland, 1914–1995* (Groningen, 1997). For industrial design related to social-economic developments in this period, see Reyer Kras, *Nederlands Fabrikaat: Industriële vormgeving* [Teleac] (Utrecht and Bussem, 1997), pp. 164–91.

2 Gert Staal and Hester Wolters, eds, *Holland in Vorm: Dutch Design, 1945–1987* (The Hague, 1987). This book was published on the occasion of a series of exhibitions in the Stedelijk Museum (Amsterdam), Museum Boijmans Van Beuningen (Rotterdam), Centraal Museum (Utrecht), Gemeentemuseum (Arnhem) and the Gemeentemuseum (The Hague). See also André Koch, ed., *Ludiek Sensueel en Dynamisch: Nederlandse jeugdcultuur en vormgeving in de jaren zestig* (Schiedam, 2002); Bert Vreeken, *Vormgeving na '60; van Pop-Art tot Postmodern*, exh. cat., Gemeentemuseum, The Hague (1987).

3 J. Bouman, P. Schuitema and P. Zwart, *Rapport inzake de richtlijnen en mogelijkheden eener technisch-kunstzinnig verantwoorde industrieele productie van gebruiksvoorwerpen op basis eener sociaal economisch verantwoorde productie* (The Hague, 1944). Few copies of the report were produced.

4 Renny Ramakers, *Tussen kunstnijverheid en industriële vormgeving: De Nederlandse Bond voor Kunst in Industrie* (Utrecht, 1985), pp 81–92; F. Huygen, 'Vechten tegen de bierkaai: De promotie van industriële vormgeving via instituten en overheid', in *Holland in Vorm*, pp. 76–86.

5 Titus Yocarini, *Vak in beweging, VANK, GKf, VRI, GVN, bNO* (Eindhoven, 1992). On the GKf, see also Mirelle Thijssen, *Het Bedrijfsfotoboek, 1945–1965* (Rotterdam, 2002), pp. 111–19.

6 For the history of the Board, the Institute and the Centre of Industrial Design, see the issues of the *Maandbericht* and its successor *IV-Nieuws*, together with the several brochures that were published by the *IIV*: Nico Verhoeven, *Doelmatigheid van Industriële Vormgeving* (1962), and Nico Verhoeven, *Raad, Instituut en Centrum voor Industriële Vormgeving* (1966).

7 Members of the IIV visited the 1951 congress of the Council of Industrial Design in London. This resulted in the publishing of the brochure *Industriële vormgeving als factor van bedrijfsvoering* (Amsterdam, 1951).

8 R. Bullhorst and R. Eggink, *Friso Kramer: Industrieel ontwerper* (Rotterdam, 1991).

9 Staal and Wolters, *Holland in Vorm*, pp. 144–7, 174; Rosalie van Egmond, *Gero, zilver voor het volk* (Rotterdam, 2002); N. Tummers and L. Strijards, eds, *Edmond Bellefroid: de wisselwerking tussen vrije kunst en design* [Bellefroid Symposium] (Maastricht, 1994); Anna Sterk, *St Maarten Porcelein*, exh. cat., Het Kruithuis, 's-Hertogenbosch (1988); Anna Sterk, *N.V. Keramische industrie Fris, Edam, 1947–1969*, exh. cat., Princessehof, Leeuwarden (1985).

10 See *Maandbericht IV* (1956) p. 97.

11 See *Maandbericht IIV*, December 1959. Peter van Dam, *Ir. Louis C. Kalff 1897–1976: Het artistieke geweten van Philips* (Eindhoven, 2006).

12 See *Maandbericht IV* (1961/2), pp. 2–6.

13 *Industriële vormgeving in Amerika: Rapport studiegroep industrie: With a Summary in English* (The Hague, 1954); Timo de Rijk, 'Een grand tour naar de Nieuwe Wereld: "Geobsedeerd door locomotieven, sex, gebakken biefstukjes en snelheid"', in R. Baarsen et al., *Het Nederlandse binnenhuis gaat zich te buiten: Internationale invloeden op de Nederlandse wooncultuur* (Leiden, 2007), pp. 369–86.

14 Strangely enough, no monograph on Wim Gilles has been written to date. His archives are kept in the Museum Boijmans Van Beuningen in Rotterdam. For Gilles, see Timo de Rijk, ed., *Designers in Nederland: een eeuw productvormgeving* (Amsterdam and Gent, 2003), p. 152; Thimo te Duits, ed., *The Origin of Things*, exh. cat., Museum Boijmans Van Beuningen, Rotterdam (2002), pp. 108–15, 126–34.

15 W. Gilles, *De produktanalyse*, Industriële Vormgeving in kort bestek, IIV (Amsterdam, 1957).

16 J. van den Heuvel, 'De opleidingen', in Staal and Wolters, *Holland in Vorm*, pp. 180–94; N. L. Prak, *Geschiedenis van het ontwerponderwijs* (De Bilt, 1979), pp. 136–7.

17 De Rijk, 'Een grand tour naar de Nieuwe Wereld'; J. Penraat, *Hoe wilt u wonen: wenken voor nieuwe inrichting* (Amsterdam, 1957).

18 Fred Vermeulen, 'Karel Suyling, portret van een allround ontwerper', *Items*, 5 (1992), pp. 44–50; Jurriaan Schrofer and Frederique Huygen, *100 Citroën-advertenties van Karel Suyling* (Rotterdam, 1987).

19 On this second trip to the USA in 1955, see the reviews in *Maandbericht*, IV of that year. See also Timo de Rijk, *Het elektrische huis: Vormgeving en acceptatie van elektrische huishoudelijke apparaten in Nederland* (Rotterdam, 1998), pp. 113–15.

20 Dick Maan, *De Maniakken: Het ontstaan en ontwikkeling van de grafische vormgeving aan de Haagse academie in de*

jaren dertig (Eindhoven, 1982); Marg van den Burgh, *Cor Alons: Binnenhuisarchitect en industrieel ontwerper* (Rotterdam, 1987), pp. 37–41.

21 Joke Hofkamp and Evert van Uitert, 'De Nieuwe Kunstschool 1933–1943', *Kunstonderwijs in Nederland*, Nederlands Kunsthistorisch Jaarboek 30 (Haarlem, 1980). For some personal memories of the Nieuwe Kunstschool, see Frederike Huygen, *Visies op Vormgeving: Het Nederlandse ontwerpen in teksten, Deel 1: 1874–1940* (Amsterdam, 2007), pp. 30–37.

22 Caroline Boot, 'Mart Stam: Kunstnijverheidsonderwijs als aanzet voor een menselijke omgeving. Dessau-Amsterdam', *Wonen TA-BK*, 11 (1982), pp. 10–21; Cor de Wit, *Johan Niegeman, 1902–1979: Bauhaus, Sowjetunie, Amsterdam* (Amsterdam 1979), pp. 113–17.

23 'Joy, circa 1948: G. Kiljan (1891–1968)', in *The Origin of Things*, ed. Te Duits (2003), pp. 101–7.

24 Ed van Hinte, *Wim Rietveld: Industrieel ontwerper* (Rotterdam, 1996); Staal and Wolters, *Holland in Vorm*, pp. 89–91, 134, 172–3.

25 G. J. van der Grinten, 'Onderzoek naar de wenselijkheid van een opleiding van industriële vormgevers in Nederland' [unpublished report] (Delft, 1960). In the 1960s and '70s the designer Piet van der Scheer taught industrial design at the Technical University in Eindhoven regularly.

26 De Rijk, *Het elektrische huis*.

27 Mayke Groffen and Sjouk Hoitsma, *Het geluk van de huisvrouw*, exh. cat., Historisch Museum, Rotterdam (2003), p. 38; Schuyt and Taverne, *Welvaart in zwart wit*, p. 279; Ruth Oldenziel and Carolien Bouw, *Schoon genoeg: Huisvrouwen en huishoudtechnologie in Nederland, 1898–1998* (Nijmegen, 1998).

28 De Rijk, *Het elektrische huis*, pp. 83–126.

29 For the design policy at the Philips firm, see De Rijk, *Het elektrische huis*, pp. 249–352; Van Dam Ir., *Louis C. Kalff*; J. Heskett, *Philips: A Study of Corporate Management of Design* (London, 1989); Frederike Huygen, 'Design bij Philips: product, strategie en identiteit', in Aad Krol and Timo de Rijk, eds, *Jaarboek Nederlandse vormgeving 03/04* (Rotterdam, 2004), pp. 50–61.

30 Sergio Derks, *Generations of Shaving Excellence: An Impression of 60 Years of Philishave* (Eindhoven, 1998).

31 On the economic developments of the furniture industry in the Netherlands after the Second World War, see Guus Vreeburg and Hadewych Martens, *UMS Pastoe: Een Nederlandse meubelfabriek, 1913–1983*, exh. cat., Centraal Museum, Utrecht (1983), pp. 42–9; Hadewych Martens, 'De Nederlandse meubelindustrie: Een korte terugblik', in Renny Ramakers, ed., *Meubelen* [vorm & industrie in Nederland 2] (Rotterdam, 1984), pp. 7–9. Also see Staal and Wolters, *Holland in Vorm*, pp. 158–68.

32 Jojanneke Clarijs, *'t Spectrum: Moderne meubelvormgeving en naoorlogs idealisme* (Rotterdam, 2002).

33 Vreeburg and Martens, *UMS Pastoe* (1983).

34 E. Jamin, L. Schwenke and S. Wijnen, *Artifort* (Rotterdam, 1990). Also see I. van Ginneke, *Kho Liang Ie: Interieurarchitect: industrieel ontwerper* (Rotterdam, 1986).

35 Rob van Holsteijn, 'Dutch Design Center: gezamenlijke promotie van het Nederlandse meubel' in Ramakers, ed., *Meubelen*, pp. 30–31. See also www.dutchdesigncenter.nl.

36 Renny Ramakers, ed., *Kantoormeubilair*, vorm & industrie in Nederland 7 (Rotterdam, 1986).

37 A. Koch, *W. H. Gispen, serieproducten, 1923–1960* (Rotterdam, 2005); A. Koch, *W. H. Gispen: A Pioneer of Dutch Design* (Rotterdam, 1998); B. Laan and A. Koch, eds, *Collectie Gispen: Meubels, lampen en archivalia in het Nai, 1916–1980* (Rotterdam, 1996); A. Koch, *Industrieel ontwerper W. H. Gispen (1890–1981): Een modern eclecticus* (Rotterdam, 1988).

38 Sylvia van Schaik et al., *Mondial: Gispen & Gerrit Th. Rietveld* (Culemborg, 2006).

39 On Ahrend, see Dirk de Wit, *60 + 40 is waarschijnlijk honderd: Ahrend passers, pennen potloden en projecten* (Zwolle, 1996); De Rijk, ed., *Designers in Nederland*, pp. 16–17. See also Bulhorst and Eggink, *Friso Kramer*; Van Hinte, *Wim Rietveld*.

40 Ineke van Ginneke, '1982. Ontwerpteam Océ van der Grinten', in *Bekroonde ontwerpers: Zes jaar Kho Liang Ie-prijs*, industrieel ontwerpen in Nederland 4 (Rotterdam, 1985), pp. 30–35.

41 J. de Lange, *Dafjes* (Rijswijk, 1997); Warna Oosterbaan, 'de wording van het dafje', *Items*, 27 (1988); J. Lammers, *Autodesign in Nederland* (Zwolle, 1993), pp. 40–47.

42 Anne-Marie van Ommen, *200 ijzersterke merken en producten NL* (Harderwijk, 2007), pp. 286–47; De Rijk, ed., *Designers in Nederland*, p. 376; T. Tummers, 'Tomado. Opkomst en ondergang van een oerhollands merk', *Items*, 29 (1989).

43 Annet Metz et al., *De eerste plastic eeuw: Kunststoffen in het dagelijks leven: Massacultuur*, exh. cat., Gemeentemuseum, The Hague (1981); Reyer and Kras, ed., *Bakeliet: Techniek/vormgeving/gebruik*, exh. cat., Museum Boijmans Van Beuningen, Rotterdam (1981). K. Sanders lectured on designing with plastics on the fifth national plastic day in 1953; see *Plastica*, 6 (1953), p. 12; *Plastica*, 7 (1954), pp. 1–2.

44 Renny Ramakers, *Huishoudelijke artikelen*, vorm & industrie in Nederland 1 (Rotterdam, 1984), pp. 18–19; Van Ommen, *200 IJzersterke merken*, pp. 80–81; www.tiger.nl; www.mepal.com.

45 De Rijk, ed., *Designers in Nederland*, pp. 309–10; E. Truijen, *Brieven van een designer* (Delft, 1987); www.robparry.nl.

46 Stijn van Diemen, *Emmapark: Het geheim van het begrijpelijke: Tel Design, 1962–2002* (The Hague, 2003); Jan Middendorp, *'Ha, daar gaat er een van mij'. Kroniek van*

het grafisch ontwerpen in Den Haag, 1945–2000 (Rotterdam, 2002), pp. 101–10; A. O. Eger, 'Jan Lucassen', *Product: Tijdschrift voor productontwikkelaars* (2005), p. 12.

47 Alain Le Quernec, ed., *Studio Dumbar* (Paris, 2006). For more on Studio Dumbar and its history, see www.studiodumbar.com.

48 F. Huygen and H. Boekraad, *Wim Crouwel: Mode en Module* (Rotterdam, 1997), pp. 126–73; Kees Broos, ed., *Ontwerp: Total Design; Design: Total Design* (Utrecht, 1983).

49 Ben Bos, *Benno Wissing* (Eindhoven, 2006); Dingenus van de Vrie, *Benno Wissing, grafisch en ruimtelijk ontwerpen*, exh. cat., Museum Boijmans Van Beuningen, Rotterdam (1999); www.nago.nl.

50 Thijsen, *Bedrijfsfotoboek*, pp. 91–100.

51 Dirk van Ginkel and Paul Hefting, *Ben Bos: Design of a Lifetime* (Amsterdam, 2000); Carry van Lakerveld, *Een keuze uit het werk van Jolijn van de Wouw, grafisch ontwerper 1942–2002* (Amsterdam, 2002); www.nago.nl.

52 Paul Mijksenaar, *Visual Function: An Introduction to Information Design* (Rotterdam, 1997); Kees Broos and Paul Hefting, *Grafische Vormgeving in Nederland: Een eeuw* (Naarden, 1995), pp. 184–5; *Jurriaan Schrofer*, exh. cat., Museum Fodor, Amsterdam (1974); Dingenus van de Vrie, *Haagse omslagen: Jurriaan Schrofer en De Ooievaarpockets, 1958–1962* (Amsterdam, 2006); www.nago.nl.

53 Wim Crouwel, *Alphabets* (Amsterdam, 2003); Wim Crouwel, *New Alphabet: A Possibility for the New Development* (Hilversum, 1967); Thimo te Duits, ed., *The Origin of Things*, exh. cat., Museum Boijmans Van Beuningen, Rotterdam (2002), pp. 146–55.

54 Wibo Bakker, 'Design in de Supermarkt', in *Jong Holland*, 2 (2006), pp. 14–23.

55 Toon Lauwen, *Holland in beeld, 1895–2008* (Bussum, 2007), pp. 192–283; Kras, *Nederlands Fabrikaat*), pp. 80–99.

56 L. van den Berg, ed., *Benno Premsela: Een vlucht naar voren*, Centraal Museum, Utrecht (1996); Martin Visser et al., *Benno Premsela onder Anderen*, exh. cat., Stedelijk Museum, Amsterdam (1981).

57 Ineke van Ginneke, '1980: José de Pauw', *Bekroonde ontwerpers* (1985), pp. 20–25.

58 See www.premsela.org.

59 Much has been written on graphic design for the Dutch government. The author used in the first place: Middendorp, *'Ha, daar er een van mij'*; Kras, *Nederlands Fabrikaat*, pp. 100–19; Staal and Wolters, eds, *Holland in Vorm*, pp. 29–68.

60 Jan Middendorp, *Dutch Type* (Rotterdam, 2004).

61 *Design* (December 1968), p. 48.

62 C. Versteeg, *100 jaar ANWB bewegwijzering* (The Hague, 1994); Kras, *Nederlands Fabrikaat*, pp. 56–79.

63 For this discussion, see 'Een ijzersterk betoog over een grote pan', in *De Vorm* (1975), and Willemijn Stokvis, 'Bewegwijzering in Nederland', in *Vrij Nederland*, 25 September 1976). Both articles are reprinted in Dingenus van de Vrie and Titus Yocarini, eds, *10 jaar designkritiek geknipt en gescheurd uit . . .* (Amsterdam, 1980).

64 J. Bolten, *Het Nederlandse bankbiljet en zijn vormgeving* (Amsterdam, 1987); Staal and Wolters, *Holland in Vorm*, pp. 38–44, 50–54; Kras, *Nederlands Fabrikaat*, pp. 100–109.

65 Jan Teunen, *Bruno Ninaber van Eyben: With Compliments* (Rotterdam, 2002).

66 Arnoud Witte and Esther Cleven, eds, *Design is geen vrijblijvende zaak: Organisatie, imago en context van de ptt-vormgeving tussen 1906 en 2002* (Breda and Rotterdam, 2006); Gerard Forde, *Design in the Public Service: the Dutch PTT, 1920–1990*, exh. cat., Design Museum, London (1991); Paul Hefting, ed., *Kunst en vormgeving bij de PTT*, special issue of *Kunstschrift/ Openbaar Kunstbezit* (October–November 1985); Staal and Wolters, *Holland in Vorm*, pp. 38–44, 50–54.

5 Design for Debate, 1960s to the Present

1 On design criticism in general in the Netherlands, see F. Huygen, *Designkritiek in Nederland: een essay* (Amsterdam and Rotterdam, 1995); C. Kuitenbrouwer and K. Sierman, *Over grafisch ontwerpen in Nederland: een pleidooivoor geschiedschrijving en theorievorming* (Rotterdam, 1996); Y. Bartholomée, *Vormgevingskritiek in de Nederlandse pers* (Rotterdam, 2003).

2 Ineke van Ginneke, *Kho Liang Ie: Interieurarchitect: industrieel ontwerper* (Rotterdam, 1986), p. 49; Ellinoor Bergvelt et al., *80 jaar wonen in het Stedelijk*, exh. cat., Stedelijk Museum, Amsterdam (1981), pp. 24–5.

3 Jan Middendorp, *'Ha, daar gaat er een van mij!': Kroniek van het grafisch ontwerpen in Den Haag, 1945–2000* (Rotterdam, 2002), pp. 88–93; B. Majorick, *Ontwerpen en verwerpen; industriële vormgeving als noodzaak* (Amsterdam, 1959).

4 *De Groene Amsterdammer*, 13 February 1960.

5 In addition to the literature already mentioned in chapter Four on Goed Wonen, mention should be made of Michel Karis, 'Modern Wonen in Nederland, 1962–1973', and Marjonne van Dijk, 'Meubels van de markt, nostalgie en sfeer in het jaren zestig interieur', both in André Koch, ed., *Ludiek Sensueel en Dynamisch: Nederlandse jeugdcultuur en vormgeving in de jaren zestig* (Schiedam, 2002), pp. 138–51, 152–61.

6 Liane Lefaivre and Ingeborg de Roode, eds, *Aldo van Eyck: The Playgrounds and the City*, exh. cat., Stedelijk

Museum, Amsterdam (2002); Francis Strauven, *Aldo van Eyck's Orphanage: A Modern Monument* (Rotterdam, 1996); Francis Strauven, *Relativiteit en verbeelding* (Amsterdam, 1994).

7 Martien de Vletter, *De kritiese jaren zeventig: Architectuur en stedenbouw in Nederland/The Critical Seventies; Architecture and Urban Planning in the Netherlands* (Rotterdam, 2004).

8 Ruud van Wezel, 'Een klus die nooit geklaard is', in Ruth Oldenziel and Carolien Bouw, *Schoon genoeg: Huisvrouwen en huishoudtechnologie in Nederland, 1898–1998* (Nijmegen, 1998), pp. 231–52.

9 Rüdiger Jungbluth, *IKEA: Het geheim van succes* (Amsterdam, 2006), pp. 100–02.

10 Frederique Huygen, 'De weg van de geleidelijkheid: interview met Jan des Bouvrie', *Items*, 21 (1986); Gert Staal, *Gijs Bakker vormgever: Solo voor een solist*, exh. cat., Centraal Museum, Utrecht (1989); Ida van Zijl and Gijs Bakker, *Objects to Use* (Rotterdam, 2000). On De Bouvrie, Bakker, Gelderland and Castelijn, see also Timo de Rijk, ed., *Designers in Nederland; een eeuw productvormgeving* (Amsterdam and Gent, 2003), pp. 33–4, 59, 76–7, 146.

11 Renny Ramakers, 'One-off Items and Mass Production', in *Holland in Vorm: Dutch Design, 1945–1987*, ed. Gert Staal and Hester Wolters (The Hague, 1987), pp. 213–28; reprinted in *Morf, tijdschrift voor vormgeving 4* (2006).

12 Caroline Boot, ed., *In het spoor van het Bauhaus: weefwerk van Kitty van der Mijll Dekker*, exh. cat., Textielmuseum, Tilburg (2007); Claudia Thunnissen and Rob Sperna Weiland, *Gerrit de Blanken, 1894–1961: pottenbakker uit Leiderdorp: virtuoze eenvoud*, exh. cat., Lakenhal, Leiden (2005); Eugène Langendijk and Mienke Simon Thomas, *Dutch Art Nouveau and Art Deco Ceramics: The Boijmans Van Beuningen Museum Collection*, Rotterdam (2001); M. Singelenberg-van der Meer, *Mobach: 100 jaar keramiek in Utrecht* (De Bilt, 1995).

13 Lucien den Arend et al., *De nieuwe vrijheid van de ambachtskunsten* (Venlo and Delft, 1981). On studio pottery, see Thimo te Duits, *Moderne Keramiek in Nederland/Modern Ceramic in the Netherlands* (The Hague, 1990); Mieke G. Spruit-Ledeboer, *Nederlandse keramiek, 1900–1975* (Amsterdam, 1976). On textile crafts, see Liesbeth Crommelin, *Textiel in het Stedelijk: Textiles in the Stedelijk*, Stedelijk Museum, Amsterdam (1993); *Textiel Nu* [1–14], exh. cat., Textielmuseum, Tilburg (1980–84). For textile crafts in the 1960s and '70s, see also the magazines *Scheppend Ambacht* and its successor *Bijvoorbeeld*. On silversmiths and jewel design, see Yvonne G.J.M. Joris, *Jewels of Mind and Mentality: Dutch Jewelry Design, 1950–2000*, exh. cat., Museum Het Kruithuis, 's-Hertogenbosch (2000); Marjan Unger, *Het Nederlandse sieraad in de 20ste eeuw* (Bussum and Utrecht, 2004). On studio glass: Titus Eliëns, *Lexicon Nederlandse glaskunst van de twintigste eeuw* (Lochem, 2004); Annette van der Kley-Blextoon, *Leerdam glas, 1878–2003: Glasfabriek Leerdam* (Lochem, 2004); Job Meihuizen, *De wereld volgens Valkema* [Glascahier Nationaal Glasmuseum] (Leerdam, 2004); K.J.H. Wasch et al., *Floris Meydam in Vorm* (Haarlem, 2003). On postwar developments in stained glass, see Carine Hoogveld, Ellinoor Bergvelt and Frans van Burkom, eds, *Glas in lood in Nederland, 1817–1968* (The Hague, 1989), pp. 163–96.

14 J.J.E. Salden (revised by Bert Rutgrink), *De kunst om van de kunst te leven: zakelijk informatie voor ambachtskunstenaars* [COSA] (Delft, 1982); A.W.H. Quaedvlieg, 'Het overheidsbeleid met betrekking tot het scheppend ambacht in Nederland: Kroniek 21 jaar COSA', *Scheppend Ambacht* (1969), pp. 110–16.

15 Bergvelt et al., *80 jaar wonen in het Stedelijk*, pp. 22–3; Titus Yocarini, *Vak in Beweging: Grafische ontwerpers en hun organisatie* (Eindhoven, 1976), p. 25.

16 Karin Gaillard et al., *Keramiekcollectie Dienst Beeldende Kunst, 's-Hertogenbosch* ('s-Hertogenbosch, 1983); André Koch, *Galerie Kapelhuis; dertig jaar vernieuwing in de toegepaste kunst, 1960–1990* (Rotterdam, 2003); Marie-José van den Hout, *Galerie & Marzee Collectie, 1979–2004* (Nijmegen, 2004); *Radiant: 30 jaar RA/30 years RA* (Amsterdam, 2006); Truus Gubbels, *Het oog voorbij: Galerie Nouvelles Images, 1960–2000: 40 jaar gedeeld galeriehouderschap, Ton Berends (1960–1988) en Erik Bos (1988–2000)* (The Hague, 2001); Spruit-Ledeboer, *Nederlandse keramiek*.

17 Karel Sanders, *De industriële ontwerper en zijn ambachtelijke collega* [Industriële vormgeving in kort bestek] (Amsterdam, 1955).

18 R. Smeets, 'Terugblikken omwille van morgen', *Scheppend Ambacht* (1964), pp. 132–5.

19 Spruit-Ledeboer, *Nederlandse keramiek*; H.J.H., 'Stichting "Keramisch Werkcentrum" in Heusden (N.Br.)', *Mededelingenblad vrienden van de Nederlandse ceramiek*, 69/70 (1973), pp. 48–51.

20 Crommelin, *Textiel in het Stedelijk*.

21 Marjan Unger, *Helly Oestreicher*, exh. cat., De Beyerd, Breda (1989).

22 Unger, *Het Nederlandse sieraad*; Joris, *Jewels of Mind and Mentality*.

23 Evert van Straaten, 'Geen angst voor potten/Don't Be Afraid of Pots', *Mededelingenblad van de vereniging van vrienden van de Nederlandse ceramiek* 3, 109/110 (1983), pp. 3–6. This magazine was, together with *Bijvoorbeeld*, an important platform for discussions on ceramics during the 1970s and '80s.

24 J. P. Smits, H. de Jong and B. van Ark, *Three Phases of Dutch Economic Growth and Technological Change*,

1815–1997, Groningen Growth and Development Centre, Rijksuniversiteit (Groningen, 1999); Jan Luiten van Zanden, *Een klein land in de 20ste eeuw: Economische geschiedenis van Nederland, 1914–1995* (Groningen, 1997).

25 Simon Mari Pruys, *Dingen vormen mensen; Een studie over produktie, consumptie en cultuur* (Bilthoven, 1972).

26 Ibid., p. 31.

27 Ibid., p. 35.

28 Ibid., pp. 102–3.

29 Ibid., p. 118.

30 Annemarie Vels Heijn et al., *Verslag van het symposium 'Massacultuur en museumbeleid', gehouden op 13 juni 1981 in het Haags Gemeentemuseum*, Nederlandse Museum Vereniging (1981).

31 Dingenus van de Vrie and Titus Yocarini, eds, *10 jaar designkritiek geknipt en gescheurd uit . . .* (Amsterdam, 1980). On developments in the graphic arts during these years, see A. Lopes Cardozo et al., 'Graphic Design', in *Holland in Vorm*, pp. 247–78. See also the surveys in Alston W. Purvis and Cees W. de Jong, *Nederlands grafisch ontwerp van de negentiende eeuw tot nu* (Laren, 2006); Jan Middendorp, *Dutch Type* (Rotterdam, 2004); Middendorp, *'Ha, daar gaat er een van mij'*; Kees Broos and Paul Hefting, *Grafische vormgeving in Nederland. Een eeuw* (Amsterdam, 1995).

32 Max Bruinsma et al., *Een leest heeft drie voeten: Dick Elffers & de kunsten* (Amsterdam, 1989).

33 Yocarini, *Vak in beweging*.

34 On this debate, see Van de Vrie and Yocarini, *10 jaar designkritiek*; F. Huygen and H. Boekraad, *Wim Crouwel: Mode en Module* (Rotterdam, 1997), pp. 159–70; Chris H. Vermaas, *Jan van Toorn* [Roots 3] (Eindhoven, 2005); Jan van Toorn, *Design's Delight* (Rotterdam, 2006).

35 Steffen Maas, ed., *Hitweek, 1965–1969*, exh. cat., De Beyerd, Centrum voor beeldende kunst, Breda (2003); Anthon Beeke, *Dutch Posters, 1960–1996* (Amsterdam, 1997).

36 Han Steenbruggen, *Swip Stolk: Master Forever*, exh. cat., Groninger Museum, Groningen (2000); Broos and Hefting, *Grafische Vormgeving in Nederland*, pp. 181–5.

37 The three first issues published by the Gerrit Jan Thieme Fund were: Wim Crouwel, *Ontwerpen en drukken* (1974); Dick Elffers, *Vorm en tegenvorm; een poging tot een portret van een ontwerper* (1976); Piet Schreuders, *Lay in, Lay out: zijn ontwerpers misdadig?* (1977). Piet Schreuders's text was reprinted in 1997 by the Buitenkant Publishers in Amsterdam and again in *Morf, tijdschrift voor vormgeving* 2 (2005).

38 Wim Crouwel, *Alphabets* (Amsterdam, 2003).

39 Huygen and Boekraad, *Wim Crouwel*, pp. 159–70; Van de Vrie and Yocarini, *10 jaar designkritiek*.

40 *De verloren jaren: Persoonlijke visies op De Bestverzorgde Boeken 1971 tot en met 1985*, exh. cat., Rijksmuseum Meermanno-Westreenianum, The Hague (1991).

41 Dingenus van de Vrie, *Kwadraat-bladen; een serie experimenten in druk op het gebied van de grafische vormgeving, beeldende kunst, literatuur, architectuur en muziek, 1955–1974* (Amsterdam, 2005).

42 Maas, *Hitweek*; Henk van Gelder and Hester Carvalho, *Gouden tijden: 50 jaar Nederlandse Popbladen* (Amsterdam, 1994), pp. 45–57.

43 Hans Oldewarris and Peter de Winter, eds, *20 jaar/years 010: 1983–2003* (Rotterdam, 2003)

44 Max Bruinsma et al., *Beeld tegen beeld: Wild Plakken*, exh. cat., Centraal Museum, Utrecht (1993); Paul Hefting et al., *Hard Werken* (Amsterdam and Rotterdam, 1995); Broos and Hefting, *Grafische Vormgeving in Nederland*, pp. 201–5; *Laatste Post* [special edition in the form of an ordinary paper dealing with the history of the Enschedese School], Stedelijk Museum, Amsterdam, and Rijksmuseum Twenthe, Enschede (1997).

45 Gijs Bakker and Evert Rodrigo, *Design from the Netherlands/Design aus den Niederlände*, Bureau Beeldende Kunst Buitenland, Ministerie CRM (Amsterdam, 1980).

46 Lidewij Edelkoort et al., *Ulf Moritz: Fascination Textiles* (Amsterdam, 2007); Jan Teunen, *Bruno Ninaber van Eyben: With Compliments* (Rotterdam, 2002); Jack Meijers, '20 jaar Vormgeversassociatie', *Items*, 8 (1994), pp. 39–44; Erik Beenker et al., *Jan van der Vaart: Ceramics* (Amsterdam, 1991); *Aldo van den Nieuwelaar: verlichting, meubelen, gordijnstoffen*, exh. cat., Metz & Co., Amsterdam (1984); Gijs Bakker, 'Ik ben vormgever, geen technoloog', *Bijvoorbeeld*, XIII/3 (1981), pp. 2–5.

47 I. Szénássy et al., *Ontwerpen voor de industrie, 1: Studies over ontwerpen voor de industrie, waarbij opgenomen de catalogus van de gelijknamige tentoonstelling over industrieel ontwerpen en vormgeven* (Groningen, 1982).

48 Ibid., p. 30.

49 Wim Crouwel was Extraordinary Professor at the Technical University in Delft from 1972 to 1978. In addition he was Professor from 1982 to 1985. Wim Crouwel, *Vormgeving – door wie* [Inaugural Lecture TH Delft] (1973); Wim Crouwel, *Vormgeving – zin en onzin* [Farewell Lecture TH Delft] (1985).

50 Ineke van Ginneke, '1979: Bruno Ninaber van Eyben', in *Bekroonde ontwerpers: zes jaar Kho Liang Ie-prijs* [Industrieel Ontwerpen in Nederland 4] (Rotterdam, 1985), pp. 14–19.

51 Jan Brand et al., *Product Design Diversity* (Arnhem [Artez] and Rotterdam, 2007); Chris Reinewald, 'We vonden onszelf toen geweldig: Gijs Bakker en oud-studenten over hun academietijd in Arnhem', *Items*, 1 (1994), pp. 38–44.

52 Rob van Holsteijn, 'Dutch Design Center: gezamenlijke promotie van het Nederlandse meubel', in Ramakers, ed., *Meubelen*, pp. 30–31; see www.dutchdesigncenter.nl.

Compare the chronological survey of furniture design in Luca Dosi Delfini et al., *The Furniture Collection, Stedelijk Museum, Amsterdam, 1850–2000: From Michael Thonet to Marcel Wanders*, museum cat., Stedelijk Museum, Amsterdam (2004).

53 Frederique Huygen, 'Nederlandse Meubelen 1980–1983', *Items*, 7 (1983), pp. 5–43.

54 Frederique Huygen, 'Tussen ambacht en industrie', *Items*, 14 (1984), pp. 4–23.

55 Meijers, '20 jaar Vormgeversassociatie', pp. 39–44.

56 Marijke Kuper, *Mart van Schijndel, kleurrijk architect* (Rotterdam, 2003).

57 De Rijk, ed., *Designers in Nederland*, pp. 119–120.

58 Jeroen Vinken et al., *Jeroen Vinken* (Tilburg, 2004).

59 The Centre for Industrial Design in Amsterdam organized in 1967 the exhibition *Nieuwe Italiaanse vormgeving* (New Italian Design) in the Beurs.

60 Jaap Huisman, 'Een enkeltje italië', *Items*, 24 (1987), pp. 31–9; Toon Lauwen, *Bob Noorda: Nederlandse nuchterheid Italiaanse elegantie*, Roots 7 (Amsterdam, 2007).

61 Ghislain Kieft, *Memphis-Design*, exh. cat., Museum Het Kruithuis, 's-Hertogenbosch (1984).

62 Titus Eliëns, *Borek Šípek ; glas, design, architectuur*, Drents Museum, Assen (2006).

63 Bart Lootsma, 'Weltschmerz: de oktoberrevolutie in Arnhem', *Items*, 31 (1989), pp. 6–11; E. Hartkamp-Jonxis, 'Het nut van het fantastische in de derde macht: Tien jaar cubic 3 Design' (1991); Jaap Huisman, 'Cubic 3 zet 17e eeuwse traditie voort met plastic', *Volkskrant*, 16 November 1991; Ed Annink, *Ed Annink Designer* (Rotterdam, 2002). Ed Annink et al., *Bright Minds, Beautiful Ideas: Parallel Thoughts in Different Times: Bruno Munari, Charles & Ray Eames, Martí Guixé and Jurgen Bey* (Amsterdam, 2003).

64 Ida van Zijl, 'Veni, vidi, vici?: Postmodernism and the Interior', in *From Neorenaissance to Postmodernism, A Hundred and Twenty-five Years of Dutch Interiors, 1870–1995*, ed. Ellinoor Bergvelt, Frans van Burkom and Karin Gaillard (Rotterdam, 1996), pp. 330–53.

65 Hein van Haaren, 'Neo-design', *Items* 12, (1984), p. 5.

66 Frederike Huygen, *Martin Visser: Oeuvreprijs 1998* (Amsterdam, 1998); Gert Staal, 'Bemiddelaar tussen kunst en design', *Items*, 33 (1990), pp. 25–9; Guus Vreeburg, 'De meubelkunst van Martin Visser', *Jong Holland*, 1/2 (1985), pp. 38–55.

67 Peter van Kester, 'Vijftien ontwerpers zoeken een industrie', *Bijvoorbeeld* (1985/6), pp. 11–15. Compare this with Gert Staal, 'The Strength of Ambivalence', *Dutch Art: Design in the Netherlands*, International Information of the Dutch Ministry of Cultural Affairs (The Hague, 1989), pp. 15–20.

68 Thalita Schoon, ed., *Het meubel verbeeld; recente tendensen in sculptuur/Furniture as Art: Recent Tendencies in Sculpture*, exh. cat., Museum Boijmans Van Beuningen, Rotterdam (1988); *Meubelsculptuur: wonderlijke tafels en stoelen . . . of beeldhouwwerken?*, exh. cat., Museum Commanderie van S-Jan/Galerie Marzee, Nijmegen (1992); Paul Donker Duyvis et al., *Schräg/Tegendraad: Parodie, humor en spot in de hedendaagse Nederlandse kunst*, exh. cat., Rheinische Landesmuseum Bonn (The Hague, 1991).

69 Lisette Thooft and Micky Otterspeer, *Goed in Vorm: gesprekken over dutch design* (Bloemendaal, 1993), pp. 77–86.

70 Jennifer Allen et al., *Atelier van Lieshout* (Rotterdam, 2007).

71 Thimo te Duits, 'Shocking: Surrealism and Fashion Now', *Surreal Things: Surrealism and Design*, exh. cat., Victoria and Albert Museum, London (2007), pp. 139–60; this article was also published in *Vreemde dingen: Surrealisme en design*, exh. cat., Museum Boijmans Van Beuningen, Rotterdam (2007), pp. 184–205. On mixing of art and fashion, see also *Droog and Dutch Design: From Product to Fashion: The Collection of the Centraal Museum Utrecht*, exh. cat., Living Design Center Ozone, Japan (Utrecht, 2000); Jan Brand and José Teunissen, eds, *The Power of Fashion: About Design and Meaning* (Arnhem, 2006).

72 Gert Staal, *De bizarre gelaagdheid van de mode: Alexander van Slobbe en Guus Beumer* (Haarlem, 2000).

73 See the interactive fashion presentation of Le Cri Néerlandais on DVD: *Défilé sans public*, Stichting Het Nederlands Vormgevingsinstituut/The Netherlands Design Institute (1995).

74 See www.viktor-rolf.com. Inez van Lamsweerde, *Viktor & Rolf* (Breda, 1998); Viktor & Rolf, *Viktor & Rolf* (Amsterdam, 1999).

75 Robert de Haas et al., *Rijksaankopen 1985: Werk van hedendaagse beeldende kunstenaars*, Rijksdienst Beeldende Kunst (Amsterdam and The Hague, 1985); Cees Strauss, 'RBK', *Bijvoorbeeld*, XX/3 (1988), pp. 15–18. In 1992 the magazine *Vormberichten* published a series of articles on collecting modern design in Dutch museums.

76 *Meubelsculptuur*; Erwin Houtenbrink et al., *Het meubelboek: Nederlands meubelontwerp, 1986–1996*, Stichting Sofa (The Hague, 1996); Marjan Unger and Christina Hosman, *Tejo Remy*, exh. cat., Centrum Beelden Kunst, Amersfoort (2001); Annink et al., *Bright Minds* (2003); Paola Antonella et al., *Wanders Wonders: Design for a New Age*, exh. cat., Het Kruithuis, Museum of Contemporary Art, 's-Hertogenbosch (1999).

77 Concerning the background, birth and development of Droog, compare Renny Ramakers and Gijs Bakker, eds, *Droog Design: Spirit of the Nineties* (Rotterdam, 1998); Ida van Zijl, 'Droog Design', *Jong Holland*, 3 (1996), pp. 47–50; Renny Ramakers, ed., *Simply Droog* (Amsterdam, 2004).

78 Piet Hein Eek et al., *Piet Hein Eek, 1990–2006* (Eindhoven, Baarn and Amsterdam, 2006).

79 Timo te Duits, 'Dick van Hoff; Prototypes keukenmachines 2003', in *The Origin of Things*, pp. 272–5; Brigitte Fitoussi et al., *Richard Hutten: Works in Use* (Oostkamp, 2006); Ed van Hinte et al., *Richard Hutten* (Rotterdam, 2002).

80 E. van Hinte and G. Williams, *Black/Ineke Hans* (Rotterdam, 2003).

81 Louise Schouwenberg, *Hella Jongerius* (London, 2003).

82 KesselsKramer, *One Hundred and One Things to Do* (Amsterdam, 2006).

83 Guus Beumer and Louise Schouwenberg, 'De Stille Kracht van vormgeving', *Metropolis M*, 1 (2004), pp. 129–40; Frederike Huygen, 'Droog serveren', *Metropolis M*, 3 (2004), pp. 47–57; Micheal Rock, 'Mad Dutch Disease: The Strange Case of Dutch Design and Other Contemporary Contagions', in *Jaarboek Nederlandse vormgeving 03/04* (Rotterdam, 2004), pp. 63–81.

84 At the moment one can observe two tendencies: some designers focus on socially engaged and/or ecological design, while others prefer to be artists and deliberately produce their products in limited editions. Some avant-garde designers, such as Marcel Wanders, Joep van Lieshout and Job Smeets, have engaged themselves with Moooi, a much more commercial organization for distributing modern design (see www.moooi.com). There have been several interesting initiatives in recent years involving cooperation with craftsmen and designers in the Third World; see, for instance, Louise Schouwenberg, 'A Dutch Perspective: the Netherlands', in *Prince Claus Journal 10a: The Future is Handmade: The Survival and Innovation of Craft*, pp. 108–21.

85 Bas van Lier, 'Nederlandse ontwerpers over hun buitenlandse escapades: "Design is gewoon business"', *Items* (July–August 2004), pp. 48–52.

86 Ed van Hinte, *Huibert Groenendijk: Down to Design* (Rotterdam, 2004); see www.welldesign.com; www.waacs.nl.

87 See www.designacademy.nl. The results of the annual final examinations of the Design Academy in Eindhoven have been published annually under the title *Graduation* since 1991.

88 In the prestigious overview *Spoon* (New York, 2002), among the hundred 'best' and 'most promising' designers of the world one could discover six Dutch artists: Ed Annink, Jurgen Bey, Tord Boontje, Richard Hutten, Hella Jongerius and Job Smeets. Five years later, six designers were again listed in *& Fork* (New York, 2007): Maarten Baas, Piet Hein Eek, Ineke Hans Chris Kabel, Joris Laarman and Wieki Somers.

89 Gert Staal, 'De grote restauratie: Gert Staal blikt terug op 10 jaar Nederlandse vormgeving', and Merel Bem, 'Hoe gaat het met startende vormgevers', in *Jaarverslag 2004 Fonds voor Heeldende Kunst, Vormgeving en Bouwkunst* (Amsterdam, 2005).

90 Ed van Hinte and Conny Bakker, *Trespassers: Inspiration for Eco-efficient Design*, Nederlands Design Instituut (Rotterdam, 1999); www.doorsofperception.com; www.ydi.nl (Young Designers and Industry).

91 *Tot hier en . . . verder: Jaarverslag Vormgevingsinstituut 2000* (Amsterdam, 2001).

92 See www.premsela.org.

93 Yocarini, *Vak in beweging*).

94 Petra Timmer, ed., *Waar kleur een specifieke rol speelt: Sikkensprijs* (Blaricum, 1997).

95 Up to 2008 the Designprijs Rotterdam (Rotterdam Design Contest) has been held nine times. The winners were: (2007) Thonik, (2003) Hella Jongerius, (2001) Jop van Bennekom, Erik Wong and redactie Forum, (1999) NL Architects, (1997) Maatschappij voor Oude en Nieuwe Media, (1996) Bob van Dijk/Studio Dumbar, (1995) Jan Erik Baars, Caroline Brouwer and Jan Paul van der Voet/Philips Corporate Design, (1994) Diek Zweegman/BRS Premsela Vonk, (1993) Roelof Mulder. See Marianne Toussaint, 'Vijf jaar Designprijs Rotterdam, een evaluatie' (Rotterdam, 1997), an unpublished report of the Rotterdamse Kunst Stichting. On each occasion the jury reports have been published, together with critical essays on design by national and international professionals.

96 A selection of entries from the first three Design Contests was shown in Bremen (Germany) in 1996; Bart Lootsma, *Mentalitäten: Niederländisch Design; prämierte Arbeiten des Designpreises Rotterdam 1993-1996/Mentalities: Dutch Design; Nominated Products of the Design Prize Rotterdam* (Amsterdam, 1996)

97 Compare the overviews and Jury reports published in *De Nederlandse Designprijzen* by the Stichting Nederlandse Designprijzen in 2003, 2005 and 2006.

Bibliography

Exhibition Catalogues

Adrichem, Jan van, et al., *Rebel, mijn hart: Kunstenaars 1940–1945*, Nieuwe Kerk, Amsterdam (Zwolle, 1995)

Aldo van den Nieuwelaar: verlichting, meubelen, gordijnstoffen, Metz & Co., Amsterdam (1984)

Antonella, Paola, et al., *Wanders Wonder: Design for a New Age*, Het Kruithuis, Museum of Contemporary Art, 's-Hertogenbosch (1999)

Baeten, Jean Paul, *Ontwerp het onmogelijke: de wereld van architect Hendrik Wijdeveld*, Nederlands Architectuurinstituut, Rotterdam (2006)

Bakker, Gijs, and Evert Rodrigo, *Design from the Netherlands: Design aus den Niederlände* [Bureau Beeldende Kunst Buitenland, Ministerie CRM] (Amsterdam, 1980) [travelling exhibition]

Beenker, Erik, and L. van den Berg, *Benno Premsela, een vlucht naar voren*, Centraal Museum, Utrecht (1996)

Beeren, Wim, et al. eds, *Het Nieuwe Bouwen in Rotterdam*, Museum Boijmans Van Beuningen, Rotterdam (1982)

Bergvelt, Ellinoor, ed., *Industry and Design in the Netherlands, 1850–1950*, Stedelijk Museum, Amsterdam (1985)

—, 'The Decorative Arts in Amsterdam, 1890–1930', in *Designing Modernity: The Arts of Reform and Persuasion, 1885–1945*, Wolfsonian, Miami Beach (1995), pp. 79–110

—, et al., *Amsterdamse School, 1910–1930*, Stedelijk Museum, Amsterdam (1975)

—, et al., *80 jaar wonen in het Stedelijk*, Stedelijk Museum, Amsterdam (1981)

Boekraad, Cees, et al., *Het Nieuwe Bouwen. De Nieuwe Beelding in de architectuur. Neo-Plasticism in Architecture De Stijl*, Gemeentemuseum, The Hague (1983)

Bois, M. de, *Chris Lebeau, 1878–1945*, Drents Museum, Assen (Zwolle, 1983)

Bois, Mechteld de, ed., *C.A. Lion Cachet, 1864–1945*, Drents Museum, Assen (Zwolle, 1994)

Bommer, Bea, ed., *Katoendruk in Nederland*, Textielmuseum, Tilburg; Gemeentemuseum, Helmond (1989)

Boot, Caroline, and Sanny de Zoete, *Artistiek damast van Brabantse bodem 1900–1960, ontwerpen van Chris Lebeau, André Vlaanderen, Jaap Gidding en tijdgenoten*, Textielmuseum, Tilburg (2005)

Boot, Caroline, ed., *In het spoor van het Bauhaus: weefwerk van Kitty van der Mijll Dekker*, Textielmuseum, Tilburg (2007)

Boot, M., 'Olanda', in *Torino 1902: le arti decorative internazionale del nuevo secolo*, exh. cat. (Turin, 1994), pp. 488–529

Brentjens, Yvonne, *G. W. Dijsselhof (1866–1924): Dwalen door het Paradijs*, Gemeentemuseum, The Hague (Zwolle, 2002)

—, *K.P.C. de Bazel (1869–1923): Ontwerpen voor het interieur*, Gemeentemuseum, The Hague (Zwolle, 2006)

—, *Rozenburg: Plateel uit Haagse kringen (1883–1917)*, Gemeentemuseum, The Hague (Zwolle, 2007)

—, *Piet Zwart: Vormingenieur*, exh. cat. Gemeentemuseum, The Hague (Zwolle, 2008)

Broos, Kees, *Piet Zwart, 1885–1977*, Gemeentemuseum, The Hague (Amsterdam, 1982)

Bruinsma, Max, et al., *Beeld tegen beeld: Wild Plakken*, Centraal Museum, Utrecht (1993)

Dam, Jan Daniël van, *Amstelhoek, 1897–1910*, Museum Het Princessehof, Leeuwarden (1986)

Donker Duyvis, Paul, et al., *Schräg/Tegendraad: Parodie, humor en spot in de hedendaagse Nederlandse kunst*, Rheinische Landesmuseum, Bonn (The Hague, 1991)

Droog and Dutch Design; From Product to Fashion:The Collection of the Centraal Museum Utrecht, Living Design Center Ozone, Japan (Utrecht, 2000)

Duits, Thimo te, 'Shocking: Surrealism and Fashion Now', in *Surreal Things: Surrealism and Design*, Victoria and Albert Museum, London (2007), pp. 139–60

Duits, Thimo te, ed., *The Origin of Things*, Museum Boijmans Van Beuningen, Rotterdam (2002)

Duits, Thimo te, *Glasfabriek Leerdam, 1915–1934: De kunstnijverheidscollectie van de Glasfabriek Leerdam, 1915–1934 Museum Boijmans Van Beuningen*, Drents Museum, Assen (Zwolle, 1998)
Dupuits, Petra, *Metz est venu*, Stedelijk Museum, Amsterdam (1992)
Edelkoort, Lidewij, et al., *Ulf Moritz: fascination textiles*, Textielmuseum, Tilburg (Amsterdam, 2007)
Eliëns, Titus, *H. P. Berlage (1856–1934): Ontwerpen voor het interieur*, Gemeentemuseum, The Hague (Zwolle, 1998)
—, *T.A.C. Colenbrander (1841–1930): Ontwerper van de Haagse Plateelbakkerij Rozenburg*, Gemeentemuseum, The Hague (Zwolle, 1999)
—, and J.J.M. Meihuizen, *K.P.C. de Bazel – de glazen van de architect*, Glasmuseum, Leerdam (2005)
Forde, Gerard, *The Dutch PTT, 1920–1990: Design in the Public Service*, Design Museum, London (1991)
Gaillard, Karin, *Labor Omnia Vincit: een idealistische meubelfabriek, 1910–1935*, Gemeentemuseum, Arnhem (1991)
Groffen, Mayke, and Sjouk Hoitsma, *Het geluk van de huisvrouw*, Historisch Museum, Rotterdam (2004)
Hefting, P. H., *Royal PTT Nederland NV – Art and Design Past and Present: A Guide*, Design Museum, London (1990)
Heslenfeld, Mariannne, *De collectie Holland: art nouveau-keramiek van de NV Faience en Tegelfabriek 'Holland', 1894–1918*, Museum Het Princessehof, Leeuwarden (2007)
Hey, Jan Jaap, *Vernieuwing en bezinning: Nederlandse beeldende kunst en kunstnijverheid ca 1885–1935 uit de collectie van het Drents Museum*, Drents Museum, Assen (Zwolle, 2004).
Hilkhuijsen, Jos, *Delftse Art Nouveau: Onderwijs en ontwerp van Adolf le Compte (1850–1921), Karel Sluyterman (1863–1931) en Bram Gips (1961–1943)*, Drents Museum, Assen (Zwolle, 2001).
Hinte, Ed van, and Timo de Rijk, eds, *Wereldverbeteraars: 100 Jaar idealen in glas*, Glasmuseum, Leerdam (2006)
Hoogenboezem, E., *Jac. Jongert, 1883–1942: Graficus tussen kunst en reclame*, Gemeentemuseum, The Hague (1982)
Huygen, Frederike, 'Zakelijk, nieuw, modern en functioneel: Some Terms Defined: Objective, New, Modern and Functional', in *1928: Beauty, Lucidity, Logica and Ingenuity*, Museum Boijmans Van Beuningen, Rotterdam (1993), pp. 42–63
Jintes, L. F., and J. T. Pol-Tyszkiewicz, *Chris van der Hoef, 1875–1933*, Rijkmuseum Het Koninklijk Penningkabinet, Leiden; Drents Museum, Assen (1994)
Jurriaan Schrofer, Museum Fodor, Amsterdam (1974)
Kho Liang Ie, *Hochschule für Gestaltung Ulm*, Stedelijk Museum, Amsterdam (1965)
Kieft, Ghislain, *Memphis-Design*, Het Kruithuis, 's-Hertogenbosch (1984)
Knol, M., *Klaas van Leeuwen, 1868–1935*, Drents Museum, Assen (Zwolle, 1988)
Kras, Reyer, ed., *Bakeliet: Techniek / vormgeving / gebruik*, Museum Boijmans Van Beuningen, Rotterdam (1981)
Krekel-Aalberse, Annelies, *Zilver: Silver: Silber: 1880–1940: Art Nouveau: Art Deco*, Gemeentemuseum, The Hague (Stuttgart, 2001)
—, *Carel J. A. Begeer, 1883–1956*, Drents Museum, Assen (Zwolle, 2001)
—, and E. Raasen-Kruimel, *Jan Eisenloeffel, 1876–1957*, Singermuseum, Laren; Drents Museum, Assen (Zwolle, 1996)
Kuper, Marijke, *Gerrit Th. Rietveld: L'oeuvre complet*, Centraal Museum, Utrecht (1993)
Lefaivre, Liane, and Ingeborg de Roode, eds, *Aldo van Eyck: The Playgrounds and the City*, Stedelijk Museum, Amsterdam (2002)
Maas, Steffen, ed., *Hitweek, 1965–1969*, De Beyerd, Centrum voor beeldende kunst, Breda (2003)
Meubelsculptuur: wonderlijke tafels en stoelen . . . of beeldhouwwerken?, Museum Commanderie van St-Jan/Galerie Marzee, Nijmegen (1992)
Metz, Annet, et al., *De eerste plastic eeuw: Kunststoffen in het dagelijks leven: Massacultuur*, Gemeentemuseum, The Hague (1981)
Molen, J.R. ter, ed., *Frans Zwollo sr, 1872–1945, en zijn tijd / Frans Zwollo sr, 1872–1945, und seine Zeit*, Museum Boijmans Van Beuningen, Rotterdam (1982)
Neerincx, Riet, ed., *T.A.C. Colenbrander (1841–1930): Plateelbakkerij 'RAM' te Arnhem (1921–1935)*, Gemeentemuseum, Arnhem (1986)
Orchard, Karin, and Isabel Schulz, eds, *Kurt Schwitters en de Avant-garde*, Museum Boijmans Van Beuningen, Rotterdam (2007)
Roode, Ingeborg de, and Marjan Groot, *Amsterdamse School textiel, 1915–1930*, Textielmuseum, Tilburg (1999)
Rijk, Timo de, *De Haagse Stijl: Art Deco in Nederland*, exh. cat. Gemeentemuseum, The Hague (Rotterdam, 2004)
Schoon, Thalita, ed., *Het meubel verbeeld: recente tendensen in sculptuur/Furniture as Art: Recent Tendencies in Sculpture*, Museum Boijmans Van Beuningen, Rotterdam (1988)
Simon Thomas, Mienke, Elly Adriaansz and Sandra van Dijk, *Jaap Gidding: Art Deco in Nederland*, Museum Boijmans Van Beuningen, Rotterdam (2006)
Sluijter-Seijffert, Nicolette, and Hans Vogels, eds, *Van decor naar design: Kunstenaars in de Goudse aardewerkindustrie, 1898–1941*, Museum het Catharina Gasthuis, Gouda (Zwolle, 2001)
Staal, Gert, *Gijs Bakker vormgever: Solo voor een solist*, Centraal Museum, Utrecht (1989)
Steenbruggen, Han, *Swip Stolk: Master Forever*, Groninger Museum, Groningen (2000)
Sterk, Anna, *N.V. Keramische industrie Fris, Edam, 1947–1969*, Princessehof, Leeuwarden (1985)
—, *St. Maarten Porcelein*, Het Kruithuis, 's-Hertogenbosch (1988)
Szénássy, I., and J. Jacobs, *Ontwerpen voor de industrie 1: Studies over ontwerpen voor de industrie, waarbij opgenomen de catalogus van de gelijknamige tentoonstelling over industrieel*

ontwerpen en vormgeven (Groningen, 1982) [travelling exhibition]
Textiel Nu [1-14], Textielmuseum, Tilburg (1980–84)
Thunnissen, Claudia, and Rob Sperna Weiland, *Gerrit de Blanken: 1894–1961: pottenbakker uit Leiderdorp: virtuoze eenvoud*, Lakenhal, Leiden (2005)
Tibbe, L., et al., *Jac. van den Bosch, 1868–1948*, Drents Museum, Assen (Zwolle, 1987)
Tibbe, Lieske, 'Gemeenschapskunst: de samenleving in symbolen', in M. Bax and C. Blotkamp, *In't diepst van mijn gedachten . . . Symbolisme in Nederland ca 1890–1930*, Drents Museum, Assen (Zwolle, 2004)
Unger, Marjan, *Helly Oestreicher*, De Beyerd, Breda (1989)
—, and Christina Hosman, *Tejo Remy*, Centrum Beelden kunst, Amersfoort (2001)
Visser, Martin, et al., *Benno Premsela onder anderen*, Stedelijk Museum, Amsterdam (1981)
Vogels, Hans, *N.V. Koninklijke Plateelbakkerij Zuid-Holland*, Stedelijke Musea, Gouda (1994)
Vreeburg, Guus, and Hadewych Martens, *UMS Pastoe: Een Nederlandse Meubelfabriek, 1913–1983*, Centraal Museum, Utrecht (1983)
Vreeken, Bert, *Vormgeving na '60; van Pop-Art tot Postmodern*, Gemeentemuseum, The Hague (1987)
Vrie, Dingenus van de, *Benno Wissing, grafisch en ruimtelijk ontwerpen*, Museum Boijmans Van Beuningen, Rotterdam (1999)
Wilk, Christopher, ed., *Modernism: Designing a New World, 1914–1939*, Victoria and Albert Museum, London (2006)
Witterholt, Madelon, ed., *De verloren jaren: Persoonlijke visies op De Bestverzorgde Boeken 1971 tot en met 1985*, Rijksmuseum Meermanno-Westreenianum, The Hague (1991)
Woud, Auke van der, *Het Nieuwe Bouwen: International: CIAM: Housing Town Planning*, Museum Kröller-Müller, Otterlo (1983)

Titles in English

& Fork (New York, 2007)
Adriaansz, Elly, et al., 'The Rotterdam Museumpark Villas', *Wiederhall*, 20 (2001)
Annink, Ed, et al., *Bright Minds, Beautiful Ideas: Parallel Thoughts in Different Times: Bruno Munari, Charles and Ray Eames, Martí Guixé and Jurgen Bey* (Amsterdam, 2003)
Bakker, Gijs, and Evert Rodrigo, *Design from the Netherlands/Design aus den Niederlanden*, Bureau Beeldende Kunst Buitenland, Ministerie CRM (Amsterdam, 1980)
Beckett, Jane, 'W. H. Gispen and the Development of Tubular Steel Furniture in the Netherlands', in *Tubular Steel Furniture*, ed. B. Campbell-Cole and T. Benton (London, 1979), pp. 28–45
Beeke, Anton, *Dutch Posters, 1960–1996* (Amsterdam, 1996)
Beenker, Erik, et al., *Jan van der Vaart: Ceramics* (Amsterdam, 1991)
Bergvelt, Ellinoor, Frans van Burkom and Karin Gaillard, eds, *From Neorenaissance to Postmodernism: A Hundred and Twenty-five Years of Dutch Interiors, 1870–1995* (Rotterdam, 1996)
Betsky, Aaron, and Adam Eeuwens, *False Flat: Why Dutch Design Is so Good* (London and New York, 2004)
Blotkamp, Carel, et al., *De Stijl: The Formative Years, 1917–1921* (Utrecht, 1982)
Brand, Jan, et al., *Product Design Diversity* (Arnhem [Artez] and Rotterdam, 2007)
—, et al., *The Power of Fashion: About Design and Meaning* (Arnhem, 2006)
Broos, Kees, ed., *Ontwerp: Total Design/Design: Total Design* (Utrecht, 1983)
Casciato, Maristella, *The Amsterdam School* (Rotterdam, 1996)
Crommelin, Liesbeth, *Textiel in het Stedelijk/Textiles in the Stedelijk* (Amsterdam, 1993)
Crouwel, Wim, *Alphabets* (Amsterdam, 2003)
—, *New Alphabet: A Possibility for the New Development*, Kwadraat-blad (Hilversum, 1967)
Day, Susan, *Art Deco and Modernist Carpets* (San Francisco, 2002)
Derks, Sergio, *Generations of Shaving Excellence: An Impression of 60 Years of Philishave* (Eindhoven, 1998)
Dosi Delfini, Luca, et al., *The Furniture Collection, Stedelijk Museum, Amsterdam, 1850–2000: From Michael Thonet to Marcel Wanders* (Amsterdam, 2004)
Duits, Thimo te, *Moderne Keramiek in Nederland/Modern Ceramic in the Netherlands* (The Hague, 1990)
Eliëns, Titus, M., Marjan Groot and Frans Leidelmeijer, *Dutch Decorative Arts, 1880–1940* (Bussum, 1997)
Fitoussi, Brigitte, et al., *Richard Hutten: Works in Use* (Oostkamp, 2006)
Ginkel, Dirk van, and Paul Hefting, *Ben Bos: Design of a Lifetime* (Amsterdam, 2000)
Heskett, J., *Philips: A Study of Corporate Management of Design* (London, 1989)
Hinte, Ed van, *Huibert Groenendijk: Down to Design* (Rotterdam, 2004)
—, ed., *Ed Annink Designer* (Rotterdam, 2002)
—, ed., *Richard Hutten: Taking Form, Making Form* (Rotterdam, 2002)
—, and Conny Bakker, *Trespassers: Inspiration for Eco-efficient Design*, Nederlands Design Instituut (Rotterdam, 1999)
—, Gert Staal and Carel Kuitenbrouwer, *Dutch Art: Design in the Netherlands*, International Information of the Dutch Ministry of Cultural Affairs (The Hague, 1989)
Industriële vormgeving in Amerika: Rapport studiegroep industrie: With a Summary in English (The Hague, 1954)
KesselsKramer, *One Hundred and One Things to Do* (Amsterdam, 2006)
Koch, André, *W. H. Gispen, a Pioneer of Dutch Design* (Rotterdam, 1998)

Krekel-Aalberse, Annelies, *Art Nouveau and Art Deco Silver* (London, 1989)
Langendijk, Eugène, and Mienke Simon Thomas, *Dutch Art Nouveau and Art Deco Ceramics: The Boijmans Van Beuningen Museum Collection* (Rotterdam, 2001)
Le Quernec, Alain, ed., *Studio Dumbar* (Paris, 2006)
Liefkes, Reino, *Andries Copier: Glass Designer, Glass Artist* (Zwolle, 2002)
Lindinger, H., ed., *Ulm Design: The Morality of Objects: Hochschule für Gestaltung Ulm, 1951–1968* (Berlin, 1990)
Lootsma, Bart, et al., *Mentalitäten: Niederländisch Design: prämierte Arbeiten des Designpreises Rotterdam, 1993–1996/Mentalities: Dutch Design: Nominated Products of the Design Prize Rotterdam* (Amsterdam, 1996)
Middendorp, Jan, *Dutch Type* (Rotterdam, 2004)
Mijksenaar, Paul, *Visual Function: An Introduction to Information Design* (Rotterdam, 1997)
Molenaar, Joris, ed., *Van Nelle, Monument in Progress* (Rotterdam, 2005)
Oldewarris, Hans, *The Covers of Wendingen* (Rotterdam, 1995)
—, and Peter de Winter, eds, *20 jaar/years 010: 1983–2003* (Rotterdam, 2003)
Petersen, Ad, *Sandberg, Designer and Director of the Stedelijk* (Rotterdam, 2004)
Purvis, Alston W., *Dutch Graphic Design, 1918–1945* (New York, 1992)
Radiant: 30 jaar RA: 30 years RA (Amsterdam, 2006)
Ramakers, Renny, ed., *Simply Droog* (Amsterdam, 2004)
—, and Gijs Bakker, eds, *Droog Design: Spirit of the Nineties* (Rotterdam, 1998)
Rock, Micheal, 'Mad Dutch Disease: The Strange Case of Dutch Design and Other Contemporary Contagions', in *Jaarboek Nederlandse vormgeving 03/04* (Rotterdam, 2004), pp. 63–81
Schaik, Sylvia van, et al., *Mondial: Gispen & Gerrit Th. Rietveld* (Culemborg, 2006)
Schouwenberg, Louise, *Hella Jongerius* (London, 2003)
Smits, J.P., H. de Jong and B. van Ark, *Three Phases of Dutch Economic Growth and Technological Change, 1815–1997*, Groningen Growth and Development Centre, Rijksuniversiteit (Groningen, 1999)
Spoon (New York, 2002)
Staal, Gert, and Hester Wolters, eds, *Holland in Vorm: Dutch Design, 1945–1987* (The Hague, 1987)
Stiphout, Wouter van, 'Stories from behind the Scenes of Dutch Moral Modernism', in *Mart Stam's Trousers: Stories from behind the Scenes of Dutch Moral Modernism* (Rotterdam, 1999)
Strauven, Francis, *Aldo van Eyck's Orphanage: A Modern Monument* (Rotterdam, 1996)
Teunen, Jan, *Bruno Ninaber van Eyben: With Compliments* (Rotterdam, 2002)
Thovez, Enrico, 'The International Exhibition of Modern Decorative Art at Turin: The Dutch Section', *The Studio*, XXVI (1902), pp. 204–13
Toorn, Jan van, *Design's Delight* (Rotterdam, 2006)
Troy, Nancy, *The Stijl Environment*, (Cambridge, MA, 1983)
Vletter, Martien de, *De kritische jaren zeventig: Architectuur en stedenbouw in Nederland/The Critical Seventies: Architecture and Urban Planning in the Netherlands* (Rotterdam, 2004)
Zijl, Ida van, and Gijs Bakker, *Objects to Use* (Rotterdam, 2000)

Titles in Dutch

Bax, Marty, *Het Web der schepping: Theosofie en kunst in Nederland van Lauweriks tot Mondriaan* (Nijmegen, 2006)
Bergvelt, Ellinoor, ed., *Goed Wonen, een Nederlandse wooncultuur 1946–1968*, special edition of *Wonen TABK* (1979) 4/5
Eliëns, Titus, and Marlite Halbertsma, eds, *Volmaakt verchroomd d3 en het avant-gardemeubel in Nederland* (Rotterdam, 2007)
Gans, L., *Nieuwe Kunst: De Nederlandse bijdrage aan de Art Nouveau* (Utrecht, 1966)
Groot, Marjan, *Vrouwen in de vormgeving, 1880–1940* (Rotterdam, 2007)
Huisman, Jaap, et al., *Honderd jaar wonen in Nederland, 1900–2000* (Rotterdam, 2000)
Huygen, Frederike, *Visies op Vormgeving: Het Nederlandse ontwerpen in teksten Deel I: 1874–1940* (Amsterdam, 2007) (Deel II will be published in October 2008)
—, and Hugh Boekraad, *Wim Crouwel: Mode en Module* (Rotterdam, 1997)
Kley-Blekxtoon, A. van der, *Leerdam glas, 1878–2003: De glasfabriek Leerdam* (Lochem, 2004)
Koch, André, ed., *Ludiek Sensueel en Dynamisch: Nederlandse jeugdcultuur en vormgeving in de jaren zestig* (Schiedam, 2002)
Kras, Reyer, *Nederlands Fabrikaat: Industriële vormgeving* [Teleac] (Utrecht and Bussem, 1997)
Lauwen, Toon, *Holland in beeld, 1895–2008* (Bussum, 2007)
—, *Dutch Design van de 20ste eeuw*, (Bussem, 2003)
Majorick, B., *Ontwerpen en verwerpen: industriële vormgeving als noodzaak* (Amsterdam, 1959)
Middendorp, Jan, *'Ha, daar gaat er een van mij'. Kroniek van het grafisch ontwerpen in Den Haag, 1945–2000* (Rotterdam, 2002)
Pruys, Simon Mari, *Dingen vormen mensen: Een studie over produktie, consumptie en cultuur* (Bilthoven, 1972)
Rijk, Timo de, ed., *Designers in Nederland: Een eeuw productvormgeving* (Amsterdam and Gent, 2003)
—, *Het elektrische huis: Vormgeving en acceptatie van elektrische huishoudelijke apparaten in Nederland* (Rotterdam, 1998)
Simon Thomas, Mienke, *De Leer van het Ornament: Versieren volgens voorschrift, 1850–1930* (Amsterdam, 1996)
Spruit-Ledeboer, Mieke G., *Nederlandse keramiek, 1900–1975* (Amsterdam, 1976)

Timmer, Petra, *Metz & Co.: De creatieve jaren* (Rotterdam, 1995)
Unger, Marjan, *Het Nederlandse sieraad in de 20ste eeuw* (Bussum and Utrecht, 2004)
Witte, Arnold, and Esther Cleven, *Design is geen vrijblijvende zaak: Organisatie, imago en context van de ptt-vormgeving tussen 1906 en 2002* (Rotterdam, 2006)
Yocarini, T., *Vak in beweging, VANK, GKf, VRI, GVN, bNO* (Eindhoven, 1992)

Websites

www.designacademy.nl
Design Academy, Eindhoven
www.designlink.nl
www.doorsofperception.com
www.dutchdesigncenter.nl
Dutch Design Center, Utrecht
www.design.startpagina.nl
www.iconenvandepost.nl
www.moooi.com
Moooi BV
www.nago.nl
Nederlands Archief Grafisch Ontwerpers (NAGO)
www.premsela.org
Premsela Stichting
www.ydi.nl
Stichting Young Designers & Industry (YD+I)
www.vivid.nl

Acknowledgements

I wish to express my gratitude to all those who read and commented on all or part of my text. First of all, I would like to thank the design historian Frederike Huygen, who was also my former colleague at the Museum Boijmans Van Beuningen, and my partner Patrick van der Kroef, who helped me to conceptualize the structure of the book. Both invested days in editing my text. Timo de Rijk, Marjan Unger, Dingenus van de Vrie and Betty Brutvan-Simon Thomas read parts and offered encouraging comments. Special thanks are due to Vivian Constantinopoulos from Reaktion Books and to Hans Oldewarris from 010 Publishers in Rotterdam, who took care of the Dutch edition.

I would also like to thank everyone who supplied me with interesting illustrations; I am especially indebted to the staff of the Museum Boijmans Van Beuningen in Rotterdam. Many illustrations in this book show objects from the collection of this fantastic museum.

The Mondriaan Foundation and the Prince Bernhard Foundation generously financed the translation of the text by Kate Williams and Lynn George. Our co-operation on this part of the project has initiated some necessary clarifications of the text, for which I am grateful.

Photo Acknowledgements

The author and publishers wish to express their thanks to the following sources of illustrative material and/or permission to reproduce it.

Archief Stedelijk Museum, Amsterdam: p. 140 foot; photo AVL Rotterdam: p. 229 (foot); photo Centraal Museum Utrecht: p. 218 (top); photo from *Dekoratieve Kunst* (1902): p. 34; Den Haags Gemeentemuseum: p. 43 top (photo courtesy of the Museum Boijmans Van Beuningen, Rotterdam); photo from *Eigen Haard* (1902): p. 28 (left); photo Jan + Robert Fock: p. 64; Gemeente Archief Amsterdam: p. 84 (photo courtesy of the Museum Boijmans Van Beuningen); photos Gemeentearchief Den Haag: pp. 90, 101, 106 left (courtesy of the Museum Boijmans Van Beuningen); photos Bob Goedewaagen: pp. 26, 30, 33 (foot), 61, 68 (foot), 69, 113, 202 (right), 203, 206; photo HEMA (courtesy of the Museum Boijmans Van Beuningen): p. 185; Instituut Collectie Nederland, Amsterdam: pp. 17 (on long-term loan from the Museum Boijmans Van Beuningen), 29 top left (photo courtesy of the Museum Boijmans Van Beuningen), 33 top (on long-term loan from the Museum Boijmans Van Beuningen), 66 (on long-term loan from the Museum Boijmans Van Beuningen); photo Bert Koenderink: p. 68 (top); photo Yves Krol: p. 11; Museum Boijmans Van Beuningen, Rotterdam: pp. 6, 21, 25, 26, 29 (foot), 30, 31, 33 (foot), 35, 37, 48, 52 (top left, foot), 61, 63, 68 (foot), 98, 108, 113, 139, 140 (top), 144, 151, 155, 162, 186, 194, 215, 217, 221, 222, 226, 227, 228; photos courtesy of the Museum Boijmans Van Beuningen: pp. 15 (from *Verslag der Centrale Commissie*, 1902), 23 (from *Elsevier's Geïllustreerd Maandschrift*, 1905), 27, 28 right (from Frans Netscher, *John Th. Uiterwijk & Co. Arts and Crafts*, 1901), 43 (foot), 52 (top right), 55 (right), 66, 75, 78, 82, 83, 91 (photo private collection), 95, 96, 97, 99, 103 (photo Jannes Linders), 105, 106 (right), 107, 111, 115, 119, 120, 121, 122, 125, 128 (reproduced from the catalogue of the 1951 exhibition 'Kunst en Kitsch' at the Gemeentemuseum, Den Haag), 130 (photo Jan Versnel), 136, 137, 138, 142, 145, 148, 153, 156, 164 (photo Lex van Pieterson), 165 (photo Jan Versnel), 168 (top), 173 (photo Luchthaven Schiphol), 176 (top), 182, 187, 188, 190, 195, 202 (left), 205, 207, 211, 219, 223 top (photo Alexander Schabraque), 223 foot (photo ANP), 229 top (photo Bianca Pilet), 231, 232, 235 (photo Fred Ernst); Nederlands Architectuurinstituut, Rotterdam: pp. 12, 16; Nederlands Textielmuseum Tilburg: p. 168 (foot); photo De Nederlandse Bank (courtesy of the Museum Boijmans Van Beuningen): p. 177; Noordbrabants Museum, 's-Hertogenbosch: p. 169; photo Océ-Nederland: p. 159; photo from *Op de Hoogte* 22 (September 1925): p. 73; Royal Collections, the Netherlands: 29 top right (photo courtesy of the Museum Boijmans Van Beuningen); Collectie Spaarnestad Photo, Haarlem: pp. 46 (photo/Het Leven/C. J. Hofker, courtesy of the Museum Boijmans Van Beuningen), 132 (photographer unknown; photo courtesy of the Museum Boijmans Van Beuningen), 161 (photo courtesy of the Museum Boijmans Van Beuningen); photos courtesy of the Stedelijk Museum, Amsterdam: pp. 39, 55 (left); photo courtesy of the Stichting Museum voor Communicatie, Den Haag: p. 86; photo Marjan Ungers: p. 178; photo courtesy of the Universiteitsbibliotheek van de Universiteit van Amsterdam: p. 19; reproduced from *Wendingen (Techniek en Kunst)* 9: 2 (1928): pp. 59, 60 top; photo Kim Zwarts (courtesy of the Museum Boijmans Van Beuningen): p. 88.

Index